T0141124

About Island Press

Island Press is the only nonprofit organization in the United States whose principal purpose is the publication of books on environmental issues and natural resource management. We provide solutions-oriented information to professionals, public officials, business and community leaders, and concerned citizens who are shaping responses to environmental problems.

In 2006, Island Press celebrates its twenty-second anniversary as the leading provider of timely and practical books that take a multidisciplinary approach to critical environmental concerns. Our growing list of titles reflects our commitment to bringing the best of an expanding body of literature to the environmental community throughout North America and the world.

Support for Island Press is provided by the Agua Fund, The Geraldine R. Dodge Foundation, Doris Duke Charitable Foundation, The William and Flora Hewlett Foundation, Kendeda Sustainability Fund of the Tides Foundation, Forrest C. Lattner Foundation, The Henry Luce Foundation, The John D. and Catherine T. MacArthur Foundation, The Marisla Foundation, The Andrew W. Mellon Foundation, Gordon and Betty Moore Foundation, The Curtis and Edith Munson Foundation, Oak Foundation, The Overbrook Foundation, The David and Lucile Packard Foundation, The Winslow Foundation, and other generous donors.

The opinions expressed in this book are those of the author(s) and do not necessarily reflect the views of these foundations.

About SCOPE

The Scientific Committee on Problems of the Environment (SCOPE) was established by the International Council for Science (ICSU) in 1969. It brings together natural and social scientists to identify emerging or potential environmental issues to address jointly the nature and solution of environmental problems on a global basis. Operating at an interface between the science and decision-making sectors, SCOPE's interdisciplinary and critical focus on available knowledge provides analytical and practical tools to promote further research and more sustainable management of the earth's resources. SCOPE's members, thirty-eight national science academies and research councils and twenty-two international scientific unions, committees, and societies, guide and develop its scientific program.

SCOPE 66

The Silicon Cycle

THE SCIENTIFIC COMMITTEE ON PROBLEMS OF THE ENVIRONMENT (SCOPE)

SCOPE Series

SCOPE 1 – 59 in the series were published by John Wiley & Sons, Ltd., U.K. Island Press is the publisher for SCOPE 60 as well as subsequent titles in the series.

SCOPE 60: *Resilience and the Behavior of Large-Scale Systems,* edited by Lance H. Gunderson and Lowell Pritchard Jr.

SCOPE 61: *Interactions of the Major Biogeochemical Cycles: Global Change and Human Impacts,* edited by Jerry M. Melillo, Christopher B. Field, and Bedrich Moldan

SCOPE 62: *The Global Carbon Cycle: Integrating Humans, Climate, and the Natural World,* edited by Christopher B. Field and Michael R. Raupach

SCOPE 63: *Invasive Alien Species: A New Synthesis,* edited by Harold A. Mooney, Richard N. Mack, Jeffrey A. McNeely, Laurie E. Neville, Peter Johan Schei, and Jeffrey K. Waage

SCOPE 64: *Sustaining Biodiversity and Ecosystem Services in Soils and Sediments,* edited by Diana H. Wall

SCOPE 65: *Agriculture and the Nitrogen Cycle: Assessing the Impacts of Fertilizer Use on Food Production and the Environment,* edited by Arvin R. Mosier, J. Keith Syers, and John R. Freney

SCOPE 66: *The Silicon Cycle: Human Perturbations and Impacts on Aquatic Systems,* edited by Venugopalan Ittekkot, Daniela Unger, Christoph Humborg, and Nguyen Tac An

SCOPE 66

The Silicon Cycle

Human Perturbations and Impacts on Aquatic Systems

Edited by
Venugopalan Ittekkot
Daniela Unger
Christoph Humborg
Nguyen Tac An

A project of SCOPE, the Scientific Committee on Problems of the Environment, of the International Council for Science

Washington • Covelo • London

Library of Congress Cataloging-in-Publication Data
The silicon cycle : human perturbations and impacts on aquatic systems / Venugopalan Ittekkot, Daniela Unger, Christoph Humborg, and Nguyen Tac An, editors.
p. cm. — (Scope ; 66)
ISBN 1-59726-114-9 (cloth : alk. paper) — ISBN 1-59726-115-7 (pbk. : alk. paper)
1. Silicon cycle (Biogeochemistry) 2. Aquatic ecology. I. Ittekkot, V. (Venugopalan), 1945- II. Series: SCOPE report ; 66.
QH344.S55 2006
577'.14—dc22 2006012462

British Cataloguing-in-Publication data available.

Printed on recycled, acid-free paper 

Manufactured in the United States of America

10 9 8 7 6 5 4 3 2 1

Contents

List of Figures and Tables

Figures

Tables

Preface

In publishing this volume, the Scientific Committee on Problems of the Environment (SCOPE) continues its assessments of the biogeochemical cycles that are essential to life on this planet. For nearly four decades SCOPE has synthesized scientific information on the complex and dynamic network of flows and interactions of major biogeochemical elements between various compartments of the global environment (see SCOPE Series Publication List). The *modus operandi* of these projects has been to bring together leading scientists working on various mechanisms, sources, sinks, and fluxes of the various cycles to critically review current knowledge and to point out gaps in knowledge and a way forward at interdisciplinary workshops. The accumulated information and results of discussions are synthesized, edited, and published as reports in the SCOPE series.

The publication of SCOPE 57 *Particle Flux in the Ocean* in 1996 brought to term the Global Carbon Cycle unit, which had concentrated on understanding the flow, interaction, and fate of carbon and other nutrients from land via rivers to lakes and to deep oceans. The summary chapter of this volume noted: "A better understanding of the interaction of the silica cycles with the cycles of other nutrients will be crucial to assess the efficiency of the oceanic biological pump which sequesters atmospheric CO-$_{2-}$".

In the intervening ten years, emphasis has been placed on obtaining much more information on the silicon cycle. It has become apparent that this work is extremely timely, as anthropogenic perturbation of the silicon cycle can seriously interfere with the functioning of water bodies, and thereby change the ability of these bodies to help sequester carbon, or of rivers to sustain fisheries. Thus, this SCOPE synthesis integrates our current understanding of these interactions and emphasizes the necessity of integrating this knowledge into efforts to mitigate anthropogenic effects in the future.

John W. B. Stewart, Editor-in-Chief
SCOPE Secretariat,
51 Boulevard de Montmorency, 75106 Paris, France
Executive Director: Véronique Plocq-Fichelet

Acknowledgments

We would like to acknowledge financial support for this project from the Deutsche Forschungsgemeinschaft through the German SCOPE committee, the Federal German Ministry for Education and Research, the European Commission, the Volkswagen Foundation, the Royal Swedish Academy of Sciences, and the Vietnam National Centre for Natural Sciences and Technology. The project has also received sponsorship from the Intergovernmental Oceanographic Commission at the United Nations Educational, Scientific and Cultural Organization through the IOC Sub-Commission for the Western Pacific (WEST PAC) and the WESTPAC office in Bangkok, as well as from the International Geosphere–Biosphere Programme via its core project on Land-Ocean Interactions in the Coastal Zone (LOICZ).

1
Introduction

Venugopalan Ittekkot, Daniela Unger, Christoph Humborg, and Nguyen Tac An

Silicon dioxide is the most abundant component of the earth's crust. It occurs as silicate minerals in association with igneous, metamorphic, and sedimentary rocks. They undergo physical and chemical weathering. The associated processes involving the release of weathered products, their transport and transformation, and their interaction with other nutrient elements form the basis of the global silicon cycle. Perturbation of this cycle by natural and anthropogenic factors and its impact on the environment and climate were at the core of the Land–Ocean Nutrient Fluxes: The Silicon Cycle project, conducted by the Scientific Committee on Problems of the Environment (SCOPE) of the International Council for Science.

Silicon rarely features in the study of nutrients at the land–sea continuum. Carbon is included when there is a need for a number to balance the carbon budget. However, most studies have concentrated on nitrogen and phosphorus because they represent the nutrient elements that are discharged by human activities. Their fluxes have dramatically gone up over the past few decades because of their increased discharges from domestic, agricultural, and industrial sources and deforestation. Accelerated algal growth in water bodies, known as eutrophication, is attributed to this increase, leading to a deterioration of water quality via oxygen depletion. Oxygen-deficient conditions in turn promote the production and emission to the atmosphere of climatically relevant gases such as nitrous oxide and methane. Toxic algal blooms also have been attributed to eutrophication, with devastating effects on fisheries and on biodiversity in general.

Not adequately taken into account is the role of silicon in all this. Diatoms form a major part of the aquatic food chain and are found to play a critical role in marine biogeochemical cycles, especially in the sequestration of carbon dioxide from the atmosphere via the biological pump. Diatoms need silicon, which is present in both cell contents and cell walls. Rock weathering is the natural source of dissolved silicate in aquatic systems. Land use changes and hydrological alterations have been changing the supply from this source to our rivers. There is evidence that the resulting changes in the ratios of nutrients (N, P, and Si) delivered to our coastal seas could effect species shifts from diatoms to nondiatoms, exacerbating the already declining quality and structure of aquatic ecosystems.

In three workshops within the project (1999, University of Linköping, Sweden; 2000, National Centre for Natural Science and Technology, Nha Trang, Vietnam; 2002, Center for Tropical Marine Ecology, Bremen, Germany), the SCOPE project brought together experts to review and synthesize the available information on the changing pattern of silicate transfer from land to sea and associated adverse impacts on the quality and structure of aquatic ecosystems. They reexamined the available data on land–ocean nutrient fluxes in light of new research on the role of silicates as nutrients and the changes in their fluxes caused by human activities. This appeared to be warranted in order to develop scientifically sound strategies to reduce the risk of ecosystem perturbation, especially in coastal waters, and to identify areas and needs for future research.

While summarizing the results of the project, the present volume also tracks the pathway of silicon from its source in the rocks and minerals on land to its burial in sediments at the sea floor. It discusses the biotic and abiotic modifications of silicon in transit and its cycling in the coastal seas. Natural geological processes in combination with atmospheric and hydrological processes are considered, as are human perturbations of the natural controls of the silicon cycle.

We hope that this book will be useful to a broad spectrum of readers, including students and researchers, coastal managers, lecturers, and national monitoring agencies, and that it will contribute to the ongoing efforts to understand the global cycling of silicon.

2

Silicate Weathering in South Asian Tropical River Basins

Vaidyanatha Subramanian, Venugopalan Ittekkot, Daniela Unger, and Natarajan Madhavan

Silicon dioxide occurs as silicate minerals in the earth's crust. Weathering of silicates can be represented as

$$2CO_2 + 3H_2O + CaSiO_3 \rightarrow Ca^{++} + 2HCO_3^- + H_4SiO_4,$$

where Ca^{++} might also be substituted by Mg^{++}.

Chemical weathering of silicates and the cycling of their products form the basis of silicon biogeochemistry and its interaction with elemental cycles such as carbon and nitrogen. Weathering of carbonates can be represented as

$$CO_2 + H_2O + CaCO_3 \rightarrow Ca^{++} + 2HCO_3^-$$

or

$$2CO_2 + 2H_2O + CaMg(CO_3)_2 \rightarrow Ca^{++} + Mg^{++} + 4HCO_3^-$$

The CO_2 consumed during these weathering reactions is supplied either directly from the atmosphere or from soils where it is produced during the degradation of plant biomass. However, in the case of silicate weathering all alkalinity produced is of atmospheric origin, whereas in the case of carbonate weathering only half of it is. Therefore, silicate weathering on land represents an important sink for atmospheric CO_2 and is of special interest in the context of controlling the concentrations of CO_2 in the atmosphere and the ocean over geological time scales (Berner et al. 1983; Wollast and Mackenzie 1983; Brady and Carrol 1994).

The dissolved products of weathering are transferred via the rivers to the oceans, where they are taken up biologically to form carbonate and biogenic silica in the tissues and skeletons of marine plankton.

Silicate weathering depends on temperature and precipitation (White and Blum 1995) and might be enhanced through global warming. Another factor affecting silicate weathering is the erosion rate, that is, the delivery of potentially weatherable material. However, estimates from the Lesser Himalayas indicate that because of the steep relief and the resulting abundant supply of eroded material, erosion is not a limiting factor in the case of Himalayan rivers (West et al. 2005). Both the dissolved silica (DSi) and alkalinity load have been shown to be related to land use changes and therefore are prone to anthropogenic alterations. DSi in pristine northern Swedish rivers was found to correlate with vegetation cover (Ittekkot et al. 2003). DSi concentration in small tropical rivers appears to decrease after the conversion of forested land into cropland when rice or crops from the Gramineae family are cultivated because these plants store silicon in phytoliths, reducing the export of Si from soil to rivers (Chapter 4, this volume). However, alkalinity increased with the proportion of cropland to forestland in the case of the Mississippi and its tributaries in the last decades (Raymond and Cole 2003). In small watersheds, as in southwestern India, extensive land use changes in the last three decades appear to have made irreversible changes in some of the major water quality parameters such as alkalinity, DSi, Ca, and Mg (Sajeev and Subramanian 2003).

In the context of CO_2-induced global warming, the role played in the carbon cycle by the possibly accelerated chemical weathering and the subsequent supply of solute components to the sea is of interest particularly in regions with enhanced anthropogenic alterations and pronounced intensity of weathering reactions.

Regional Background

The geology of the Indian subcontinent is diverse, representing Precambrian hard rocks (shield area) to the most recent alluvial terrain in the Indo-Gangetic belt (Table 2.1). The Himalayan region has a mixture of hard igneous and metamorphic rocks associated with Tertiary sediments, and their proportion varies in the drainage basins of rivers.

However, even in watersheds characterized by the predominance of silicate rocks it was found that water chemistry was determined to a large extent by the weathering of carbonates present in veins and thin layers (Blum et al. 1998; Quade et al. 2003). Silicate weathering appears to be much more intense in the huge Ganges Basin than in the high Himalayan region (West et al. 2002). The peninsular region is characterized by basaltic Deccan traps in the central and hard rocks in the southern parts. Ancient and recent sediments occur in the northern and coastal regions. Climatically, south Asia faces extreme conditions, from humid tropics in the southern part to cold arid temperatures in the Ladakh Himalayan region, which affect

Table 2.1. Extension of river basins and dominant basin lithology for major south Asian rivers.

River System	Basin Area (km^2)	Basin Lithology
Indus	470,000	Upper reaches: granite and other hard rocks 35% Middle reaches: sedimentary rocks 35% Lower reaches: sedimentary rocks and alluvium 30%
Ganges–Padma	970,000	Upper reaches: granite and other hard and soft rocks 20%, sedimentary rocks 20%, alluvium 60%
Brahmaputra	924,000	India: granite and other hard rocks 50%, sedimentary rocks 40%, alluvium 10% Bangladesh: alluvium 90%, metamorphic rocks 10%
Megna	36,200	India: granite and other hard rocks 80% Bangladesh: sedimentary rocks 10%
Narmada	87,900	Deccan basaltic traps 70%, sedimentary rocks 30%
Tapti	49,150	Deccan basaltic traps 70%, sedimentary rocks 30%
Mahanadi	88,320	Granite 34%, sandstone 22%, limestone and shale 17%, charnokite and other hard rocks 20%, coastal sediments 5%
Godavari	313,00	Granite and other hard rocks 20%, Deccan basaltic traps 56%, sedimentary rocks 24%
Krishna	251,360	Crystalline rocks 78%, Deccan basaltic traps 20%, alluvium 2%
Cauvery	66,250	Gneiss 60%, charnokite 25%, sandstone and limestone 10%, other sedimentary rocks 5%
West-flowing small rivers	24,000	Granite and other hard rocks 80%, alluvium 20%

weathering intensity. The mountainous drainage basins of Himalayan rivers experience high chemical and physical erosion rates. The same is true for small west-flowing rivers on the Indian subcontinent, as in Kerala (Ittekkot et al. 1999). Dense populations on either side of the Himalayas in China and south Asia lead to severe land use alteration throughout the region. Thus south Asian rivers are particularly suited for study of the link between land use changes, weathering, alkalinity, DSi, and atmospheric CO_2 drawdown.

Data and Discussion

For this study, pCO_2 was estimated using the pH and HCO_3^{-1} values for individual rivers. Alkalinity from silicate weathering (where all values are in milliequivalent units), following the method of Raymahasay (1986), is calculated as

$$\text{Alkalinity}_{\text{carbonate weathering}} = 0.74\ \text{Ca}_{\text{total}} + 0.4\ \text{Mg}_{\text{total}}$$

and

$$\text{Alkalinity}_{\text{silicate weathering}} = \text{Alkalinity}_{\text{total}} - \text{Alkalinity}_{\text{carbonate weathering}}.$$

Generally, the observed concentration of alkalinity and Ca and Mg in river water is the net balance from possible diverse sources (carbonate versus silicate rocks; Quade et al. 2003) and removal by precipitation of Ca–Mg–carbonate minerals in the river basin (Stallard and Edmond 1981).

According to Probst et al. (1998), weathering of both silicate and nonsilicate rocks contributes 70 percent of the river alkalinity, and the remaining 30 percent is accounted for by river-based biological processes caused by decomposition of organic matter. For large rivers in the subcontinent, the relative weathering contribution in the beginning of monsoon season is likely to be even larger because of the large discharge of various river systems flushing out weathered topsoil in the initial stages of the rain, when biochemical reactivity is inhibited by the rapid flow of water.

Figure 2.1 shows the plot of carbonate alkalinity against the total concentration of Ca and Mg for major rivers in south Asia as seasonally adjusted values.

This plot shows that HCO_3 nearly balances Ca + Mg, which is indicative of mainly carbonate weathering and was observed earlier by Sarin et al. (1989) for highland tributaries of the rivers Ganges and Brahmaputra. But at higher alkalinities, a deficit of Ca + Mg occurs, which would have to be balanced by alkalis (i.e., Na, Ka) from silicate weathering.

Because of the monsoon-related precipitation pattern, all the rivers in the subcontinent show extreme seasonality in their flows. Whereas the rivers in the southern region transport 95 percent of their annual runoff in three monsoon months (June to August), the Himalayan rivers—Ganges, Indus, and Brahmaputra—carry up to 50 percent of their annual runoff during the four-month monsoon period (June to September). Seasonal variability of river water composition as a result of weathering is well documented for a number of river systems in the world; see Raymond and Cole (2003) for the Mississippi River, Stallard and Edmond (1981) and Gibbs (1970) for the Amazon, and Galy and France-Lanord (1999) and Subramanian (2000) for several rivers in the Indian subcontinent. Major south Asian rivers listed in Table 2.2 reveal strong seasonality in their solute transport, reflecting variable export of DSi, Ca, Mg, and alkalinity in response to seasonal changes in the weathering zone (i.e., precipitation and temperature). In the downstream region of the Indus River in Pakistan, for example, the winter values for alkalinity, Ca, and Mg are generally higher than the corresponding summer values (data from Karim and Veizer 2000). Similar variation was observed for the Indus from the Ladakh Himalayas and for the Ganges and the Brahmaputra. The Cauvery, which is one of the peninsular Indian rivers draining hard Precambrian shield areas, shows low

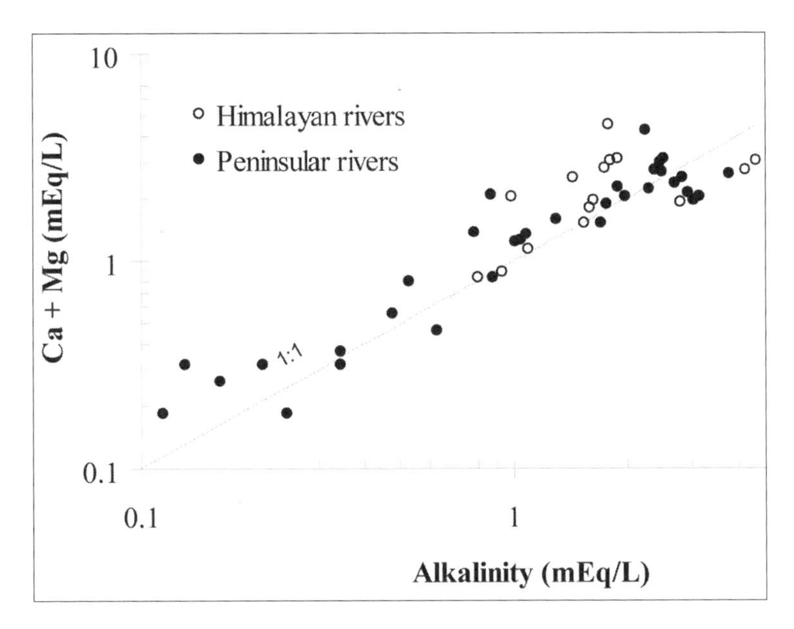

Figure 2.1. Correlation of alkalinity and total Ca and Mg.

HCO_3^{-1}/SiO_2 ratios compared with the Indus, reflecting the relevance of silicate weathering (Quade et al. 2003). This implies that the source of excess Mg and Ca in Cauvery waters are the silicate minerals of the shield rocks. Data presented in Table 2.2 also show that the amount of DSi in the lower Ganges–Brahmaputra system is generally larger than reported by many earlier workers because the data here are based long-term observation.

Because silicate and carbonate weathering differ in their significance for the carbon cycle, it is essential to separate the components of alkalinity derived from silicate and nonsilicate weathering. The use of $^{87}Sr/^{86}Sr$ led to the conclusion that accelerated silicate weathering especially in the Himalayas contributed significantly to reduced atmospheric CO_2 levels during the Cenozoic (Raymo and Ruddiman 1992). However, recent work has shown that weathering of Himalayan carbonates might well produce similarly high isotopic ratios (Blum et al. 1998; Quade et al. 2003).

In Figure 2.2 we plot alkalinity from silicate weathering against total alkalinity in order to estimate the relative contribution of silicate weathering to alkalinity.

Dessert et al. (2001) calculated the alkalinity from silicate weathering in the basaltic terrain in central India and found it to be higher than in other regions such as the Himalayan watershed. Accordingly, the pCO_2 consumed during weathering to generate the alkalinity, shown in Figure 2.3, exhibits great variability because of the diverse lithology in the subcontinent.

Table 2.2. Seasonal variation of pH, alkalinity, Ca^{+2}, Mg^{+2}, and SiO_2 for Brahmaputra, Ganges, Indus, and Cauvery.

River and Locations	Season	pH	HCO_3^{-1} (mEq/L)	Ca^{+2} (mEq/L)	Mg^{+2} (mEq/L)	SiO_2 (mmol/L)
Brahmaputra						
Bhahadurabad ghat	Winter	7.3	1.549	0.297	0.164	0.168
	Summer	7.9	0.942	0.141	0.091	0.070
Ganges						
Harding bridge	Winter	8.0	2.811	0.509	0.319	0.196
	Summer	8.3	2.122	0.343	0.231	0.137
Indus						
Humayun bridge	Winter	8.1	1.271	0.734	0.304	0.009
	Summer	8.3	1.031	—	—	0.064
Thakot bridge	Winter	8.0	1.360	0.756	0.317	0.036
	Summer	8.3	0.939	0.550	0.171	0.056
Sarki	Winter	8.3	1.697	0.835	0.397	0.101
	Summer	8.2	1.298	0.599	0.199	0.069
Dadu-Moro bridge	Winter	8.4	2.528	1.068	0.506	0.124
	Summer	8.1	1.662	0.694	0.247	0.122
Cauvery						
Shimoga	Winter	7.8	0.155	7.719	2.464	0.406
	Summer	7.5	0.100	11.355	1.971	0.406
Erode	Winter	7.8	0.295	7.470	3.696	0.250
	Summer	7.6	0.020	7.968	2.136	0.906
Tirichy	Winter	7.7	0.219	8.566	2.793	0.422
	Summer	8.1	0.154	11.454	2.546	0.453
Chidambaram	Winter	7.8	0.344	8.267	3.450	0.484
	Summer	7.2	0.190	14.841	2.218	0.625

Sources: Ganges data, Datta and Subramanian (1997), Subramanian (2000, 2002); Indus data, Karim and Veizer (2000); Cauvery data, Ramanathan et al. (1993).

The lower part of the Indus in Pakistan (data calculated from Karim and Veizer 2000) and the combined Ganges–Brahmaputra–Megna system in Bangladesh (data calculated from Datta and Subramanian 1997) show consumption of higher levels of pCO_2 than in the upper reaches of these rivers in the Himalayas. This observation is in line with the results of West et al. (2002), who found silicate weathering to be much more intense in the huge Ganges Basin than in the high Himalayan region. Additionally, tributaries might also affect the solute load composition of the Ganges in its lower reaches, as pointed out by Galy and France-Lanord (1999) and Sarin et al. (1989).

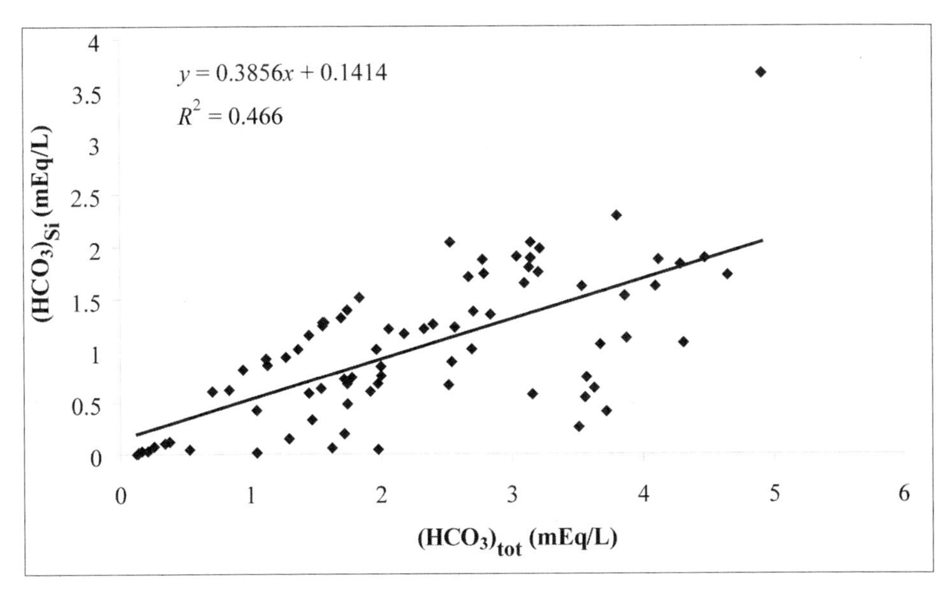

Figure 2.2. Correlation of total alkalinity and silicate alkalinity.

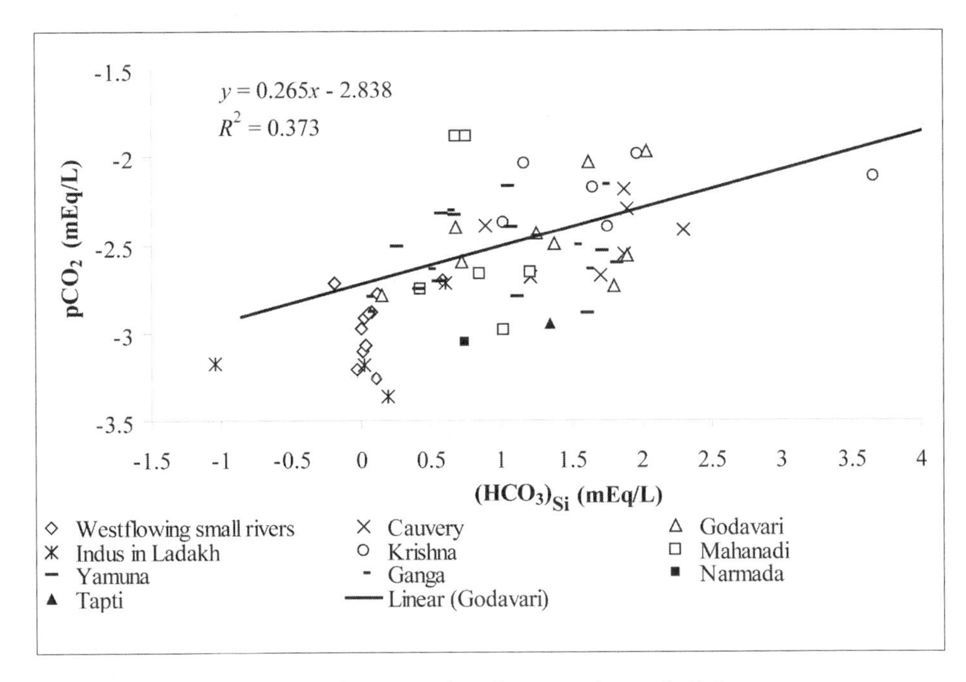

Figure 2.3. Computed values for pCO_2 plotted against silicate alkalinity.

Ca/Na and Mg/Na observed by Quade et al. (2003) result in seasonally variable fractions of CO_2 consumed by silicate weathering. Their calculations for the rivers Ganges, Brahmaputra, and Indus result in a silicate weathering contribution of 15–19 percent and 36 percent during the rainy and nonrainy seasons, respectively, an estimate slightly higher than that reported by Galy and France-Lanord (1999).

Based on the data in Table 2.3, the export of silicate alkalinity to the adjoining oceans is estimated to be 2.94×10^{13} g C/year. It is interesting to note that for the Mississippi drainage basin, Raymond and Cole (2003) report an increase of total bicarbonate export from 1.1 to 1.75×10^{13} g C/year during the second half of the last century, of which 60 percent is estimated to originate from atmospheric CO_2. The comparison highlights the importance of south Asian rivers with respect to alkalinity supply to the ocean. Further research is needed to assess the variation and changes resulting from anthropogenic impact and climate change.

Table 2.3. Discharge, basin area, and computed values of silicate alkalinity and pCO_2 for south Asian rivers.

Rivers	Discharge Q (km^3)	Basin Area (km^2)	HCO_3 (Si) (mEq/L)	pCO_2 (mEq/L)
Indus (Ladakh)	38	470,000	0.056	–3.11
Indus (Pakistan)	398	1,165,000	1.036	–0.04
Ganges (India)	393	970,000	0.914	–2.55
Ganges–Brahmaputra–Megna (Bangladesh)	1,200	1,640,000	0.900	0.55
Brahmaputra (India)	654	924,000	0.392	–2.42
Mahanadi	55	88,320	0.820	–2.48
Narmada	47	87,889	0.740	–3.05
Tapti	10	49,136	1.350	–2.95
Godavari	92	313,000	1.280	–2.45
Krishna	33	251,360	1.870	–2.19
Cauvery	12	66,243	1.680	–2.46
West-flowing small rivers	5	24,000	0.070	–2.96

Conclusion

Weathering of silicate rock and its products is the primary source of silicates and a major source of alkalinity in river water and coastal oceans. This process of weathering depends on temperature, precipitation, and, as recently shown, human activities in the watersheds. The alkalinity export is higher in cropland than in forestland, and in most south Asian river basins forestland is diminishing because of urbanization or conversion to cropland. Because of the geological and anthropogenic setting, south Asian river basins are important in the transfer of DSi and alkalinity to rivers and oceans. Therefore, the relative importance of the two mechanisms ultimately affecting atmospheric CO_2—weathering and anthropogenic impact—must be quantified, particularly for rivers in these densely populated regions with large-scale land use changes in historical and modern times.

Acknowledgments

The manuscript was prepared while V. Subramanian was a fellow at the Hanse Institute of Advanced Study, Delmenhorst, Germany.

Literature Cited

Berner, R. A., A. C. Lasaga, and R. M. Garrels. 1983. The carbonate–silicate geochemical cycle and its effect on atmospheric carbon dioxide over the past 100 million years. *American Journal of Science* 283:641–683.

Blum, J. D., C. A. Gazis, A. D. Jacobson, and C. P. Chamberlain. 1998. Carbonate versus silicate weathering in the Raikhot watershed within the high Himalayan crystalline series. *Geology* 26:411–414.

Brady, P. V., and S. A. Carrol. 1994. Direct effects of CO_2 and temperature on silicate weathering: Possible implications for climate control. *Geochimica Cosmochimica Acta* 58:1853–1856.

Datta, D., and V. Subramanian. 1997. Nature of solute load in the rivers of Bengal Basin, Bangladesh. *Journal of Hydrology* 198:196–208.

Dessert, C., B. Dupré, L. M. François, J. Schott, J. Gaillardet, G. Chakrapani, and S. Bajpai. 2001. Erosion of Deccan traps determined by river geochemistry: Impact on the global climate and the $^{87}Sr/^{86}Sr$ ratio of seawater. *Earth and Planetary Science Letters* 188:459–474.

Galy, A., and C. France-Lanord. 1999. Weathering processes in the Ganges–Brahmaputra Basin and the riverine alkalinity budget. *Chemical Geology* 159:31–60.

Gibbs, R. J. 1970. Mechanisms controlling world water chemistry. *Science* 170:1088–1090.

Ittekkot, V., C. Humborg, L. Rahm, and N. Tac An. 2003. Carbon–silicon interactions. Pp. 311–322 in *Interactions of the major biogeochemical cycles: Global change and human impacts*, edited by J. M. Mellilo, C. B. Field, and B. Moldan. SCOPE 61. Washington, DC: Island Press.

Ittekkot, V., V. Subramanian, and S. Annadurai (eds.). 1999. Biogeochemistry of rivers in tropical South and Southeast Asia. *Mitteilungen aus dem Geologisch-Paläontologischen Institut der Universität Hamburg* 82, SCOPE special issue.

Karim, A., and J. Veizer. 2000. Weathering processes in the Indus River Basin: Implications from riverine carbon, sulfur, oxygen, and strontium isotopes. *Chemical Geology* 170:153–177.

Probst, A., A. El Gh' Mari, B. Fritz, P. Stille, and D. Aubert. 1998. Weathering processes in the granitic Strengbach catchment (north-eastern France): Laboratory experiments under acidic conditions and field assessment using Sr isotopes. *Mineralogical Magazine* 62A:1214–1215.

Quade, J., N. English, and P. G. Decelles. 2003. Silicate versus carbonate weathering in the Himalaya: A comparison of the Arun and the Seti River watersheds. *Chemical Geology* 202:275–296.

Ramanathan, A. L., P. Vaithiyanathan, V. Subramanian, and B. K. Das. 1993. Nature and transport of solute load in the Cauvery River Basin, India. *Water Research* 28:1585–1593.

Raymahashay, B. C. 1986. Geochemistry of bicarbonate in river water. *Geological Science India* 27:114–118.

Raymo, M. E., and W. F. Ruddiman. 1992. Tectonic forcing of Late Cenozoic climate. *Nature* 359:117–122.

Raymond, P., and J. Cole. 2003. Increase in the export of alkalinity from North America's largest river. *Science* 302:985.

Sajeev, R., and V. Subramanian. 2003. Land use/land cover changes in Asthamudi wetland region of Kerala: A study using remote sensing and GIS. *Journal of Geological Society of India* 61:573–580.

Sarin, M. M., S. Krishnaswami, K. Dilli, B. L. K. Somayajulu, and W. S. Moore. 1989. Major ion chemistry of the Ganga–Brahmaputra river system: Weathering processes and fluxes to the Bay of Bengal. *Geochimica et Cosmochimica Acta* 53:997–1009.

Stallard, R. F., and J. M. Edmond. 1981. Geochemistry of the Amazon: precipitation chemistry and the marine contribution to the dissolved load at the time of peak discharge. *Journal of Geophysical Research* 86:9844–9858.

Subramanian, V. 2000. *Water: Quantity–quality perspective in South Asia.* Surrey, UK: Kingston International Publishers.

Subramanian, V. 2002. *A text book on environmental sciences.* New Delhi Narosa Publishing House.

West, A. J., M. J. Bickle, R. Collins, and J. Brasington. 2002. Small-catchment perspective on Himalayan weathering fluxes. *Geology* 30:355–358.

West, A. J., A. Galy, and M. Bickle. 2005. Tectonic and climatic controls silicate weathering. *Earth and Planetary Science Letters* 235:211–228.

White, A. F., and A. E. Blum. 1995. Effects of climate on chemical weathering in watersheds. *Geochimica et Cosmochimica Acta* 59:1729–1747.

Wollast, R., and F. T. Mackenzie. 1983. The global cycle of silica. Pp. 39–76 in *Silicon geochemistry and biogeochemistry*, edited by S. E. Aston. London: Academic Press.

3

Silicon in the Terrestrial Biogeosphere

Daniel J. Conley, Michael Sommer, Jean Dominique Meunier, Danuta Kaczorek, and Loredana Saccone

Silicon (Si) is the second most common element of the earth's crust, with a mean content of 28.8 wt% (Wedepohl 1995). Large transfers of Si occur in the geosphere with plate tectonics: Mountains are built and eroded, new sea floor is formed, and sediments are deposited and subsequently subducted and recycled. Diagenesis occurs, with plate tectonics changing the chemical composition of Si-containing minerals (Siever 1979). Si can be melted and reformed with the intense heat and pressures in the crust. All large-scale geological transformations, such as the building of new continents and subduction of crust, involve movement of Si from one geological reservoir to another.

The Si contained in the crust is also broken down by physical, chemical, and biological processes known as weathering. On geological time scales geologists consider that all products of weathering are eventually moved from the continents to the oceans. On shorter time scales, the products of weathering are used in biological processes at the surface of the earth. The products of weathering include both particulate components (e.g., clays) and dissolved silica (DSi).

It is well known that part of the DSi produced in weathering is used in the biogeosphere in aquatic ecosystems by diatoms, sponges, chrysophytes, radiolarians, and silicoflagellates (Conley et al. 1993). What is being increasingly appreciated is that terrestrial plants take up a significant portion of DSi produced during weathering (Sommer et al. 2006). Conley (2002) suggested that the flux of silicic acid transferred from the land to the oceans is primarily leakage out of the internal terrestrial biogeochemical cycle. Recent evidence supports this paradigm, with a large fraction of silicic acid going through plants before its transfer to rivers (Kurtz and Derry 2004; Derry et al. 2005). The paradigm is that silicic acid is produced and recycled on the continents before its eventual transfer to the oceans through rivers.

Relative to that in aquatic ecosystems, the biogeochemical cycling of Si in terrestrial ecosystems is poorly studied. The portion of the biogeochemical Si cycle that is cycled by geochemical reactions has been the subject of numerous investigations for many years in a variety of fields, including geology (Siever 1991; Clarke 2003), agriculture (Hall and Morrison 1906), and soil science (McKeague and Cline 1963). However, the influence of biological processes on the biogeochemical cycle of Si is being increasingly investigated by various fields (Datnoff et al. 2002; Conley 2002).

Clarke (2003) reviewed the occurrence of biogenic silica (BSi) and the potential importance to the biogeosphere, but a quantitative assessment of the actual amount of amorphous silica, whether of biological or inorganic origin, in the biogeosphere is difficult to obtain. Earlier workers have qualitatively identified various biological and chemical forms of amorphous Si in soils (Wilding et al. 1977), and there are some estimates of the percentage composition of phytoliths in soils (see Clarke 2003). Conley (2002) estimated the amount of DSi fixed by plants and deposited as amorphous silica in phytoliths by scaling DSi uptake to global primary production and found that DSi uptake amounted to 60–200 Tmol/yr, making plant uptake a significant component of the global biogeochemical cycle.

Here we would like to summarize the existing knowledge on Si pools in soils and transformations and fluxes into and out of terrestrial biogeosystems. This chapter starts by describing the different inorganic pools of Si in the biogeosphere, building from the classic soil literature. The next section discusses the biogeochemical transformations and fluxes that occur in the terrestrial biogeosphere. A number of studies of Si cycling in catchments and forest ecosystems are reviewed and the general features of the cycles highlighted. The final section of this chapter treats the anthropogenic alteration of the terrestrial Si cycle, focusing on the transformations in soils with agriculture and changing vegetation, including the anthropogenic use of Si as a fertilizer. The conclusion focuses on the link between terrestrial and aquatic ecosystems and gives some future perspectives on research needed on Si in the terrestrial biosphere.

Types of Silicon in the Terrestrial Biogeosphere

Although Si is the second most common element in the earth's crust (Wedepohl 1995), in the pedosphere Si shows a huge span, ranging from values less than 1 wt% Si in histosols to 45 wt% Si in very old podzols of quartzitic sands (Skjemstad et al. 1992). Some Si-enriched horizons in soils are composed almost entirely of Si (SiO_2 more than 95 wt%; Summerfield 1983). The Si content in soils is controlled primarily by two factors: the parent material (inherited Si) and the actual stage of soil development. The latter can be viewed as a function of climate and the duration of soil development, which itself is related to landscape stability. The Si content of soils is also controlled by the duration of soil development and climatic, hydrological, and mineralogical conditions.

Weathering of primary minerals is one of the major soil-forming processess worldwide and the ultimate source of Si in the terrestrial biogeosphere (see Chapter 1, this volume).

Nevertheless, Si dynamics in terrestrial biogeosystems cannot be understood solely as mineral weathering. This is because of the dynamic character of biogeosystems over time, especially in soil formation. Very different pools of Si in both inorganic and biogenic forms develop over time (Sauer et al. in press). These pools have different solubilities in water and reaction rates with soil solutions because of their reactive surface areas and chemical and mineralogical composition. Consequently, fluxes from terrestrial to aquatic ecosystems differ according to the relative proportion of these pools.

Silicon Pools and Transformations in Soils

Soils are the main reactor of terrestrial biogeosystems in which chemical processes interact with biological processes. Vertical and lateral transport processes and temporary or permanent immobilization processes operate at different scales, which lead to very different Si pools in soils (McKeague and Cline 1963). They can be subdivided into inorganic and biogenic pools. At the mineral scale inorganic Si pools consist of three major phases: primary minerals inherited from parent material (see Chapter 2); secondary crystalline phases developed through soil formation, mainly clay minerals (see Chapter 2); and secondary low crystalline (authigenic quartz, opal-CT, chalcedon) to amorphous phases (opal-A), which are also a result of soil formation (Chadwick et al. 1987b; Drees et al. 1989; Monger and Kelly 2002).

DSi, which is liberated from minerals by weathering through hydrolysis or complexing agents, is not necessarily included in clay mineral formation but might also interact with soluble Al to build "short range order minerals" such as imogolite or allophanes (Harsh et al. 2002). These products of soil formation are poorly crystalline aluminosilicates with a high specific surface area and variable charge. Allophanes are likely to occur in acid soils showing a molar Si:Al ratio of 0.5 to 1.0. Si may also be precipitated from soil solution as almost pure, amorphous silica phases at mineral surfaces (Drees et al. 1989). This phenomenon often can be observed in very acid soils, such as podzols of temperate and boreal climates or acid topsoils of luvisols as a result of ultimate (clay) mineral destruction.

In addition to the inorganic pools found in soils, Si is chemically adsorbed at surfaces of soil constituents such as carbonates, aluminum hydroxides, and iron hydroxides (Beckwith and Reeve 1963; Iler 1979; Dietzel 2002). Iron hydroxides play a key role in the interaction between the solid and liquid phases in soils (Beckwith and Reeve 1964). At Fe oxide surfaces, polysilicic acid might be built from DSi (Dietzel 2002), the latter of which is the dominant species in soil solutions with pH less than 8. In redoximorphic soils showing cyclic patterns of repeated waterlogging and drying, strong variations of redox potentials occur, with additional Si released as a result of H^+ production during oxidation of ferrous to ferric compounds (ferrolysis; Brinkman 1970). Although there is a long debate about the significance of this process in soils (reviewed in van Ranst and De Coninck 2002), laboratory results confirmed an enhanced release of DSi even from quartz surfaces after the oxidation of ferrous silicate surface coatings (Morris and Fletcher 1987). Their results are in accordance with greater in situ DSi concentrations of soil solutions in redoximorphic horizons (Sommer 2002; Sommer et al. 2006). There is strong evidence that Si dynamics in soils are linked to redox processes if iron oxides are present.

DSi in soil solutions might also be reprecipitated in soils, which can lead to substantial accumulations of Si (Milnes et al. 1991). In climates with a pronounced dry period, subsoil horizons might be hardened by amorphous silica infillings and subsequent cementation of grains. These processes lead to hardpans or duripans in soils (Chartres 1985; Chadwick et al. 1987a, 1987b; Monger and Kelly 2002). Surplus DSi might stem from weathering of dust inputs (external flux), soil horizons above (internal flux), groundwater (external flux), or a lateral transport from higher parts in the landscape (external flux). Surface encrustation of soils in semiarid climates is also related in part to a redistribution and precipitation of DSi (hardsetting soils; Breuer and Schwertmann 1999). On the landscape scale DSi can be precipitated in areas of groundwater soils (Pfisterer et al. 1996). When desiccation takes place over a longer time, hardened layers develop, the silcretes (Summerfield 1983). These continental Si accumulation layers vary from 1 to 5 m in thickness and show SiO_2 contents more than 95 wt% (Summerfield 1983). Thus, Si in these different forms accumulates and does not leave the continents (Thiry 1997).

BSi pools in soils can be subdivided into phytogenic (including phytoliths), bacterial, and protozoic Si. In general, knowledge of these pools is scarce for almost all soils. For bacterial and protozoic Si we have only qualitative evidence that these pools exist (Clarke 2003). Most information has been gained for phytogenic silica, which is defined here as Si precipitated in roots, stems, branches, leaves, and needles in plants. DSi that has been taken up from soil solution (actively or passively) is precipitated primarily as amorphous silica (opal-A; $SiO_2 \bullet nH_2O$) in cell walls, lumens, and the intercellular voids called phytoliths (Piperno 1988). Infrared spectroscopy analyses show that the structure of phytoliths is characterized by elementary spherical units ranging from 5,000 to 10,000 nm, similar to those forming silica gel and other forms of amorphous Si or BSi (diatoms, sponge spicules). Although the thermodynamic properties of phytoliths are not well known, we can use amorphous silica as an analogue.

The size of Si precipitates in plants ranges over four orders of magnitude, from less than 0.2 mm (Watteau and Villemin 2001) to 200 mm (Piperno 1988), as seen in Figure 3.1.

The relative proportion of clay size phytogenic Si (less than 2 mm) varies from 18 to 65 percent of total phytogenic Si, depending on the plant species (Wilding and Drees 1974; Bartoli and Wilding 1980). The biosilicification mechanisms that take place in plants are still poorly known. Oversaturation is needed to form silicate polymerization and amorphous precipitates. Precipitation of phytogenic Si therefore is favored as water

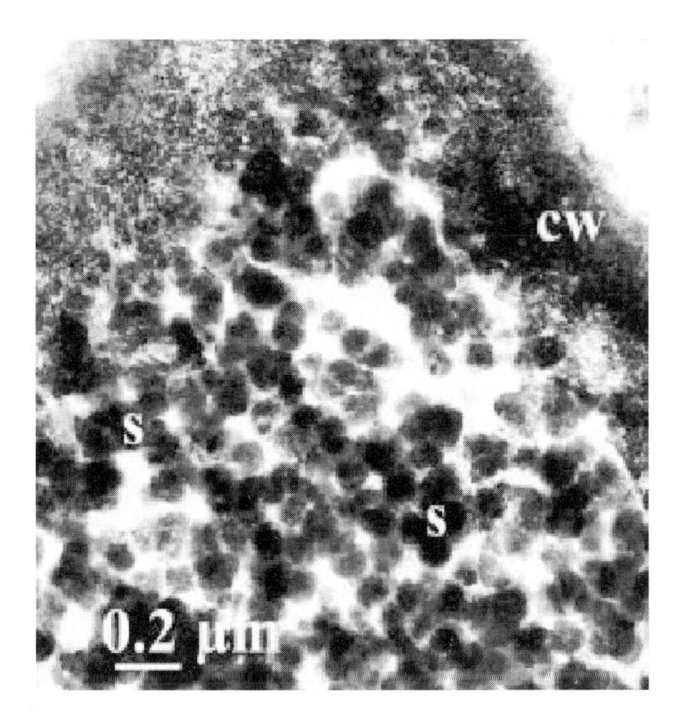

Figure 3.1. Small amorphous silica spheres (denoted by S) in beech leaves (Watteau and Villemin 2001). cw = cell wall.

evaporates from the leaves. We still don't know whether biochemical processes play a role during silicification, although the active uptake of DSi by rice plants through biochemical processes has been shown (Ma et al. 2001a, 2006).

On the basis of the existing limited data, phytoliths can be regarded as the major component of the biogenic-derived amorphous silica pool in soils before diatoms and sponge spicules (Clarke 2003). Phytolith content in soils ranges from 0.01 wt% to 50 wt% (Jones and Beavers 1964; Alexandre et al. 1997; Clarke 2003). However, most soil horizons show phytolith contents ranging from 0.1 wt% to 3 wt% of soil's fine fraction (less than 2 mm). Total phytogenic Si in soils must be even higher because phytolith analysis normally is limited to silt-sized particles and discards the clay fraction of soil horizons. The depth profile of phytoliths in soils is variable (Clarke 2003). Although the highest contents in undisturbed soils should be in surface horizons (Gol'eva 1999), vertical translocation into deeper soil, turbation processes (e.g., mixing by burrowing animals), and erosion and deposition events modify the normative depth profile. Single sedimentary soil layers (in colluvial soils) may show up to 90 wt% SiO_2, derived almost exclusively from phytoliths (Meunier et al. 1999). We have some evidence that phytogenic amorphous silica content in soils is related to soil drainage. Poorly drained, redoximorphic soils contain more phytoliths than non-redoximorphic soils (Jones and Beavers 1964; Clarke 2003). However, no systematic survey of the occurrence of phytogenic amorphous silica as a function of vegetation and soil type exists.

If phytolith productivity is higher than the dissolution rate, a BSi sink can form in soil. Meunier et al. (1999) studied a 15-cm-thick, phytolith-rich soil formation located on the west side of the Piton des Neiges (La Réunion Island, Indian Ocean) between 1,600 and 1,800 m in elevation. Phytoliths extracted from soils and from endemic plants growing in the area show that the phytolith-rich formation originated from the turnover of parent rock Si by the leaves and stems of *Nastus borbonicus*, an endemic bamboo species. The Si analyses show that the bamboos are the plants with the highest Si content in the area (from 41 to 58 mg/g of dry organic matter). Charcoal ^{14}C dating demonstrates that the phytoliths were deposited between 335 ± 90 and 3,820 ± 85 yr BP. An estimation of the biogenically derived amorphous silica deposited in the soil by bamboos gives values of about 453–649 kg/ha/yr. These values are higher than the values presented before because they result from the combination of Si accumulator plants, parent rocks enriched in easily weatherable Si, and a warm and wet climate.

Studies on the chemical properties of phytogenic silica and its solubility under different environmental conditions are limited to phytoliths (Jones and Beavers 1963; Wilding et al. 1979; Bartoli and Wilding 1980). Phytoliths contain accessory elements (e.g., Al, Fe, Ti, C) and can be up to 6 wt% of the composition, most of which is Al. Experimental studies on phytolith dissolution in distilled water (Bartoli and Wilding 1980) showed equilibrium DSi concentrations of 100–360 μmol Si/L, which was between those of quartz (36 μmol Si/L) and synthesized pure silica gels (2,000 μmol Si/L). Clear differences between plant species were observed. Phytoliths from pine needles showed lower DSi concentrations (100 μmol Si/L) than those from beech leaves, with higher DSi concentrations in solution (300 μmol Si/L). This may correspond to a higher Al content in the phytoliths of pine needles compared with beech leaves (Hodson and Sangster 1999). Experiments of increased Al substitution in silica gels leading to a reduced solubility in water support this causal relationship. Finally, Bartoli (1985) concluded that a higher Al substitution at phytoliths' surfaces (Si:Al ratio of 1.5–8) is responsible for the reduction in their solubility. Although systematic investigations must be carried out, one can assume that phytogenic Si in soil, especially that found in the clay size fraction (less than 2 mm) with its high surface:volume ratio, plays an important role in the biogeochemical Si cycle because amorphous silica is the most soluble form of Si in the system.

The dissolution and recycling of phytolith Si in soils, as with the dissolution of diatoms in aquatic ecosystems, influence its use for reconstruction of paleoenvironments. Here the morphology of phytoliths is regarded as a stable property that can be related to taxonomic units (Piperno 1988; Gol'eva 1999). Despite the complex processes that may affect the phytolith particle incorporated in the soil surface (translocation, dissolution, denudation, and colluviation) Alexandre et al. (1999) have shown that soil phytoliths may be good tracers of vegetation changes. The study of phytolith assemblages of soils has also been used to trace Holocene grassland dynamics and forest–grassland shifts (Piperno 1988; Delhon et al. 2003). Recent studies have used both oxygen (Webb and Longstaffe 2002) and carbon isotopic composition of phytoliths (Kelly et al. 1998a) to deduce paleoenvironmental conditions.

Silicon in the Biosphere

In the global biosphere, Si is not the most abundant element after oxygen (as it is in the lithosphere), with an average concentration of 0.03 wt%, ranking after H, O, C, N, S, Ca, and K (Exley 1998). The Si content of terrestrial plants is higher than the average value for soils, ranging from 0.1 wt% to 10 wt% (dry weight basis; Raven 1983; Epstein 1999), although some plants have higher Si contents. No Si–O–C or Si–C bonds have been observed so far (Exley 1998). Interactions between Si and other elements such as Al, Mn, P, or N are common. A decrease of Al toxicity in ecosystems subjected to acid rain is one of the important possible roles of Si in the biogeosphere.

The elements taken up by plants are either stored in the biomass or returned to the soil through litter fall and subsequent decomposition. Since the work of Lovering and Engel (Lovering 1967; Lovering and Engel 1967), it has been well established, although often forgotten, that plants can play a major role in the rate of silicate weathering and soil formation. Lovering and Engel (1967) remarked, "A forest of silica-accumulator plants averaging 2.5 percent of Si and 16 tons dry weight new growth per year would extract about 2000 tons of Si per acre in 5000 years—equivalent to the Si in 1 acre foot of basalt." Experimental and in situ studies have confirmed the role of plants in weathering processes (Kelly et al. 1998b, Hinsinger et al. 2001) and its link to the carbon cycle (Andrews and Schlesinger 2001).

Biogeochemical Transformations

With the exception of areas showing substantial dust inputs (semiarid to arid climates), atmospheric Si inputs into terrestrial biogeosystems have been considered negligible (Sommer 2002). Because most fluxes of Si in terrestrial biogeosystems are mediated through water, solubilities of the different Si compounds are important. In general they are a function of temperature, particle size, chemical composition (accessory elements), and the presence of disrupted surface layers (Drees et al. 1989). Solubility of amorphous silica and crystalline Si is essentially constant between pH 2.5 and 8.5. Quartz has a solubility of 36 to 250 μmol/L in water (Iler 1979; Bartoli and Wilding 1980) depending on particle size and temperature. The lower value is more realistic for quartz in soils because of surface coatings (Fe hydroxides, organic matter) of quartz in soils. Somewhat higher solubilities can be observed for biogenic amorphous silica (20–360 μmol/L) and cristobalite, whereas (synthetic) amorphous silica shows much higher solubilities (1,800–2,100 μmol/L; Drees et al. 1989). Si concentrations in soil solutions range from 0.4 to 2,000 μmol/L (Gerard et al. 2002; Sommer 2002) depending on the parent material, stage of soil development, soil horizon, and temperature (seasonal effects). Gerard et al. (2002) showed capillary water to differ in DSi concentration and dynamics from free percolating soil water.

Wherever there is a positive water balance, Si leaches from soils, a soil process known as desilication. Contrary to former textbook knowledge, this process is not limited to tropical areas. Recent desilication rates from soils in humid climates range from 20 to 370 μmol Si/m/yr (Cornu et al. 1998; Sommer 2002), although still higher values probably result from parent material effects. Recent rates and absolute losses over total time of soil development vary as a function of parent material, hydrological conditions, and stage of soil development (White and Blum 1995). With increasing age of the land surface, more Si is leached from soils. Consequently, highest recent desilication is to be expected from young soils on ultrabasic rocks in tropical humid climates.

Not all of the DSi in soil solutions is leached and enters the aquatic ecosystem. A part of it is recycled through vegetation. The magnitudes of the fluxes of Si in terrestrial ecosystems is still not well established because published studies on biomass and litter fall including Si data are scarce. The annual productivity of the main natural ecosystems is intimately linked to climate. The most productive ecosystem is represented by tropical rain forests with an average primary productivity of 20 t/ha/yr of dry matter. In the Amazon rain forest, Lucas et al. (1993) showed that each year, 41 kg/ha/yr of plant-derived Si is deposited into soil, whereas in the experimental ecosystem of Calhoun (South Carolina, USA) the DSi flux between plant and soil is 12 kg/ha/yr (Markewitz and Richter 1998). The cultivation of Si-rich plants, such as rice and sugarcane, can lead to even larger values of Si accumulation in plants. More than 100 kg/ha/yr of Si can be recycled each year by sugarcane (Berthelsen et al. 2001). In Australia, 30 years of sugarcane cultivation have led to a decrease of the Si easily available in soil (using 0.01 M $CaCl_2$) to about half of the original content of Si (respectively 5.3 and 13.1 mg/kg) (Berthelsen et al. 2001). In order to balance the loss of Si, the use of Si fertilizers has been common since the 1950s (Korndofer and Lepsch 2001). Therefore, these studies demonstrate that plants are a major component of silicon mobility in the soil. Plants can also affect the flux of DSi to the rivers (Fulweiler and Nixon 2005). At the watershed scale the flux of Si exported from areas covered by vegetation can be two to eight times higher than the flux of Si from bare areas (Moulton et al. 2000). Two mechanisms can explain this difference: Soils covered by plants are the loci of CO_2, and various acids are released, which favors the chemical weathering of silicates; and the amorphous silicate in plant litter fall is more easily weathered than the primary mineral silicates in soils (Lucas 2001).

A biogeochemical cycle approach at the ecosystem scale has been used in a limited number of studies to determine the biogeochemical Si cycle in the biogeosphere, especially regarding the importance of biogenically derived amorphous silica (phytolith) dissolution and dissolution of primary soil minerals (Bartoli 1983; Alexandre et al. 1999). With this approach, we assume that the ecosystem is at a steady state and calculate the flux of DSi (F_m) from the soil minerals as

$$F_m = (F_b + F_a) - (F_p + F_s),$$

where

F_b = flux of DSi from the BSi particles (phytoliths),
F_a = Si atmospheric input,
F_p = flux of Si taken up by plants, and
F_s = flux of Si exported from soil solution to groundwater and rivers.

Bartoli (1983) compared two forested ecosystems developed above Triassic sandstones located at 600 m above sea level in a temperate climate: a cambisol covered by a deciduous forest and a podzol covered by conifers. The model shows that the deciduous forest is characterized by a low F_m (2.5 kg/ha/yr) and a high F_s (22 ± 7 kg/ha/yr). Conversely, the conifer ecosystem is characterized by a high F_m (29.5 ± 13 kg/ha/yr) and a low F_s (4 ± 2 kg/ha/yr). A similar approach has been used by Alexandre et al. (1997) for a rain forest developed above a latosol in Congo: About 74 percent of the DSi in soil solution originates from the dissolution of phytoliths (F_b = 54 ± 5 to 70 ± 7 kg/ha/yr). The authors also show that 5 kg/ha/yr of phytoliths is stored in the soil. This result shows that all the phytolith particles do not dissolve at the same rate.

The biogeochemical Si cycle in forested ecosystems, as presented by Bartoli (1983), Lucas et al. (1993), and Meunier et al. (1999), is summarized in Conley (2002). To demonstrate the utility of this approach in analyzing the pools and fluxes of Si, we have used the data from Markewitz and Richter (1998) to present the biogeochemical cycle through a 40-year-old stand of loblolly pine planted at the Calhoun Experimental Forest, South Carolina over depleted agricultural soils used for cotton production for 150 years (Figure 3.2).

Features of the terrestrial biogeochemical Si cycle are observed in other studies of bio geochemical cycles in both terrestrial and aquatic ecosystems (Tréguer et al. 1995). For example, the soil pool of amorphous silica can be 500–1,000 times larger than that found in plant biomass, with the majority of Si needed by plants supplied from internal recycling in the forest ecosystem, especially the forest floor, and not from the weathering of new minerals (Bartoli 1983). Comparison of these different studies also demonstrates that the flux of DSi out of terrestrial ecosystems varies depending on the type of plant community. Further quantification of terrestrial ecosystems is needed, especially studies from grassland ecosystems, which are known to accumulate vast quantities of amorphous silica (Wilding et al. 1977).

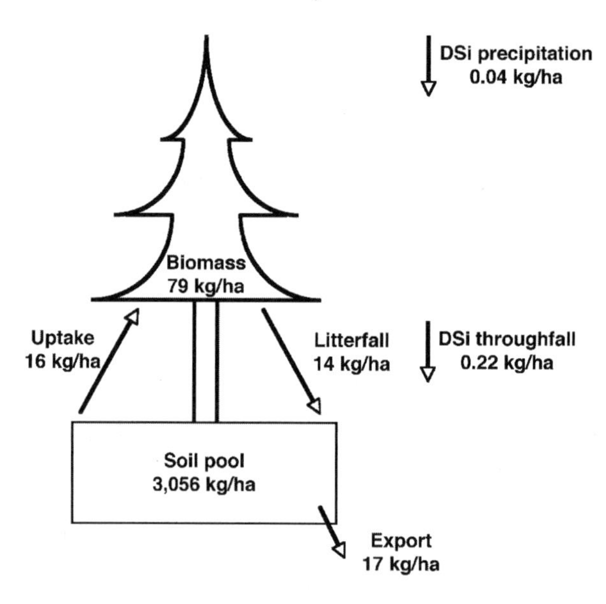

Figure 3.2. The biogeochemical Si cycle in a loblolly pine forest (from data reported in Markewitz and Richter 1998).

Anthropogenic Alteration

Landscape Effects

Human activities have altered the biogeochemical Si cycle in both terrestrial and aquatic systems. The anthropogenic alteration of the aquatic biogeochemical Si cycle has been the subject of a number of reviews (e.g., Officer and Ryther 1980; Conley et al. 1993) demonstrating the role of nutrient enrichment in increasing the sedimentation and storage of BSi in sediments, the role of dams in reducing the transport of DSi, and the role of benthic organisms in retaining particulate BSi in estuaries. Fundamental changes have also occurred in the landscape with anthropogenic activities. Throughout history humans have cut down trees, cleared land through biomass burning, and mechanically altered soil for agriculture. These activities have changed the pools and pathways of the biogeochemical Si cycle in terrestrial ecosystems. In this section we focus primarily on alteration of the biogeochemical Si cycle in the biogeosphere.

Increasing losses of Si and other elements from terrestrial organic matter reservoirs (phytomass and humus) have occurred mainly through deforestation and consequently increased humus demineralization, erosion, and transport to the coastal margins by rivers and runoff. Examination of runoff patterns after deforestation reveals that increases in DSi also follow forest clearance (Likens et al. 1970). With our current knowledge of the biogeochemical Si cycle in the biogeosphere, it is clear that much of this increase can be attributed to the dissolution of phytoliths in dead plant biomass on the soil surface, in the upper soil horizons, and the transport of DSi after the dissolution of the amorphous silica contained in phytoliths and the particulate transport of phytolith amorphous silica after forest clearance.

When forest systems are converted to pastures, the change from forest to pasture vegetation alters the biological, chemical, and mineralogical properties of the soil and changes the compartmentalization of Si (Kelly et al. 1998b). In a study of soils in Hawaii, forest soils appeared to have two Si pools, reflecting a greater proportion of plant-available Si in soluble silicate stores as an oxalate-extractable form of Si under the forest and a greater proportion of phytolith-stabilized Si under the pasture (Kelly et al. 1998b). This biological transformation can have a dramatic effect on the weathering of soils, the leakage of DSi to aquatic ecosystems, and the overall biogeochemical Si cycle in these systems (Kelly et al. 1998b). These results imply that fundamental changes have occurred in the terrestrial cycle with anthropogenic alteration of the biogeosphere. Opportunities for studying the subtleties of human influence through the process of forest clearance and development of the landscape with agriculture (plowing) are now available given our new understanding of this paradigm.

Si as Fertilizer

In the early nineteenth century Si was regarded as an essential nutrient for many plants because of its regular occurrence in the ashes of vegetation, sometimes in large amounts (Webb 1991). This view came to an end when other studies showed that plants, usually containing large proportions of Si, could achieve the same complete maturity in its absence. Even though Si is not found to be involved in plant metabolism, its availability in the soil is thought to have a beneficial effect on crop plants such as rice, wheat, barley, and cucumber (Ma et al. 2001b, Richmond and Sussman 2003) as it makes plants resistant to diseases, diminishes metal toxicity, and reduces lodging of cereal crops (Figure 3.3). As a result, it is often used as fertilizer for crops such as rice and sugarcane that have large demands for Si (Matichencov and Bocharnikova 2001).

Observations involving the impact of experimental additions of sodium silicate have been made since the mid-nineteenth century at Rothamsted Experimental Station, England, highlighting the importance of the function of Si on plants (Webb 1991). The results indicate that the function of Si is to allow the plant to draw more efficiently on the reserve of phosphoric acid in the soil (Hall and Morrison 1906). In addition, the role of Si in the nutrition of many plants was described by Brenchley et al. (1927), who focused on the enhancement of growth and interactions with P in soils.

The use of Si as a fertilizer, especially on crops with a large need for Si, is increasing (Datnoff et al. 2001). In a recent study undertaken in the greenhouse industry in the Netherlands, it was found that Si contents in plant tissue of crops grown in soilless

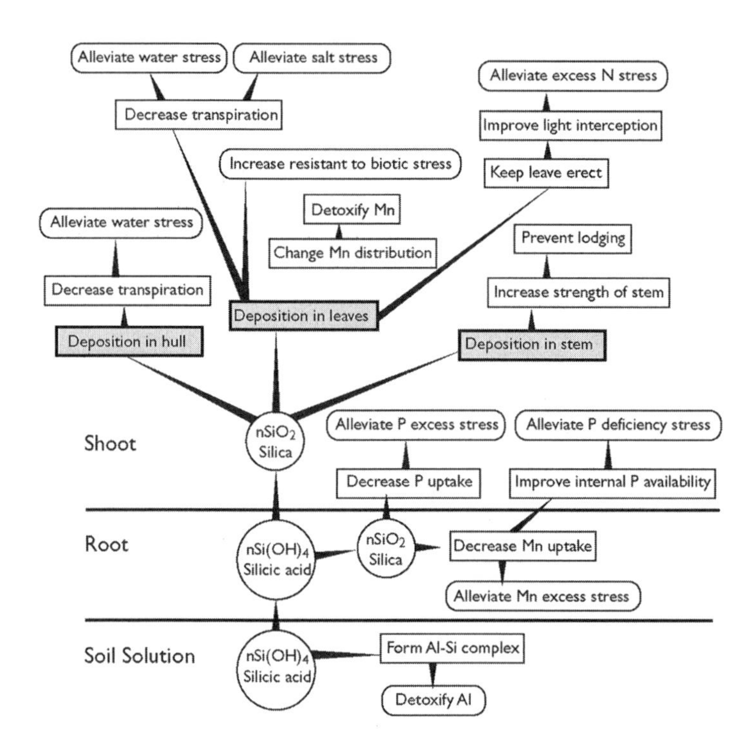

Figure 3.3. The benefits of Si for plants under various stresses (Ma et al. 2001b).

culture were much lower than in crops grown in soil (Voogt and Sonneveld 2001). Results from this study showed that cucumber, rose, and zucchini could benefit from greater DSi concentrations in the root environment. In lettuce, it was found that Si uptake affects Mn distribution and alleviates toxicity in the plant. Beneficial effects are also observable in crops such as tomatoes that do not accumulate Si (Ma et al. 2001b). Additional studies concerning the use and impact of Si-based fertilizers are needed to determine their potential widespread use (Datnoff et al. 2002).

Conclusion

Our understanding of the biogeochemical Si cycle in the biogeosphere is in its infancy. Only recently has it been recognized that the amorphous pool of Si in phytoliths in vegetation and soils has an impact on the global biogeochemical cycle of Si (Conley 2002). More studies are needed to elucidate the pathways, pools, and mechanisms of Si transfer in the biogeosphere (Sommer et al. 2006). Grasslands, which are one of the major biomes trapping phytolith amorphous silica in soils, have never been studied regarding their biogeochemical Si cycle. We need a more complete understanding of the role of different plant communities in the biogeochemical Si cycle.

Gaining a better understanding of the terrestrial biogeochemical Si cycle would be an advance for science in general, but one of the biggest challenges will be to develop the link between the terrestrial biogeochemical Si cycle and the aquatic biogeochemical Si cycle. How is Si transferred from terrestrial ecosystems to the aquatic realm? How has the flux of DSi from the biogeosphere changed with anthropogenic alteration? How has the flux of Si from the biogeosphere changed with global climate change? We have always considered DSi fluxes to be stable throughout geologic time depending on the intensity of weathering and orogeny (mountain building). We now know that this flux changes significantly through time and must be quantified.

Literature Cited

Alexandre, A., J. D. Meunier, F. Colin, and J. M. Koud. 1997. Plant impact on the biogeochemical cycle of silicon and related weathering processes. *Geochimica et Cosmochimica Acta* 61:677–682.

Alexandre, A., J. D. Meunier, A. Mariotti, and F. Soubies. 1999. Late Holocene paleoenvironmental record from a latosol at Salitre (southern central Brazil): Phytolith and carbon isotope evidence. *Quaternary Research* 51:187–194.

Andrews, J. A., and W. H. Schlesinger. 2001. Soil CO_2 dynamics, acidification, and chemical weathering in a temperate forest with experimental CO_2 enrichment. *Global Biogeochemical Cycles* 15:149–162.

Bartoli, F. 1983. The biogeochemical cycle of silicon in two temperate forest ecosystems. In *Environmental biogeochemistry*, edited by R. Hallberg. *Ecological Bulletin* 35:469–476.

Bartoli, F. 1985. Crystallochemistry and surface properties of biogenic opal. *Journal of Soil Science* 36:335–350.

Bartoli, F., and L. P. Wilding. 1980. Dissolution of biogenic opal as a function of its physical and chemical properties. *Soil Science Society of America Journal* 44:873–878.

Beckwith, R. S., and R. Reeve. 1963. Studies on soluble silica in soils. I. The sorption of silicic acid by soils and minerals. *Australian Journal of Soil Research* 1:157–168.

Beckwith, R. S., and R. Reeve. 1964. Studies on soluble silica in soils. II. The release of monosilicic acid from soils. *Australian Journal of Soil Research* 2:33–45.

Berthelsen, S., A. D. Noble, and A. L. Garside. 2001. Silicon research down under: Past, present and future. Pp. 241–256 in *Silicon in agriculture*, edited by L. E. Datnoff, G. H. Snyder, and G. H. Korndörfer. Studies in Plant Science 8. Amsterdam: Elsevier.

Brenchley, W. E., E. J. Maskell, and K. Warington. 1927. The inter-relation between silicon and other elements in plant nutrition. *Annals of Applied Biology* 14:45–82.

Breuer, J., and U. Schwertmann. 1999. Changes to hardsetting properties of soil by addition of metal hydroxides. *European Journal of Soil Science* 50:657–664.

Brinkman, R. 1970. Ferrolysis, a hydromorphic soil forming process. *Geoderma* 3:199–206.

Chadwick, O. A., D. M. Hendricks, and W. D. Nettleton. 1987a. Silica in duric soils: I. A depositional model. *Soil Science Society of America Journal* 51:975–982.

Chadwick, O. A., D. M. Hendricks, and W. D. Nettleton. 1987b. Silica in duric soils: II. Mineralogy. *Soil Science Society of America Journal* 51:982–985.

Chartres, C. J. 1985. A preliminary investigation of hardpan horizons in north-west New South Wales. *Australian Journal of Soil Research* 23:325–337.

Clarke, J. 2003. The occurrence and significance of biogenic opal in the regolith. *Earth Science Reviews* 60:175–194.

Conley, D. J. 2002. Terrestrial ecosystems and the global biogeochemical silica cycle. *Global Biogeochemical Cycles* 16:1121.

Conley, D. J., C. L. Schelske, and E. F. Stoermer. 1993. Modification of silica

biogeochemistry with eutrophication in aquatic systems. *Marine Ecology Progress Series* 101:179–192.
Cornu, S., Y. Lucas, J. P. Ambrosi, and T. Desjardins. 1998. Transfer of dissolved Al, Fe and Si in two Amazonian forest environments in Brazil. *European Journal of Soil Science* 49:377–384.
Datnoff, L. E., G. H. Snyder, and G. H. Korndörfer (eds.). 2001. *Silicon in agriculture.* Studies in Plant Science 8. Amsterdam: Elsevier.
Delhon, C., A. Alexandre, J. F. Berger, S. Thiébault, J. L. Brochier, and J. D. Meunier. 2003. Phytolith assemblages as a promising tool for reconstructing Mediterranean Holocene vegetation. *Quaternary Research* 59:48–60.
Derry, L. A., A. C. Kurtz, K. Ziegler, and O. A. Chadwick. 2005. Biological control of terrestrial silica cycling and export fluxes to watersheds. *Nature* 433:728–731.
Dietzel, M. 2002. Interaction of polysilicic and monosilicic acid with mineral surfaces. Pp. 207–235 in *Water–rock interaction,* edited by I. Stober and K. Bucher. Dordrecht, The Netherlands: Kluwer.
Drees, L. R., L. P. Wilding, N. E. Smeck, and A. L. Sankayi. 1989. Silica in soils: Quartz and disordered silica polymorphs. Pp. 913–974 in *Minerals in soil environments,* edited by J. B. Dixon and S. B. Weed. Book Series 1. Madison, WI: Soil Science Society of America.
Epstein, E. 1999. Silicon. *Annual Review of Plant Physiology and Plant Molecular Biology* 50:641–664.
Exley, C. 1998. Silicon in life: A bioinorganic solution to bioorganic essentiality. *Journal of Inorganic Biochemistry* 69:139–144.
Fulweiler, R. W., and S. W. Nixon. 2005. Terrestrial vegetation and the seasonal cycle of dissolved silica in a southern New England coastal river. *Biogeochemistry* 74:115–130.
Gerard, F., M. Francois, and J. Ranger. 2002. Processes controlling silica concentration in leaching and capillary soil solutions of an acidic brown forest soil (Rhone, France). *Geoderma* 107:197–226.
Gol'eva, A. A. 1999. The application of phytolith analysis for solving problems of soil genesis and evolution. *Eurasian Soil Science* 32:884–891.
Hall, A. D., and C. G. T. Morrison. 1906. On the function of silica in the nutrition of cereals. Part I. *Proceedings of the Royal Society B* 77:455–477.
Harsh, J. B., J. Chorover, and E. Nizeyimana. 2002. Allophane and imogolite. Pp. 291–322 in *Soil mineralogy with environmental applications,* edited by J. B. Dixon and D. G. Schulze. Book Series 7. Madison, WI: Soil Science Society of America.
Hinsinger, P., O. N. F. Barros, M. F. Benedetti, Y. Noack, and G. Callot. 2001. Plant-induced weathering of a basaltic rock: Experimental evidence. *Geochimica et Cosmochimica Acta* 65:137–152.
Hodson, M. J., and A. G. Sangster. 1999. Aluminium/silicon interactions in conifers. *Journal of Inorganic Biogeochemistry* 76:89–98.
Iler, R. K. 1979. *The chemistry of silica.* New York: Wiley.
Jones, R. L., and A. H. Beavers. 1964. Aspects of catenary and depth distribution of opal phytolites in Illinois soils. *Soil Science Society of America Proceedings* 28:413–416.
Kelly, E. F., S. W. Blecker, C. M. Yonker, C. G. Olson, E. E. Wohl, and L. C. Todd. 1998a. Stable isotope composition of organic matter and phytoliths as paleoenvironmental indicators. *Geoderma* 82:59–81.
Kelly, E. F., O. A. Chadwick, and T. E. Hilinski. 1998b. The effect of plants on mineral weathering. *Biogeochemistry* 42:21–53.
Korndörfer, G. H., and I. Lepsch. 2001. Effect of silicon on plant growth and crop yield.

Pp. 133–147 in *Silicon in agriculture*, edited by L. E. Datnoff, G. H. Snyder, and G. H. Korndörfer. Studies in Plant Science 8. Amsterdam: Elsevier.

Kurtz, A. C., and L. A. Derry. 2004. Tracing silicate weathering and terrestrial silica cycling with Ge/Si ratios. Pp. 833–836 in *Proceedings of the 11th International Symposium on Water Rock Interaction*, edited by R. B. Wanty and R. R. Seal. Rotterdam: A.A. Balkema Publishers.

Likens, G. E., F. H. Bormann, N. M. Johnson, D. W. Fisher, and R. S. Pierce. 1970. Effects of forest cutting and herbicide treatment on nutrient budgets in the Hubbard Brook watershed-ecosystem. *Ecological Monographs* 40:23–47.

Lovering, T. S. 1967. Significance of accumulator plants in rock weathering. *Bulletin of the Geological Society of America* 70:781–800.

Lovering, T. S., and C. Engel. 1967. Translocation of silica and other elements from rock into *Equisetum* and three grasses. *U.S. Geological Survey Professional Paper* 594-B:1–16.

Lucas, Y. 2001. The role of plants in controlling rates and products of weathering: Importance of biological pumping. *Annual Review Earth Planet Science* 29:135–163.

Lucas, Y., F. J. Luizao, A. Chauvel, J. Rouiller, and D. Nahon. 1993. The relation between biological activity of the rain forest and mineral composition of soils. *Science* 260:521–523.

Ma, J. F., S. Goto, K. Tamai, and M. Ichii. 2001a. Role of root hairs and lateral roots in silicon uptake by rice. *Plant Physiology* 127:1773–1780.

Ma, J. F., Y. Miyake, and E. Takahashi. 2001b. Silicon as a beneficial element for crop plants. Pp. 17–39 in *Silicon in agriculture*, edited by L. E. Datnoff, G. H. Snyder, and G. H. Korndörfer. Studies in Plant Science 8. Amsterdam: Elsevier.

Ma, J. F., K. Tamai, N. Yamaji, N. Mitani, S. Konishi, M. Katsuhara, M. Ishigure, Y. Murata, and M. Yano. 2006. A silicon transporter in rice. *Nature* 440:688–691.

Markewitz, D., and D. Richter. 1998. The *bio* in aluminum and silicon geochemistry. *Biogeochemistry* 42:235–252.

Matichencov, V. V., and E. A. Bocharnikova. 2001. The relationship between silicon and soil physical and chemical properties. Pp. 209–219 in *Silicon in agriculture*, edited by L. E. Datnoff, G. H. Snyder, and G. H. Korndörfer. Studies in Plant Science 8. Amsterdam: Elsevier.

McKeague, J. A., and M. G. Cline. 1963. Silica in soils. *Advances in Agronomy* 15:339–396.

Meunier, J. D., F. Colin, and C. Alarcon. 1999. Biogenic silica storage in soils. *Geology* 27:835–838.

Milnes, A. R., M. Thiry, and M. J. Wright. 1991. Silica accumulations in saprolites and soils in South Australia. Pp. 121–149 in *Occurrence, characteristics, and genesis of carbonate, gypsum, and silica accumulation in soils*, edited by W. D. Nettleton. Special Publication 26. Madison, WI: Soil Science Society of America.

Monger, H. C., and E. G. Kelly. 2002. Silica minerals. Pp. 611–636 in *Soil mineralogy with environmental applications*, edited by J. B. Dixon and D. G. Schulze. Book Series 7. Madison, WI: Soil Science Society of America.

Morris, R. C., and A. B. Fletcher. 1987. Increased solubility of quartz following ferrous–ferric iron reactions. *Nature* 330:558–561.

Moulton, K. L., J. West, and R. A. Berner. 2000. Solute flux and mineral mass balance approaches to the quantification of plant effects on silicate weathering. *American Journal of Science* 300:539–570.

Officer, C. B., and J. H. Ryther. 1980. The possible importance of silicon in marine eutrophication. *Marine Ecology Progress Series* 3:83–91.

Pfisterer, U., H.-P. Blume, and M. Kanig. 1996. Genesis and dynamics of an oxic dystrochrept and a typic haploperox from ultrabasic rock in the tropical rain forest climate of south-east Brazil. *Zeitschrift für Pflanzenernährung und Bodenkunde* 159:41–50.

Piperno, D. R. 1988. *Phytoliths: An archaeological and geological perspective.* London: Academic Press.
Raven, J. A. 1983. The transport and function of silicon in plants. *Biological Reviews* 58:179–207.
Richmond, K. E., and M. Sussman. 2003. Got silicon? The non-essential beneficial plant nutrient. *Current Opinion in Plant Biology* 6:268–272.
Sauer, D., L. Saccone, D. J. Conley, L. Hermann, and M. Sommer. In press. Review of methodologies for extracting plant-available and amorphous Si from soils and aquatic sediments. *Biogeochemistry.*
Siever, R. 1979. Plate-tectonic controls on diagenesis. *Journal of Geology* 87:127–155.
Siever, R. 1991. Silica in the oceans: Biological–geochemical interplay. Pp. 287–295 in *Scientists on Gaia.* Cambridge, MA: MIT Press.
Skjemstad, J. O., R. W. Fitzpatrick, B. A. Zarcinas, and C. H. Thompson. 1992. Genesis of podzols on coastal dunes in southern Queensland: II. Geochemistry and forms of elements as deduced from various soil extraction procedures. *Australian Journal of Soil Science* 30:615–644.
Sommer, M. 2002. *Biogeochemie bewaldeter Einzugsgebiete und ihr pedogenetischer Kontext.* Stuttgart-Hohenheim: Hohenheimer Bodenkundliche Hefte 65.
Sommer, M., D. Kaczorek, Y. Kuzyakov, and J. Breuer. 2006. Silicon pools and fluxes in soils and landscapes: A review. *Journal of Plant Nutrition and Soil Science* 169:294–314.
Summerfield, M. A. 1983. Silcrete. Pp. 59–93 in *Chemical sediments and geomorphology,* edited by A. S. Goudie and K. Pye. London: Academic Press.
Thiry, M. 1997. Continental silicifications: A review. Pp. 191–221 in *Soils and sediments,* edited by H. Paquet and N. Clauer. Berlin: Springer.
Tréguer, P., D. M. Nelson, A. J. van Bennekom, D. J. DeMaster, A. Leynaert, and B. Quéguiner. 1995. The silica balance in the world ocean: A reestimate. *Science* 268:375–379.
van Ranst, E., and F. De Coninck. 2002. Evaluation of ferrolysis in soil formation. *European Journal of Soil Science* 53:513–519.
Voogt, W., and C. Sonneveld. 2001. Silicon in agricultural crops in soilless culture. Pp. 115–131 in *Silicon in agriculture,* edited by L. E. Datnoff, G. H. Snyder, and G. H. Korndörfer. Studies in Plant Science 8. Amsterdam: Elsevier.
Watteau, F., and G. Villemin. 2001. Ultrastructural study of the biogeochemical cycle of silicon in the soil and litter of a temperate forest. *European Journal of Soil Science* 52:385–396.
Webb, E. A., and F. J. Longstaffe. 2002. Climatic influence on the oxygen isotopic composition of biogenic silica in prairie grass. *Geochimica et Cosmochimica Acta* 66:1891–1904.
Webb, M. 1991. *Rothamsted: The classical experiments.* Watton, Norfolk, UK: Rapid Printing.
Wedepohl, K. H. 1995. The composition of the continental crust. *Geochimica et Cosmochimica Acta* 59:1217–1232.
White, A. F., and A. E. Blum. 1995. Effects of climate on chemical weathering in watersheds. *Geochimica et Cosmochimica Acta* 59:1729–1747.
Wilding, L. P., and L. R. Drees. 1974. Contributions of forest opal and associated crystalline phases to fine silt and clay fractions of soils. *Clays and Clay Minerals* 22:295–306.
Wilding, L. P., C. T. Hallmark, and N. E. Smeck. 1979. Dissolution and stability of biogenic opal. *Soil Science Society of America Journal* 43:800–802.
Wilding, L. P., N. E. Smeck, and L. R. Drees. 1977. Silica in soils: Quartz, cristobalite, tridymite, and opal. Pp. 471–552 in *Minerals in soil environments.* Madison, WI: Soil Science Society of America.

4

Factors Controlling Dissolved Silica in Tropical Rivers

Tim C. Jennerjahn, Bastiaan A. Knoppers, Weber F. L. de Souza, Gregg J. Brunskill, E. Ivan. L. Silva, and Seno Adi

The role of dissolved silica (DSi) as an essential nutrient has gained attention in the discussion on nutrient load and composition of rivers and receiving coastal water bodies, which previously focused largely on N and P (Kristiansen and Hoell 2002; Turner et al. 2003). Riverine DSi, generally controlled by natural processes, contributes about 80 percent of the annual input into the ocean (Tréguer et al. 1995). However, in recent decades human activities in river catchments have gained more importance. Although damming generally leads to the retention of sediments and nutrients, downstream human activities lead to additional inputs of N and P but hardly of DSi, which comes mainly from natural sources. Zeolites replacing polyphosphates in detergents are an additional anthropogenic silicon source but make up only a small portion of the total riverine DSi (Chapter 10, this volume). Recent studies have shown that it is not merely the load of N and P but also the composition of the nutrient mix including DSi that affect food webs and elemental cycles of the coastal ocean (e.g., Conley et al. 1993; Humborg et al. 1997; Turner et al. 2003). Knowledge of these processes comes mostly from temperate regions. Tropical coastal regions may face similarly high inputs of N and P through human activities. For example, the average annual fertilizer use of developing countries (100 kg/ha of cropland) exceeds that of developed countries (81 kg/ha of cropland). Among continents, Asia has the highest rate of fertilizer application, followed by North America, Europe, and South America (World Resources Institute 1998). Also, hydrological alterations are as significant (if not more) in tropical regions (Vörösmarty et al. 1997). However, climate- and source rock–related weathering and runoff largely controlling DSi inputs (Turner et al. 2003) are significantly different in tropical and nontropical regions. In weathering-limited systems transport processes remove substances derived from weathering faster than weathering processes can produce new material. In contrast, in transport-limited systems river load depends on the availability of water as a transporting agent rather than on weathering rates (Drever 1997). This probably will vary significantly between low and high latitudes and thereby affect DSi.

Although a vast set of N and P data from tropical coastal regions has been collected in the frame of Land–Ocean Interactions in the Coastal Zone (LOICZ) in recent years, the amount of existing data on DSi still is astoundingly small except for the data on major world rivers collected in the GEMS database (Global Environmental Monitoring System 2000). GEMS data were collected between 1976 and 1995, for some rivers representing monitoring of only two years. In this chapter we examine the natural and anthropogenic factors controlling concentration, yield, and load of DSi in tropical rivers. Besides those from GEMS, we use as yet unpublished data from medium and small rivers and from various other sources. We compare these with data on nontropical rivers from the GEMS database.

Definition of a Tropical River

The term *tropics* is often used but has no exact meaning. For the purpose of quantifying riverine DSi fluxes in the tropics we therefore need to define the region covered. The term *tropics* is derived from the Tropics of Capricorn and Cancer at northern and southern latitudes of 23.5°. It provides a geographic definition that has been considered too simplistic for many studies, and therefore several definitions of the tropics exist (e.g., McGregor and Nieuwolt 1998; Buddemeier et al. 2002). Because temperature and precipitation exert major control on the processes governing the DSi load of rivers, we think that these are appropriate parameters to define the tropics in our case. Commonly used boundaries are the 18°C (air temperature) sea level isotherm and precipitation, which do not exactly match but largely fall into the range between 23.5° of northern and southern latitude, respectively (McGregor and Nieuwolt 1998). Our "tropical" rivers debouch into the sea between the Tropics of Cancer and Capricorn. With a few exceptions their catchments are also located entirely or partly between these borders.

Use of Data and Calculations

For reasons of compatibility with other studies we not only separate rivers into tropical and nontropical but also subdivide them into three groups. Two of them cover the GEMS database (tropical GEMS, twenty-eight rivers; nontropical GEMS, fifty rivers), and the third is our set of unpublished data and data from various other sources not included in the GEMS database (tropical non-GEMS, thirty-seven rivers). For the discussion of processes we use all data sets. The tropical rivers not included in the GEMS database with a few exceptions are smaller in terms of catchment size than GEMS rivers. For reasons of clarity the groups of rivers discussed will be named as follows. Tropical rivers (TRs) include the large tropical rivers (LTRs) included in the GEMS database and the small tropical rivers (STRs) which are not included in it. The nontropical rivers (NRs) drain 27.3 percent of the total world landmass of 133.29×10^6 km^2 (World Resources Institute 1998), whereas the LTRs and the STRs drain 16.0 percent and 0.6 percent, respectively. In terms of discharge, NRs contribute 17.6 percent of the global total of 38,200 km^3/yr (Vörösmarty et al. 1997), whereas LTRs and STRs contribute 34.7 percent and less than 0.1 percent, respectively. For the calculation of global-scale budgets we use rivers from the GEMS database only. For the discussion of anthropogenic factors affecting silicon biogeochemistry we use data on population, land use, and land cover from the World Resources Institute (1998), which are available for almost all rivers included in the GEMS database.

DSi features discussed include concentration, yield, and load. The measured DSi concentration (μM) in combination with river discharge and catchment size allows the calculation of DSi yield ($kg/km^2/yr$, 10^3 $mol/km^2/yr$), that is, the rate of DSi released per unit of area and time; and DSi load (10^3 t/yr, 10^9 mol/yr), that is, the total amount of DSi released per catchment or region per year.

Results and Discussion

In general, natural factors such as geology, climate, hydrology, and vegetation are first-order controls of the amount and composition of substances transported by rivers. Formerly, human activities in coastal areas were a second-order control in this respect. This relationship has reversed in many regions of the world in the twentieth century because of population growth and the increasing importance of coastal zones as areas of settlement and exploitation of their natural resources. The attendant environmental changes have altered the biogeochemical fluxes and composition of rivers (Humborg et al. 1997; Turner et al. 2003). Tropical coastal zones are critical in terms of fluvial inputs of nutrients and sediments into the ocean and in terms of human modifications of the coastal zone altering these inputs (Milliman and Meade 1983; Elvidge et al. 1997). On a global scale they provide about three quarters of freshwater and sediment input into the ocean annually. However, less is known about the changes of biogeochemical fluxes and composition of rivers after environmental changes in the tropics than in temperate regions, particularly in terms of Si.

Natural Factors Controlling Dissolved Silica in Tropical Rivers

The range of DSi concentrations is almost similar but the average concentration is significantly higher in tropical than in nontropical rivers (Table 4.1).

The average DSi yield (expressed in $kg/km^2/yr$) is two to three times higher in tropical than in nontropical rivers. Similarly, the average DSi load (the total amount in 10^3 t/yr) and the range of loads are about three times higher in LTRs as in NRs. Among the tropical rivers the STRs have higher concentrations and yields of DSi than the LTRs but a much smaller annual load.

The large scatter of DSi concentrations and a missing relationship with catchment size, river discharge, and runoff indicate that the DSi concentration alone does not provide clues on the sources and processes affecting riverine fluxes of DSi (Figure 4.1).

Table 4.1. Hydrological and hydrochemical data of large tropical (LTR), small tropical (STR), and nontropical (NR) rivers.

		Area (10^3 km^2)	Discharge (km$^{3/yr}$)	Runoff (mm/yr)	DSi (μM)	DSi:N	DSi (kg/km^2/yr)	Yield (10^3 mol/ km^2/yr)	DSi (10^3 t/yr)	Load (10^9 mol/yr)	Total Suspended Solids (10^6 t/yr)	Total Dissolved Solids (10^6 t/yr)
LTR	Average	762	473	566	179	20	2,514	89	1,825	65	154	35
	Range	61–6,112	2–6,450	18–2,203	37–317	2–148	127–9,708	5–347	14–20,769	1–742	<1–1,200	<1–280
	Sum	21,342	13,247	15,852					47,450	1,694	4,017	908
	Percentage of total	37	66	56					64	64	64	42
STR	Average	23	13	296	190	75	2,834	101	62	2	5	1
	Range	<1–76	<1–84	14–2,500	18–467	<1–1,073	65–16,100	2–575	1–542	1–19	<1–40	<1–4
NR	Average	729	135	247	128	12	1,026	37	640	23	53	26
	Range	14–3,270	<1–873	1–934	3–473	<1–97	16–12,382	1–442	2–5,613	1–200	<1–900	1–193
	Sum	36,439	6,732	12,333					26,883	960	2,216	1,234
	Percentage of total	63	34	44					36	36	36	58

Sources: Data for STR are from Bolaji et al. (2002), Carneiro (1998), Carvalho (1997), Global Environmental Monitoring System (2000), Hungspreugs and Utoomprurkporn (1999), Lesack et al. (1984), Martins (1987), Mwashote and Jumba (2002), Petr (1983), Qunying et al. (1987), Viers et al. (2000), Whitten et al. (1996), Zhang (1996).

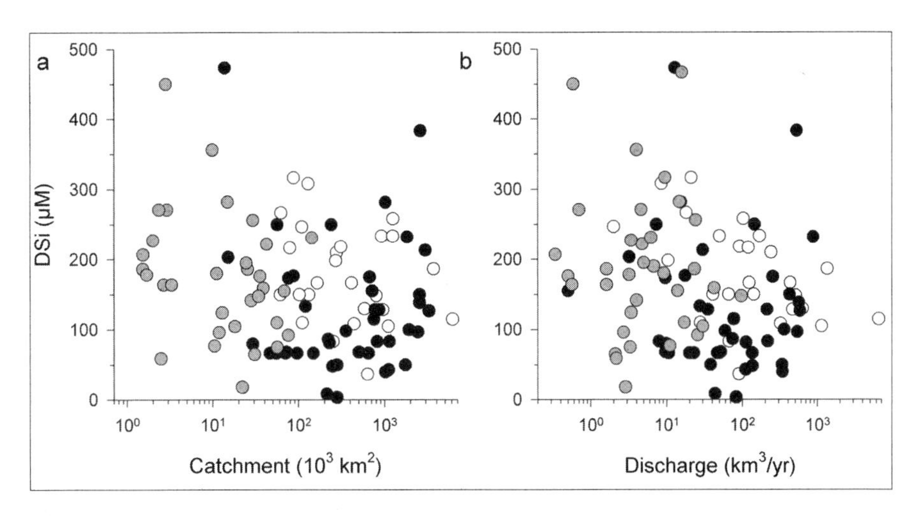

Figure 4.1. DSi concentrations in large tropical rivers *(white circles)*, nontropical rivers *(black circles)*, and small tropical rivers *(gray circles)* related to *(a)* catchment size and *(b)* discharge. Note the logarithmic scale for catchment size and river discharge.

For all groups of rivers the DSi yield (expressed in $kg/km^2/yr$) is positively correlated with runoff (Figure 4.2). The slope of the correlation near 1 for the LTRs and the NRs simply shows that the DSi released by weathering is almost constant with runoff. Interestingly, the slope for the STRs is less steep and the scatter higher than for the LTRs.

Weathering and Tectonic Activity

Gaillardet et al. (1999) found chemical weathering of silicates to be generally tightly coupled to physical denudation and runoff temperature. Despite some regional variability it demonstrates an intimate link between climatic and tectonic weathering controls. We subdivided our data set of tropical rivers by continent and found distinct differences in DSi concentration, yield, and load (Table 4.2, Figure 4.3).

Comparing DSi yields with weathering rates provides insight into the regional variability of factors controlling DSi in rivers (Table 4.3). For this purpose we used dimensionless relative silicate weathering rates calculated by Gaillardet et al. (1999) for rivers from the GEMS database and a few others and normalized these to the weathering rates of the Amazon, the largest single freshwater and DSi source on Earth. These relative silicate weathering rates have the advantage of being unbiased by regional variability in lithology and texture and thereby facilitate assessment of regional variability in silicate weathering. High DSi yields match with high relative silicate weathering rates for LTRs from tropical Asia and Oceania.

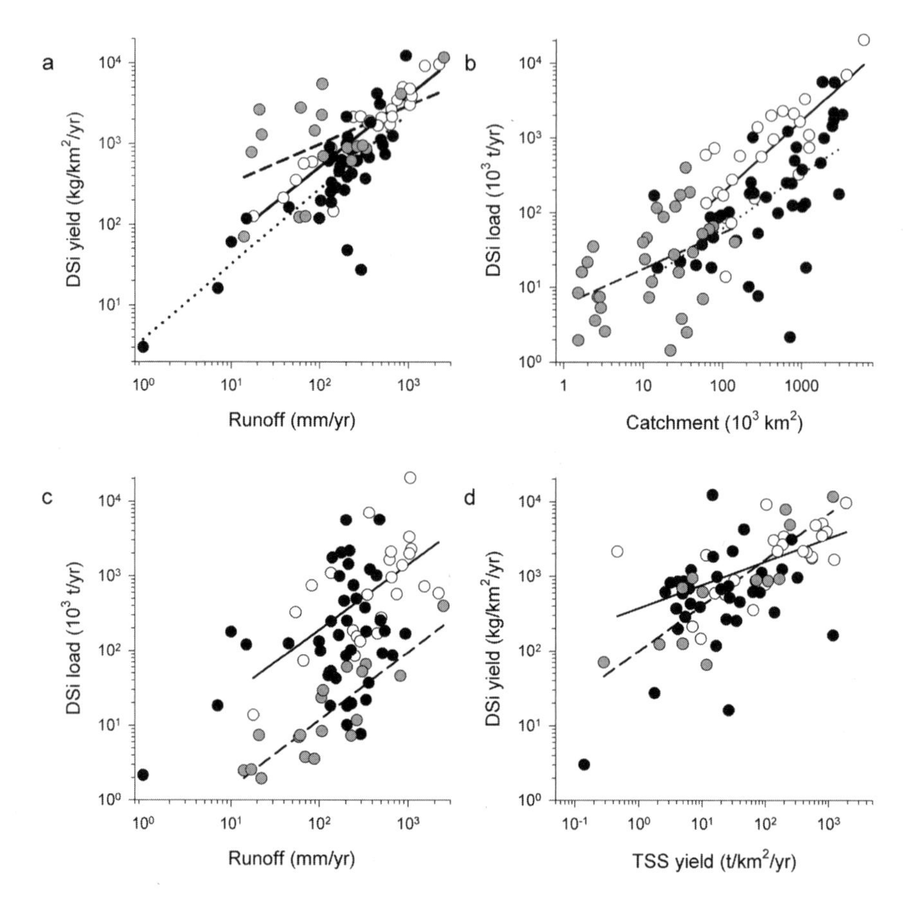

Figure 4.2. DSi yield and load versus catchment features: *(a)* DSi yield versus runoff, *(b)* DSi load versus catchment size, *(c)* DSi load versus runoff, and *(d)* DSi yield versus total suspended solid (TSS) yield. Note the logarithmic scale of all axes. Data are from large tropical rivers (LTRs) *(white circles, solid regression)*, nontropical rivers (NRs) *(black circles, dotted regression)*, and small tropical rivers (STRs) *(gray circles, dashed regression)*. Regression equations are given in the same order except where indicated. *(a)* $y = 0.88x + 0.94$, $r^2 = 0.85$, $n = 28$; $y = 0.93x + 0.58$, $r^2 = 0.64$, $n = 50$; $y = 0.51x + 1.98$, $r^2 = 0.24$, $n = 25$; *(b)* $y = 0.93x + 0.35$, $r^2 = 0.50$, $n = 28$; $y = 0.72x + 0.34$, $r^2 = 0.37$, $n = 50$; $y = 0.48x + 0.76$, $r^2 = 0.19$, $n = 25$; *(c)* LTR $y = 0.92x + 0.40$, $r^2 = 0.46$, $n = 28$; STR $y = 0.90x - 0.74$, $r^2 = 0.75$, $n = 25$; *(d)* LTR $y = 0.31x + 2.57$, $r^2 = 0.34$, $n = 28$; STR $y = 0.61x + 1.99$, $r^2 = 0.71$, $n = 25$.

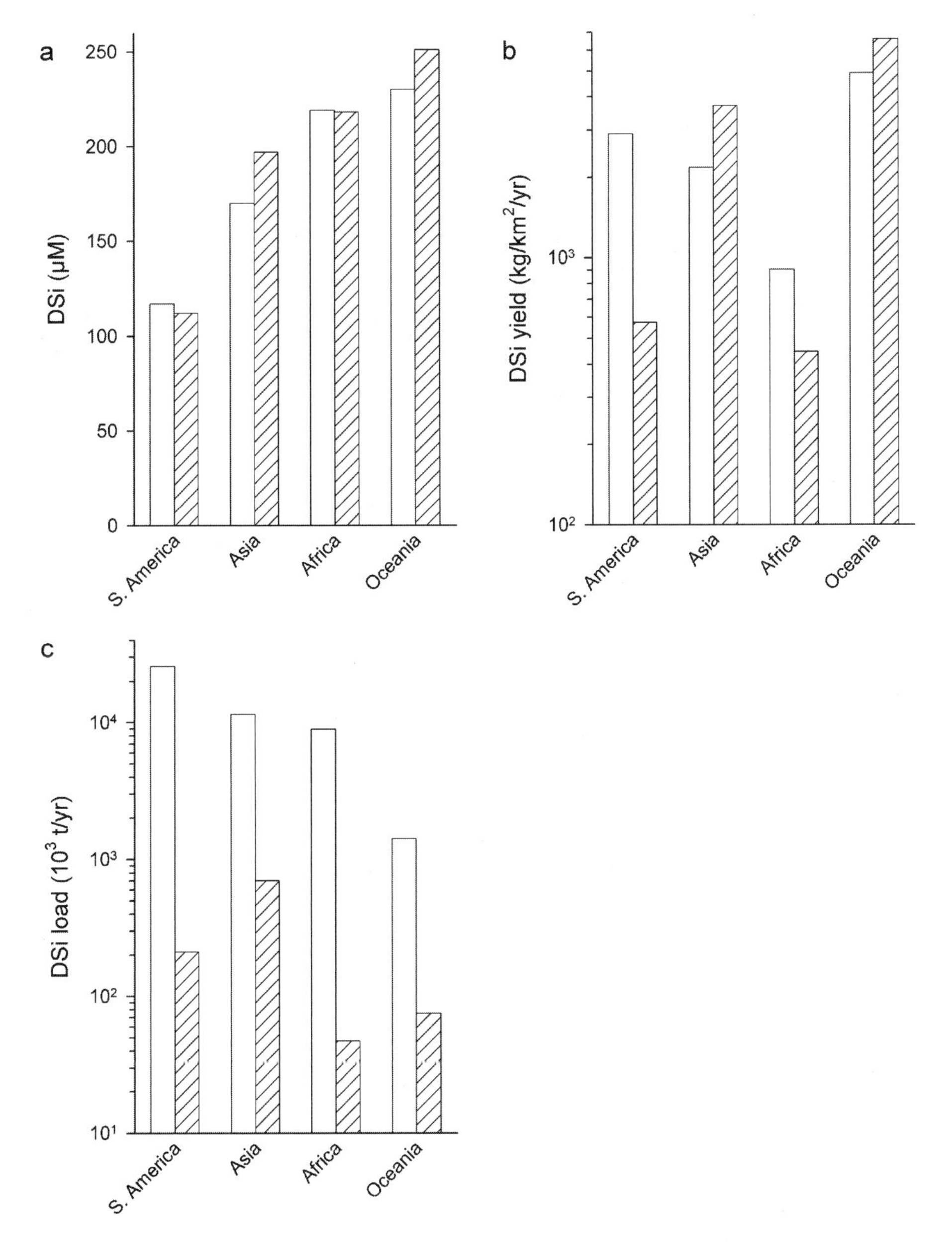

Figure 4.3. *(a)* DSi concentration, *(b)* yield, and *(c)* load of tropical rivers by continent, subdivided into large tropical rivers *(white bars)* and small tropical rivers *(hatched bars)*. Note the logarithmic scale for DSi yield and load.

Table 4.2. Hydrological and hydrochemical data of tropical rivers subdivided per continent and into large tropical rivers (LTRs) and small tropical rivers (STRs).

		Area (10^3 km^2)	Discharge (km^3/yr)	Runoff (mm/yr)	DSi (µM)	DSi:N	DSi (kg/km^2/yr)	Yield (10^3 mol/km^2/yr)	DSi (10^3 t/yr)	Load (10^9 mol/yr)	Total Suspended Solids (10^6 t/yr)	Total Dissolved Solids (10^6 t/yr)
South	Average	1,774	1,657	717	117	13	2,907	104	6,398	229	330	86
America	Range	276–6,112	90–6,450	143–1,055	37–210	10–18	147–5,049	5–180	92– 20,769	3–742	6–1,200	6–281
LTRs	Sum	8,868	8,284	3,585	—	—	—	—	25,592	914	1,651	343
South	Average	43	9	186	112	47	572	20	26	1	2	—
America	Range	12–76	1–26	14–336	65–176	6–187	70–947	3–34	3–66	<1–2	<1–8	—
STRs	Sum	346	68	1,484	—	—	—	—	211	8	16	—
Asia	Average	376	204	522	170	21	2,177	78	821	29	152	32
LTRs	Range	62–950	18–630	54–1,087	83–267	2–148	355–4,825	13–172	86–2,293	3–82	<1–540	5–86
	Sum	5,646	3,061	7,833	—	—	—	—	11,464	409	2,135	451
Asia	Average	12	11	155	197	17	3,702	132	58	2	3*	1
STRs	Range	2–38	1–42	21–820	59–450	2–56	784–5,912	28–211	2–188	<1–7	—	<1–4
	Sum	143	138	1,241	—	—	—	—	699	25	3*	11
Africa	Average	1,613	408	156	219	28*	902	32	2,239	80	26	21
LTRs	Range	270–3,690	10–1,350	39–366	187–258	—	214–1,912	8–68	58–7,056	2–252	2–43	1–56
	Sum	6,450	1,631	622	—	—	—	—	8,957	320	105	82

Table 4.2. Hydrological and hydrochemical data of tropical rivers subdivided per continent and into large tropical rivers (LTRs) and small tropical rivers (STRs). (continued)

		Area (10^3 km^2)	Discharge (km^3/yr)	Runoff (mm/yr)	DSi (µM)	DSi:N	DSi (kg/ km^2/yr)	Yield (10^3 mol/ km^2/yr)	DSi (10^3 t/yr)	Load (10^9 mol/yr)	Total Suspended Solids (10^6 t/yr)	Total Dissolved Solids (10^6 t/yr)
Africa	Average	18	4	110*	218	278	445	16	16	1	<1*	<1*
STRs	Range	<1–42	3–5	—	18–429	1–1,073	65–700	2–25	1–29	1	—	—
	Sum	92	12	—	—	—	—	—	47	2	<1*	<1*
Oceania	Average	94	68	953	230	—	4,914	176	352	13	42	8
LTRs	Range	61–129	2–141	18–2,203	150–308	—	127–9,708	5–347	14–728	1–26	3–115	<1–16
	Sum	378	272	3,812	—	—	—	—	1,407	50	126	32
Oceania	Average	26	16	2,500*	251	—	6,592	235	75	3	14	—
STRs	Range	2–143	1–97	—	105–356	—	281–15,013	10–536	5–401	<1–14	1–40	—
	Sum	238	179	2,500*	—	—	—	—	673	24	55	—

*From one or two rivers only.

Table 4.3. Average relative weathering rates and DSi yield per continent. Weathering rates are normalized to the rate of the Amazon, which is set to 1. Numbers in parentheses denote number of rivers. Weathering rates are from Gaillardet et al. (1999). Average DSi yields presented are only from the rivers for which weathering rates are given. Because of the extreme differences of weathering rates of the Magdalena and the other South American rivers, we show both values.

Rivers per Continent	Relative Silicate Weathering Rates			DSi Yield		DSi Load	
	Chemical	Cationic	Physical	($t/km^2/yr$)	(10^3 $mol/km^2/yr$)	(10^6 t/yr)	(10^9 mol/yr)
Large tropical rivers							
South America (3)	0.62	0.70	0.60	2.19	78.21	24.20	864.29
South America (4, including Magdalena)	1.81	1.60	5.95	2.90	103.57	25.59	913.93
Africa (3)	0.18	0.19	0.10	1.13	40.36	8.96	320.00
Asia (10)	1.60	3.62	3.19	2.36	84.29	11.49	410.36
Oceania (2)	2.80	6.75	6.18	9.48	338.57	1.41	50.36
Small tropical rivers							
South America (8)	—	—	—	0.57	20.36	0.21	7.50
Africa (3)	—	—	—	0.45	16.07	0.05	1.79
Asia (12)	—	—	—	3.70	132.14	0.70	25.00
Oceania (2)	15.50	17.50	13.00	13.10	467.86	0.72	25.71
Nontropical rivers							
South America (2)	0.37	0.68	0.10	3.19	113.93	6.63	236.79
Africa (1)	0.10	0.36	0.09	0.06	2.14	0.30	10.71
Asia (9)	0.52	1.06	1.99	0.79	28.21	12.84	458.60
Oceania (1)	0.03	0.13	0.07	0.02	0.71	0.19	6.79
North America (7)	0.71	1.66	0.70	0.75	26.79	5.34	190.71
Europe (7)	1.42	3.84	2.93	0.79	28.21	1.59	56.79

In contrast, nontropical rivers generally display much lower relative silicate weathering rates and DSi yields; the lowest are encountered in semiarid Africa. Among the STRs, maximum DSi yields are also found in rivers from Asia and Oceania, and they are higher than those of the respective LTRs. In the case of the Papua New Guinean Purari and Kikori rivers, the maximum DSi yield also matches the maximum relative weathering rates.

The young geology in Asia and Oceania, particularly of high-standing islands, and active tectonism in the so-called circum-Pacific ring of fire in combination with high precipitation and runoff are responsible for the worldwide highest yields and loads of suspended sediments transported by rivers (Milliman and Meade 1983; Milliman and Syvitski 1992). The combination of seismicity and runoff variability exerting major control on erosion and sediment delivery to the ocean has recently been demonstrated for the island of Taiwan, for example, which is located at the northern edge of the tropics and drained by a number of small-scale rivers (Dadson et al. 2003). With respect to the tight coupling of chemical silicate weathering with physical weathering and runoff temperature, the high physical denudation together with our finding of maximum DSi yield suggest tropical Asia and Oceania to be the most efficient sources of DSi to the ocean. In this context, the fact that the nontropical South American rivers Paraná and Uruguay drain large areas of a high mountain region may explain the disproportionately high DSi yield when compared with all other nontropical rivers. Furthermore, parts of their catchments are also embedded in the humid subtropical region. The fact that the DSi yield of STRs from Asia and Oceania is higher than that of the respective LTRs can also be attributed to exceptionally high physical denudation because most of these rivers drain high mountainous areas (Milliman and Syvitski 1992). The larger scatter of DSi yield versus runoff for the STRs as a whole may be attributed to the fact that South American and African rivers with a few exceptions drain low-lying areas and therefore have lower physical denudation, resulting in lower total suspended solid (TSS) yields. Additionally, small catchments generally have less storage capacity and respond more strongly to episodic events (Milliman and Syvitski 1992; Milliman et al. 1999; Dadson et al. 2003).

Eolian Deposition and Vegetation

Deposition of atmospheric dust and vegetation cover may be additional sources for riverine DSi in some tropical regions. Dust fluxes, mainly via dry deposition, are highest near the sources: the African, Arabian, Asian, and Australian deserts (Tegen et al. 2002). In remote areas, however, wet deposition is the dominant mechanism. Whereas dry deposition is dominated by larger particles consisting of, say, quartz and feldspar, wet deposition fluxes are dominated by smaller particles such as clays, which have a higher surface to volume ratio and therefore may be a more efficient silicon source with respect to mineral weathering. Wet deposition is the major pathway of dust fluxes over remote ocean regions and on the continents over the Amazon and Congo basins and the

Southeast Asian and western Pacific islands (Chapter 7, this volume). Moreover, the Amazon Basin is covered by the largest single tropical rain forest on Earth, and the Congo drains the equatorial greenbelt of Africa. A high supply of organic protons lowering pH in topsoils and thereby enhancing chemical weathering may compensate for otherwise low weathering rates. This process appears to be the major factor controlling DSi in weathering-limited boreal and subarctic watersheds (Chapter 5, this volume). It is conceivable that silicon release from atmospheric dust, vegetation, and soil is responsible for the high DSi yields despite low weathering rates of the Amazon and Congo rivers which, because of their size (catchment area, discharge), determine to a large extent the average DSi yields of South America and Africa (Table 4.3).

Combination of Factors Controlling Regional Variability of Dissolved Silica

The total DSi load of rivers expressed in mass units per time generally is a function of catchment size and runoff (Figure 4.2). It simply reflects the significance of total surface area drained and water available as a solvent and transporting agent for DSi input into the oceans. In a log–log plot DSi load is positively correlated with catchment area. Interestingly, STRs and NRs plot almost on the same line, whereas the slope of the linear regression is much steeper and average load of LTRs is higher than that of NRs by a factor of three (Figure 4.2, Table 4.1). For the STRs the scatter in DSi load of more than three orders of magnitude at a given catchment area can be attributed to the differences in yields. Rivers with high yields from Asia and Oceania generally plot above the line, and those with lower yields from Africa and South America plot below the line. Similarly, for all tropical rivers the DSi yield is correlated with the TSS yield, but the slope of the regression is much steeper for the STRs than for the LTRs (Figure 4.2). For Asia and Oceania the DSi yield is higher in STRs than in LTRs, and the opposite is observed for Africa and South America (Table 4.2).

Optimum weathering and transport conditions are found along the active continental margins of tropical Asia and Oceania. High tectonic uplift rates up to approximately 4 cm/yr, as in Papua New Guinea (Pigram and Davies 1987), for instance, lead to high physical denudation rates and consequently to the high TSS yields of tropical rivers. Additionally, high runoff and high runoff temperature provide ideal conditions for high chemical weathering rates and high DSi transport (Gaillardet et al. 1999). The combination of these factors causes the observed maximum DSi yields related to high TSS yields in tropical rivers from Asia and Oceania. The lower DSi yields of LTRs among the tropical rivers may result from the fact that the larger rivers generally have large floodplain grasslands, which act as a filter that reduces the river supply to the coastal sea. Dissolved and particulate river loads are deposited there, and DSi can be extracted by grasses and stored in phytoliths (Wüst and Bustin 2003; Ding et al. 2004). An opposite pattern is observed for African and South American rivers, with DSi yields higher in LTRs than in STRs. Because of the tight coupling of chemical and physical weathering, less tectonic activity and lower physical weathering rates in tropical Africa and South America result in lower chemical weathering rates (Table 4.3). Tropical regions generally have favorable weathering conditions, making them transport-limited regimes (Drever 1997). However, it appears that factors that otherwise apply in the weathering-limited regions gain importance in the control of riverine DSi in tropical Africa and South America. Vegetation and soils are major factors for chemical weathering in boreal and subarctic watersheds (Chapter 5, this volume). Therefore, it is conceivable that the much larger vegetation and soil cover in large catchments is responsible for the higher DSi yield in LTRs than in STRs of Africa and South America.

Anthropogenic Factors Controlling Dissolved Silica in Tropical Rivers

Human activities in river catchments greatly affect nutrient and sediment loads of rivers (Milliman 1997; Turner et al. 2003). Major environmental changes of river catchments and the coastal zone are also reported from tropical regions, particularly from Asia (Elvidge et al. 1997; Dudgeon 2000; Vörösmarty et al. 2003). This could also have implications for the DSi load of rivers because silicon is the second most abundant element in the lithosphere. For most of the rivers included in the GEMS database, data are available on factors describing effects of human activities on hydrological and hydrochemical characteristics of the rivers. We chose population density, the number of dams, and the percentages of developed land, cropland, and forest loss as parameters representative of urban and industrial use and waste disposal, regulation of watershed hydrology, and land use change. These are supposed to be among the most important human activities in river catchments affecting river fluxes and composition.

Urbanization, Deforestation, and Agriculture

Despite large variability, the DSi load of the tropical rivers (except for the São Francisco and Senegal rivers) and the NRs is negatively correlated with the percentage of developed land (Figure 4.4).

This is probably related to urbanization, which leads to the sealing of natural land cover and hence to a decrease of weathering and erosion and DSi input into rivers. The DSi load of LTRs is generally higher than that of NRs at a given developed land area because of the originally higher yields in LTRs. However, the following two examples illustrate how combinations of natural and anthropogenic factors on a regional scale can lead to a deviation of DSi characteristics from global trends. For example, the Brazilian São Francisco River and the African Senegal River have extremely low DSi loads despite a low percentage of developed land. Large parts of the areas drained have a semiarid to arid climate and hence lower physical and chemical weathering rates. Additionally, six large dams in the case of the São Francisco River are responsible for a reduction in river discharge and loads of dissolved nutrients and sediments in the past decades (Jennerjahn et al. in press; Knoppers et al. 2006).

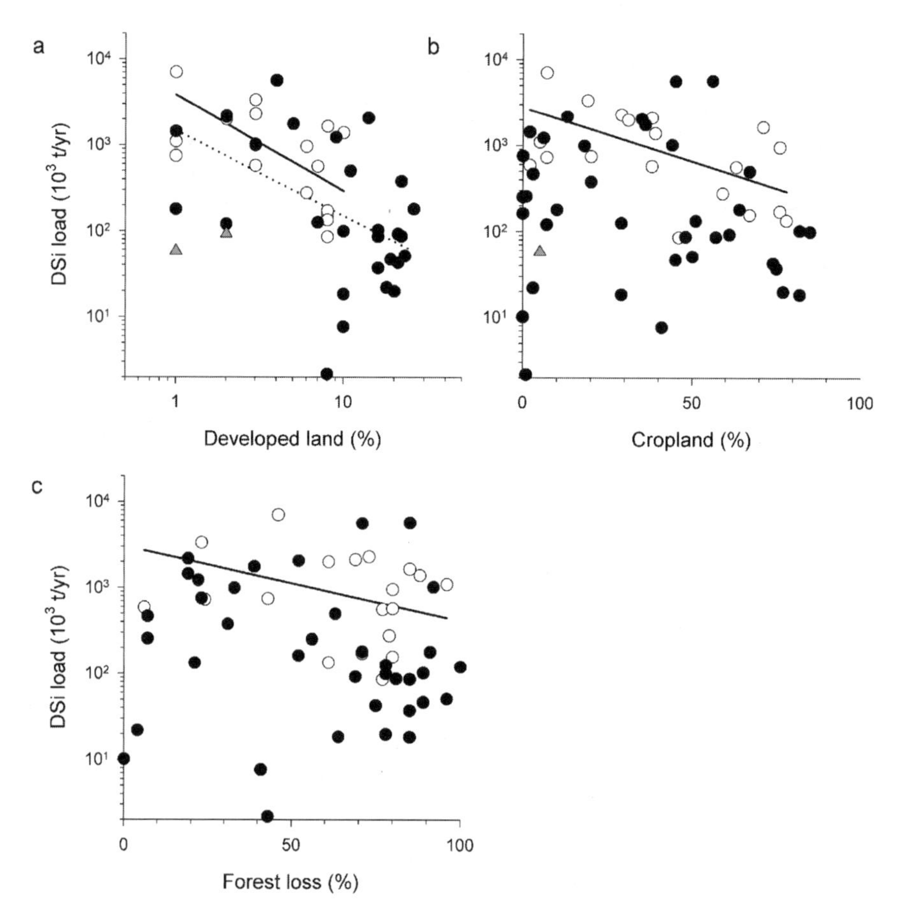

Figure 4.4. DSi load versus land use features in the catchment: *(a)* DSi load versus developed land, *(b)* DSi load versus cropland, and *(c)* DSi load versus forest loss. Note the logarithmic scale of both axes in *(a)* and of DSi load in *(b)* and *(c)*. Data are from large tropical rivers (LTRs) *(white circles, solid regression)* and nontropical rivers (NRs) *(black circles, dotted regression)*. The tropical São Francisco and Senegal rivers *(gray triangles)* have not been included in regression analysis in *(a)* and the Senegal in *(b)* for reasons given in the text. Regression equations: *(a)* LTR $y = -1.12x + 3.58$, $r^2 = 0.44$, $n = 15$; NR $y = -0.99x + 3.17$, $r^2 = 0.23$, $n = 50$; *(b)* LTR $y = -0.01x + 3.45$, $r^2 = 0.33$, $n = 19$; *(c)* LTR $y = -0.01x + 3.54$, $r^2 = 0.20$, $n = 21$.

Surprisingly, the DSi load of LTRs is negatively correlated with percentages of cropland and forest loss, and there is no statistically significant trend for the NRs (Figure 4.4). Again, the São Francisco and Senegal rivers deviate from this pattern for the reasons given earlier. Deforestation should enhance erosion and weathering and thereby increase the DSi load. Instead the observed trend suggests lower release of silicon with increasing cropland for tropical rivers. In weathering-limited regimes vegetation and soils are the most important factors for the release of DSi (Chapter 5, this volume). However, this process is much less efficient in supplying DSi than the high physical and chemical weathering rates of tropical regions, particularly in Asia and Oceania. It is illustrated by the much lower DSi concentrations, yields, and loads in nontropical regions (Table 4.1). The cultivation of crops in tropical regions therefore may lead to a lower supply of DSi as compared to natural weathering conditions. Agriculture is a major land use generating income to the population in tropical regions. Rice is among the most important crops on Earth, about three quarters of which is grown under intensive irrigation. This is particularly important in Asia, which has 70 percent of the global irrigated agricultural land of 273×10^6 ha and in 2002 produced 91 percent of the global rice harvest of 576×10^6 t (Food and Agriculture Organization of the United Nations 2003). Plants can store silicon in the form of phytoliths (see Chapter 3, this volume), which appear to be geochemically stable for millennia and are being used in late Quaternary vegetation reconstruction (e.g., Lu et al. 2002). They are abundant particularly in grasses (Gramineae), the plant family to which rice belongs (Piperno and Pearsall 1998). Deforestation and conversion to agricultural fields usually include a shift from a mature ecosystem such as a tropical forest with efficient recycling and little net gain in biomass to an immature ecosystem, which is actively extracting nutrient elements to produce crop biomass. Therefore, it is conceivable that forest loss and the cultivation of rice or other crops from the Gramineae family may reduce DSi in tropical rivers because of silicon storage in phytoliths despite high weathering rates.

Damming and Land Use Change

The DSi:N ratio is critical in terms of phytoplankton community development, a change in which is one of the factors affecting the food web in coastal waters off temperate rivers suffering from environmental change, such as the Danube and the Mississippi (Turner and Rabalais 1994; Humborg et al. 1997). Silicon is an essential nutrient besides nitrogen and phosphorus for plankton producing silica tests. Diatoms, the most abundant and hence most important phytoplankton on Earth, take up dissolved Si and N in a ratio of 1:1 (Redfield et al. 1963; Brzezinski 1985). Although variability is high, DSi:N ratios of tropical and nontropical rivers on average are above the critical ratio of 1 (Table 4.1). However, in a number of nontropical rivers this ratio falls below 1. On a global scale the Mississippi is the most prominent and quantitatively most significant river, where human activities in the catchment have altered nutrient fluxes and composition so that the receiving coastal water body, the northern Gulf of Mexico, is DSi limited and suffering from the consequences of cultural eutrophication. Dramatically increased N input from fertilizer use in agriculture is the major reason for the observed environmental changes. An attendant decrease in DSi concentration is attributed mainly to increased freshwater production and sedimentation of diatoms caused by increased P input (Turner and Rabalais 1994; Rabalais et al. 2000).

In our global data set the DSi:N ratio does not display any relationship to catchment size, discharge, runoff, or latitude. For the NRs the DSi:N ratio is negatively correlated with population and the percentages of developed land, cropland, and forest loss; although the tropical rivers display similar trends, it is correlated only with population density (Figure 4.5).

This is in agreement with the findings of Turner et al. (2003), who found the effects of land use and other human influences on the nitrogen cycle to be the major control of the DSi:N ratio rather than changes in DSi. Interestingly, for the DSi:N ratio and the DSi concentration, yield, and load of TRs and NRs there is no relationship to the number of large dams (Figure 4.5).

From these findings we infer that anthropogenically induced environmental changes do not have such a significant impact on the DSi:N ratio of tropical rivers as has been observed for many NRs, such as the Danube and the Mississippi. Despite the significant environmental changes in the coastal zone of tropical regions, it appears that the two- to three-fold higher DSi yield and load of tropical rivers maintain the DSi:N ratio at levels higher than the critical threshold of 1:1 below which significant changes in plankton community entailing further ecological and biogeochemical consequences are expected.

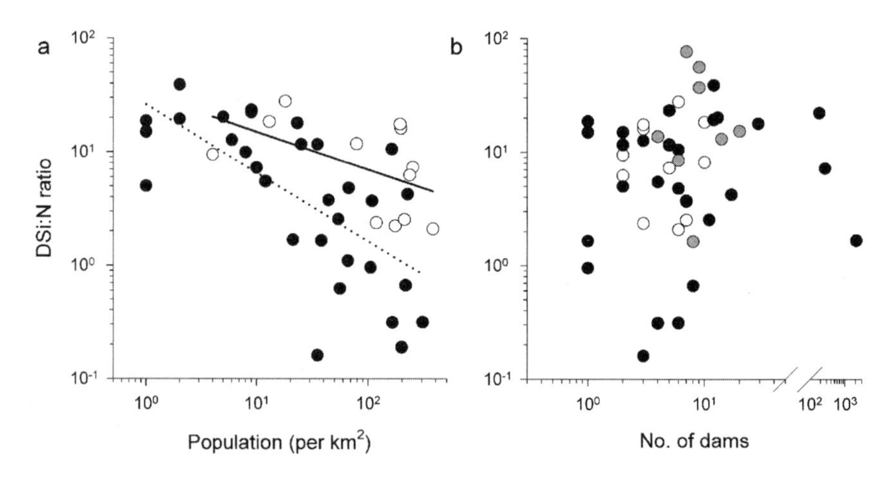

Figure 4.5. Ratios of DSi to N in large tropical rivers (LTRs) *(white circles, solid regression)*, nontropical rivers (NRs) *(black circles, dotted regression)*, and small tropical rivers (STRs) *(gray circles)* related to *(a)* population density and *(b)* number of dams. Note the logarithmic scale for all axes. Regression equations in *(a)* are LTR $y = -0.33x + 1.51$, $r^2 = 0.26$, $n = 12$; NR $y = -0.60x + 1.42$, $r^2 = 0.47$, $n = 50$.

Table 4.4. Hydrological and hydrochemical data of the Brazilian São Francisco River. Data from 1984–1985 represent an annual cycle before construction of the Xingo Dam in 1994. Data from 2000–2001 represent an annual cycle after the construction of the Xingo Dam. Discharge of this cycle was 15% below the average of 58 km^3/yr for the period 1995–2000.

	Discharge (km^3/yr)	Runoff (mm/yr)	Total Suspended Solids (10^6 t/yr)	DSi (µM)	DSi Yield (kg/km^2/yr)	DSi Load (10^3 t/yr)	N Load (10^3 t/yr)	DSi:N
1984–1985	98	157	2.7	239	1,045	658	72	5
2000–2001	49	79	0.2	317	693	437	3	77
Change (%)	–50	–50	–93	33	–43	–34	–96	1,540

However, on a regional scale human activities in river catchments may dramatically affect the ecology and biogeochemistry of tropical rivers and receiving coastal waters, as, for instance, in the case of the Brazilian São Francisco River. It is impounded by a number of large dams, the last of which, the Xingo Dam, has been in operation since 1994. The operation of these dams since the late 1970s led to a drastic reduction in discharge and a change in the seasonal discharge pattern from formerly unimodal (range 800–8,000 m^3/s) to less than 2,000 m^3/s throughout the year; the attendant ecological and biogeochemical changes led to a drastic reduction of fish catch in coastal waters (Marques et al. 2002; Knoppers et al. 2006). Predam and postdam studies showed that although discharge was reduced by about 50 percent, TSS and N were almost completely retained in the reservoirs. However, the DSi load of the river was reduced by one third, raising the DSi:N ratio from 5 to 77; nitrogen retention by the dams therefore was the main feature responsible for the raising of the DSi:N ratio (Table 4.4; Jennerjahn et al. in press; Knoppers et al. 2006).

In the Indonesian Brantas River, which drains a densely populated region of eastern Java and is regulated by eight large dams, the DSi:N ratio is less than 2 (Jennerjahn et al. 2004). This low ratio appears to be related to high nitrogen inputs from fertilizer use in the catchment, where about 50 percent of the land use is agriculture, rather than to a reduction of DSi caused by the so-called artificial lake effect, which is supposed to decrease the riverine DSi concentration through increased deposition and preservation of biogenic silica in sediments behind dams and reported from many rivers in nontropical regions (see review by Conley et al. 1993). Despite flow regulation by dams, the DSi yield (149 × 10^3 mol/km^2/yr, 4,173 kg/km^2/yr) and the DSi load (1.6 × 10^9 mol/yr, 46.1 × 10^3 t/yr) of the Brantas River are very high in relation to its catchment area (11 × 10^3 km^2; Tables 4.1 and 4.2; Figure 4.2).

Generally the TSS concentration, yield, and load are higher by a factor of three to five in tropical than in nontropical rivers (Table 4.1). This difference indicates lower light penetration, which limits plankton growth and hence biogenic removal of DSi despite generally rapid settling of particles in reservoir lakes. Also, wind-induced turbulence mixing may result in prolonged periods of high TSS, particularly in shallow reservoirs during monsoon. In addition to the initially higher DSi in TRs than in NRs, the biogenic removal of DSi in reservoir lakes may be less efficient in tropical than in nontropical regions. This is supported by the fact that neither DSi concentration, yield, and load nor the DSi:N ratio correlate with the number of large dams (Figure 4.5).

Our examples suggest that the DSi of tropical rivers is much less affected by the artificial lake effect than reported from rivers in nontropical regions. Therefore, it is much less likely that DSi becomes a limiting nutrient in tropical rivers and adjacent coastal waters, as has been observed in nontropical rivers and coastal zones. The conclusions about the significance of anthropogenic factors exerting control on riverine DSi were drawn mostly from observations made in nontropical regions. Our findings indicate that natural factors control DSi in tropical rivers and hence ecosystem features related to silicon rather than anthropogenically induced changes of catchment characteristics and hydrology. The latter are much more significant in nontropical rivers, where DSi yields and loads are initially lower. However, it can be expected that anthropogenically induced environmental changes will gain importance in the control of riverine DSi, particularly in the densely populated coastal regions of south and Southeast Asia in the future.

Climate Change Effects

Climate change related to fossil fuel emissions also is a consequence of human activities, but in contrast to the effects described earlier it operates on a global scale. Therefore, global climate change may affect riverine DSi mainly through changes in the natural control factors. An increase in atmospheric CO_2 via the production of carbonic acid may increase chemical weathering on a global scale. However, regional changes in the distribution of runoff and runoff temperature are the major factors for possible changes of riverine DSi in tropical regions. Recent findings indicate a change in the freshwater balance of the Atlantic Ocean, for example, with a freshening of high-latitude waters and increasing salinity of tropical and subtropical surface waters over the past 40 years. The observed changes are ascribed to altered circulation and precipitation patterns and intensified trade winds and probably related to global warming (Curry et al. 2003; Bryden et al. 2005). Projections of future climate by the Intergovernmental Panel on Climate Change with a few exceptions indicate a decrease in precipitation over tropical land areas and an increase in precipitation over the tropical Pacific Ocean and high-latitude regions (Houghton et al. 2001). A decrease in precipitation probably would result in reduced chemical weathering and hence less DSi in tropical rivers. In conjunction with reduced river runoff, the DSi input into the ocean from tropical rivers could be significantly lower than today. It appears that in terms of global change it will be large-scale variations of temperature and moisture distribution rather than regional-scale human activities that affect DSi in tropical rivers.

Global Relevance of Dissolved Silica Fluxes from Tropical Rivers

Of the $6.1 \pm 2.0 \times 10^{12}$ mol DSi introduced into the world ocean annually, $5.0 \pm 0.6 \times 10^{12}$ mol, 82 percent of the global total, is calculated to come from rivers (Tréguer et al. 1995). Because all major world rivers are included in the GEMS database, we assume that extrapolations made from our data for the GEMS rivers are representative of the respective tropical and nontropical regions. The total DSi load of the rivers from our data set amounts to 74.3×10^6 t/yr or 2.7×10^{12} mol/yr, making up about 54 percent of the calculated global total input from rivers. The total DSi load for the STRs, which are not included in the GEMS database, is on the order of $1{,}630 \times 10^3$ t/yr or 58×10^9 mol/yr and accounts for 2.2 percent of the total for tropical regions in our data set. Tropical rivers contribute about 70 percent of the annual total DSi input to the world ocean, even when small rivers are not included. On the continent scale, South America is the largest contributor of DSi, a major part of which comes from the Amazon (Figure 4.3). With 21×10^6 t/yr or 742×10^9 mol/yr, it is the largest single DSi source on Earth, accounting for 15 percent of the calculated global total. Interestingly, the DSi load of African rivers is in the same range as that of Asian rivers, although TSS yield and relative silicate weathering rates are lower by an order of magnitude (Figure 4.3, Tables 4.2 and 4.3). This appears to be related to the larger area drained, the high supply of reactive atmospheric dust, the reduced vegetation cover, and hence the larger mineral surface area exposed to weathering in Africa.

Conclusion

In general, tropical rivers have higher concentrations, yields, and loads of DSi than nontropical rivers and contribute about 70 percent of the total annual DSi input by rivers into the ocean. Our findings indicate that natural factors such as climate, geology, and geomorphology rather than anthropogenic factors such as river regulation, urban and industrial waste disposal, and land use change control DSi in tropical rivers. Asian and Oceanian rivers display the highest DSi yields. They appear to be the most efficient sources of DSi because of optimum weathering and transport conditions, the combination of a young geology and active tectonism in the circum-Pacific ring of fire, high precipitation and runoff, and the fact that silicate weathering rates are highest in that area.

When compared with the larger rivers from the GEMS database, small tropical rivers display higher DSi concentrations and yields in Asia and Oceania and lower DSi concentrations and yields in Africa and South America. Weathering and DSi supply generally are higher in Asia and Oceania. There, extensive floodplains of the large rivers, often covered by grasslands, can store a significant portion of riverine DSi in soils and sediments and as phytoliths in vegetation. Because of the lack of floodplain grasslands, smaller rivers display higher DSi concentrations and yields. As in the weathering-limited boreal and arctic regions (Chapter 5, this volume), vegetation and soils are more important for the supply of DSi in Africa and South America because of generally lower weathering rates. Therefore, the larger vegetation and soil cover in catchments of larger African and South American rivers result in higher DSi concentrations and yields when compared with small rivers.

Although anthropogenic controls may be of regional importance on a global scale, the far more important natural controls determine the DSi concentration, yield, and load of tropical rivers. It is conceivable that large-scale processes of climate change and associated changes of temperature and moisture distribution will affect silicon cycling in tropical rivers and adjacent coastal waters more than human activities. This will be the case particularly for Asia and Oceania, which are the most efficient sources of DSi to the ocean. Nevertheless, increasing human activities entailing further environmental changes may gain importance in controlling DSi in tropical rivers, particularly in South America and Africa, which are less efficient in transferring silicon from the lithosphere into the dissolved form. At present it appears that ecological changes in coastal zones related to changes in nutrient fluxes and composition generally are less affected by changes in the silicon cycle than by changes of the nitrogen and phosphorus cycles in tropical regions.

Literature Cited

Bolaji, G. A., O. S. Awokola, A. M. Ghadebo, and O. Martins. 2002. Movement of minerals and nutrients from river basins in south-western Nigeria into the Atlantic Ocean. In *African basins: LOICZ global change assessment and synthesis of river catchment–coastal sea interaction and human dimensions*, edited by R. S. Arthurton, H. H. Kremer, E. Odada, W. Salomons, and J. I. Marshall Crossland. *LOICZ Reports & Studies* 25:163–167. Texel, The Netherlands: LOICZ.

Bryden, H., H. R. Longworth, and S. A. Cunningham. 2005. Slowing of the Atlantic meridional overturning circulation at 25°N. *Nature* 438:655–657.

Brzezinski, M. A. 1985. The Si:C:N ratio of marine diatoms: Interspecific variability and the effect of some environmental variables. *Journal of Phycology* 21:347–357.

Buddemeier, R. W., S. V. Smith, D. P. Swaney, and C. J. Crossland (eds.). 2002. The role of the coastal ocean in the disturbed and undisturbed nutrient and carbon cycles. *LOICZ Reports & Studies* 24. Texel, The Netherlands: LOICZ.

Carneiro, M. E. R. 1998. *Origem, transporte e destino da materia organica no estuario do Rio Paraiba do Sul, RJ.* Ph.D. thesis, Departamento de Geoquimica, Universidade Federal Fluminense, Niterói, Brazil.

Carvalho, C. E. U. 1997. *Distribuicao espacial, temporal e fluxo de metais pesados na porcao inferior da bacia de drenagem do Rio Paraiba do Sul, RJ.* Ph.D. thesis, Departamento de Geoquimica, Universidade Federal Fluminense, Niterói, Brazil.

Conley, D. J., C. L. Schelske, and E. F. Stoermer. 1993. Modification of the biogeochemical cycle of silica with eutrophication. *Marine Ecology Progress Series* 101:179–192.

Curry, R., B. Dickson, and I. Yashayaev. 2003. A change in the freshwater balance of the Atlantic Ocean over the past four decades. *Nature* 426:826–829.

Dadson, S. J., N. Hovius, H. Chen, W. B. Dade, M.-L. Hsieh, S. D. Willett, J.-C. Hu, M.-J. Horng, M.-C. Chen, C. P. Stark, D. Lague, and J.-C. Lin. 2003. Links between erosion, runoff variability and seismicity in the Taiwan orogen. *Nature* 426:648–651.

Ding, T., D. Wan, C. Wang, and F. Zhang. 2004. Silicon isotope compositions of dissolved silicon and suspended matter in the Yangtze River, China. *Geochimica et Cosmochimica Acta* 68:205–216.

Drever, J. I. 1997. *The geochemistry of natural waters.* Upper Saddle River, NJ: Prentice Hall.

Dudgeon, D. 2000. Large-scale hydrological changes in tropical Asia: prospects for riverine biodiversity. *BioScience* 50:793–806.

Elvidge, C. D., K. E. Baugh, E. A. Kihn, H. W. Kroehl, and E. R. Davis. 1997. Mapping city lights with nighttime data from the DMSP operational linescan system. *Photogrammetric Engineering and Remote Sensing* 63:727–734.

Food and Agriculture Organization of the United Nations. 2003. www.fao.org.

Gaillardet, J., B. Dupre, P. Louvat, and C. J. Allegre. 1999. Global silicate weathering and CO_2 consumption rates deduced from the chemistry of large rivers. *Chemical Geology* 159:3–30.

Global Environmental Monitoring System. 2000. www.gemswater.org.

Houghton, J. T., Y. Ding, D. J. Griggs, M. Noguer, P. J. van der Linden, and D. Xiaosu (eds.). 2001. *Climate change 2001: The scientific basis. Contribution of Working Group I to the Third Assessment Report of the Intergovernmental Panel on Climate Change (IPCC).* Cambridge: Cambridge University Press.

Humborg, C., V. Ittekkot, A. Cociasu, and B. von Bodungen. 1997. Effect of Danube River dam on Black Sea biogeochemistry and ecosystem structure. *Nature* 386:385–388.

Hungspreugs, M., and W. Utoomprurkporn. 1999. Trace metal contamination in Thai rivers. In *Biogeochemistry of rivers in tropical South and Southeast Asia,* edited by V. Ittekkot, V. Subramanian, and S. Annadurai. SCOPE Sonderband, *Mitteilungen aus dem Geologisch-Paläontologischen Institut der Universität Hamburg* 82:139–157.

Jennerjahn, T. C., V. Ittekkot, S. Klöpper, Seno Adi, Sutopo Purwo Nugroho, Nana Sudiana, Anyuta Yusmal, Prie Hartanto, and B. Gaye-Haake. 2004. Biogeochemistry of a tropical river affected by human activities in its catchment: Brantas River estuary and coastal waters of Madura Strait, Java, Indonesia. *Estuarine, Coastal and Shelf Science* 60:503–514.

Jennerjahn, T. C., B. Knoppers, W. F. L. de Souza, C. E. V. Carvalho, G. Mollenhauer, M. Hübner, and V. Ittekkot. In press. Factors controlling the production and accumulation of organic matter along the Brazilian continental margin between the equator and 22°S. In *Carbon and nutrient fluxes in continental margins: A global synthesis,* edited by K. K. Liu, R. Quinones, L. Talaue-McManus, and L. Atkinson. New York: CMTT/IGBP, Springer.

Knoppers, B., P. R. P. Medeiros, W. F. L. de Souza, and T. Jennerjahn. 2006. The São Francisco estuary, E-Brazil. Pp. 51–70 in *Water pollution in estuaries. Handbook of environmental chemistry 5,* edited by P. Wangersky. Berlin: Springer Verlag.

Kristiansen, S., and E. E. Hoell. 2002. The importance of silicon for marine production. *Hydrobiologia* 484:21–31.

Lesack, L. F. W., R. E. Hecky, and J. M. Melack. 1984. Transport of carbon, nitrogen, phosphorus, and major solutes in the Gambia River, West Africa. *Limnology and Oceanography* 29:816–830.

Lu, H., Z. Liu, N. Wu, S. Berne, Y. Saito, B. Liu, and L. Wang. 2002. Rice domestication and climatic change: Phytolith evidence from east China. *Boreas* 31:378–385.

Marques, M., B. Knoppers, and M. Machmann-Oliveira. 2002. The São Francisco River Basin: Environmental impacts and the causal chain analysis. In *Proceedings of the First International Symposium on Transboundary Waters Management. Associacíon Mexicana de Hidrologia* 1:1–17.

Martins, O. 1987. The Ogun River: Geochemical characteristics of a small drainage basin. In *Transport of carbon and minerals in major world rivers* Part 4, edited by E. T. Degens, S. Kempe, and Gan Weibin. SCOPE Sonderband, *Mitteilungen aus dem Geologisch-Paläontologischen Institut der Universität Hamburg* 64:475–482.

McGregor, G. R., and S. Nieuwolt. 1998. *Tropical climatology*, 2nd ed. Chichester, England: Wiley.

Milliman, J. D. 1997. Blessed dams or damned dams? *Nature* 386:325–327.

Milliman, J. D., K. L. Farnsworth, and C. S. Albertin. 1999. Flux and fate of fluvial sediments leaving large islands in the East Indies. *Journal of Sea Research* 41:97–107.

Milliman, J. D., and R. H. Meade. 1983. World-wide delivery of river sediment to the oceans. *Journal of Geology* 91:1–21.

Milliman, J. D., and J. P. M. Syvitski. 1992. Geomorphic/tectonic control of sediment discharge to the ocean: The importance of small mountainous rivers. *Journal of Geology* 100:525–544.

Mwashote, B. M., and I. O. Jumba. 2002. Quantitative aspects of inorganic nutrient fluxes in the Gazi Bay (Kenya): Implications for coastal ecosystems. *Marine Pollution Bulletin* 44:1194–1205.

Petr, T. 1983. Dissolved chemical transport in major rivers of Papua New Guinea. In *Transport of carbon and minerals in major world rivers* Part 2, edited by E. T. Degens, S. Kempe, and H. Soliman. SCOPE Sonderband, *Mitteilungen aus dem Geologisch-Paläontologischen Institut der Universität Hamburg* 55:477–481.

Pigram, C. J., and H. L. Davies. 1987. Terranes and accretion history of the New Guinea orogen. Bureau of Mineral Resources. *Journal of Australian Geology and Geophysics* 10:193–211.

Piperno, D. R., and D. M. Pearsall. 1998. The silica bodies of tropical American grasses: Morphology, taxonomy, and implications for grass systematics and fossil phytolith identification. *Smithsonian Contribution to Botany* 85:1–40.

Qunying, Z., F. Lin, X. Li, and M. Hu. 1987. Major ion chemistry and fluxes of dissolved solids with rivers in southern coastal China. In *Transport of carbon and minerals in major world rivers* Part 4, edited by E. T. Degens, S. Kempe, and Gan Weibin. SCOPE Sonderband, *Mitteilungen aus dem Geologisch-Paläontologischen Institut der Universität Hamburg* 64:243–249.

Rabalais, N. N., R. E. Turner, D. Justic, Q. Dortch, W. J. Wiseman Jr., and B. K. Sen Gupta. 2000. Gulf of Mexico biological system responses to nutrient changes in the Mississippi River. Pp. 241–268 in *Estuarine science: A synthetic approach to research and practice*, edited by J. Hobbie. Washington, DC: Island Press.

Redfield, A. C., B. H. Ketchum, and F. A. Richards. 1963. The influence of organisms on the composition of sea-water. Pp. 26–77 in *The sea*, edited by M. N. Hill. New York: Interscience Publishers.

Tegen, I., S. P. Harrison, K. E. Kohfeld, I. C. Prentice, M. C. Coe, and M. Heimann. 2002. The impact of vegetation and preferential source areas on global dust aerosols: Results from a model study. *Journal of Geophysical Research* 107(D21):4576.

Tréguer, P., D. M. Nelson, A. J. van Bennekom, D. J. DeMaster, A. Leynaert, and B. Queguiner. 1995. The silica balance in the world ocean: A reestimate. *Science* 268:375–379.

Turner, R. E., and N. N. Rabalais. 1994. Coastal eutrophication near the Mississippi River delta. *Nature* 368:619–621.

Turner, R. E., N. N. Rabalais, D. Justic, and Q. Dortch. 2003. Global patterns of dissolved N, P and Si in large rivers. *Biogeochemistry* 64:297–317.

Viers, J., B. Dupre, J.-J. Braun, S. Deberdt, B. Angeletti, J. N. Ngoupayou, and A. Michard. 2000. Major and trace element abundances, and strontium isotopes in the Nyong Basin rivers (Cameroon): Constraints on chemical weathering processes and element transport mechanisms in humid tropical environments. *Chemical Geology* 169:211–241.

Vörösmarty, C. J., M. Meybeck, B. Fekete, K. Sharma, P. Green, and J. P. M. Syvitski. 2003. Anthropogenic sediment retention: Major global impact from registered river impoundments. *Global and Planetary Change* 39:169–190.

Vörösmarty, C. J., K. P. Sharma, B. M. Fekete, A. H. Copeland, J. Holden, J. Marble, and J. A. Lough. 1997. The storage and aging of continental runoff in large reservoir systems of the world. *Ambio* 26:210–219.

Whitten, T., R. E. Soeriaatmadja, and S. A. Afiff. 1996. *The ecology of Java and Bali. The Ecology of Indonesia Series* Vol. II. Halifax, NS: Periplus Editions, Dalhousie University.

World Resources Institute. 1998. www.wri.org.

Wüst, R. A. J., and R. M. Bustin. 2003. Opaline and Al–Si phytoliths from a tropical mire system of West Malaysia: Abundance, habit, elemental composition, preservation and significance. *Chemical Geology* 200:267–292.

Zhang, J. 1996. Nutrient elements in large Chinese estuaries. *Continental Shelf Research* 16:1023–1045.

5

Dissolved Silica Dynamics in Boreal and Arctic Rivers: Vegetation Control over Temperature?

Christoph Humborg, Lars Rahm, Erik Smedberg, Carl-Magnus Mörth, and Åsa Danielsson

The Siberian and Canadian rivers contribute some 17 percent (6,000 km^3/yr) and 10 percent (0.5 Tmol/yr) to the total riverine water and dissolved silica (DSi) input, respectively, into the ocean (Gordeev et al. 1996; Shiklomanov et al. 2000). During the Quaternary, their contributions to the water discharge and thereby to the DSi fluxes have undergone enormous variations, and especially during deglaciation periods these inputs probably were much higher. For example, it was estimated that a single river, the St. Lawrence, discharged some 6,000 km^3/yr during the Younger Dryas (Broecker et al. 1989; Clark et al. 2002). In contrast, the tropical rivers were subjected to less climate variability; that is, it can be assumed that these rivers showed much more constant input patterns of both water and DSi. Therefore, we suggest that boreal and arctic rivers that drain glaciated environments might have been mainly responsible for wide temporal variability of DSi inputs into the ocean. Still, the biogeochemical significance of glacial periodicity to such river systems and their biogeochemical influence on the marine environment are unclear. Both DSi and dissolved inorganic nitrogen (DIN) inputs into the contemporary ocean are estimated at 5–6 Tmol/yr (Jacobsen et al. 2000; Tréguer et al. 1995). Thus, the Si:N ratio of total annual input, that is, loads from rivers and the atmosphere (including N fixation), correspond to the molar uptake demand ratio of diatoms, which is about 1 (Brzezinski 1985). An improved understanding of changes in inputs of DSi will help explain the variation of diatom production over geological time scales.

The pioneering work by Meybeck (1979) showed that DSi concentrations in major world rivers are a function of temperature; that is, the DSi content in river waters commonly varies from about 100 µM in arctic regions to 700 µM in the tropics. Other authors stress the combined effect of runoff temperature and physical denudation as the controlling variables for chemical silicate weathering rates on a global scale and emphasize the significance of active physical denudation of continental rocks, as for example in the Himalayas, for elevated DSi concentrations (Gaillardet et al. 1999). However, watersheds of boreal and arctic rivers are mostly located in central plains, and physical denudation is caused mainly by ice and wind; that is, observed variations in DSi concentrations probably cannot be related to a young geology and active tectonism, as in the tropics (Chapter 4, this volume). DSi concentrations observed at the mouths of major arctic rivers such as the Ob (164 μM), Yenesei (107μM), Lena (70 μM), and Kolyma (67 μM) show a trend of decreasing values toward the east along the Eurasian plate (Gordeev et al. 1996), that is, with decreasing temperature but also with a change from taiga to tundra-dominated vegetation.

Is temperature the leading environmental variable or possibly the effect of vegetation cover that also covaries with temperature? In fact, there are strong links between vegetation and weathering. It has been shown both in experiments (Hinsinger et al. 2001) and in investigations of small streams (Drever and Zobrist 1992; Oliva et al. 1999) that silicate weathering is a function of vegetation cover, which lowers the pH in the topsoils, and this increases silicate weathering (Brantley and Stillings 1996). Among plausible chemical mechanisms by which vascular plants may enhance the weathering rate, three in particular deserve further attention: root exudation of organic acids (Grayston et al. 1997), the activity of ectomycorrhizal fungi (Landeweert et al. 2001), and mineral dissolution by bacteria (Bennett et al. 2001). In a large-scale perspective that covers entire watersheds and even subcontinents (Figure 5.1), total organic carbon (TOC) that consists of more than 80 percent of dissolved organic carbon (DOC) in high-latitude rivers (Wetzel 2001) can be regarded as a proxy for these various organic acids and ligand-forming organic molecules, and a positive correlation between TOC and DSi appears for the Swedish rivers, the rivers entering the Laptev Sea, and the Mackenzie. The TOC and DSi concentrations for the other Siberian rivers shown in Figure 5.1 fall in the same range as the Swedish rivers but are less correlated. A possible explanation might be that these rivers are undersampled, and the values given often consist of only a few measurements. A recent study in the Mackenzie River Basin showed that mean DOC and DSi concentrations were 1088 μM and 78 μM, respectively, and a positive relationship between DOC and DSi concentrations (Figure 5.1, $R^2 = 0.31$, $p = 0.007$) and weathering rates have also been reported (Millot et al. 2003). These relationships between DSi concentrations and vegetation cover and their significance for DSi land–sea fluxes within glacial cycles and perturbations of boreal and arctic rivers by human activities such as hydrological alterations and eutrophication are discussed in this chapter.

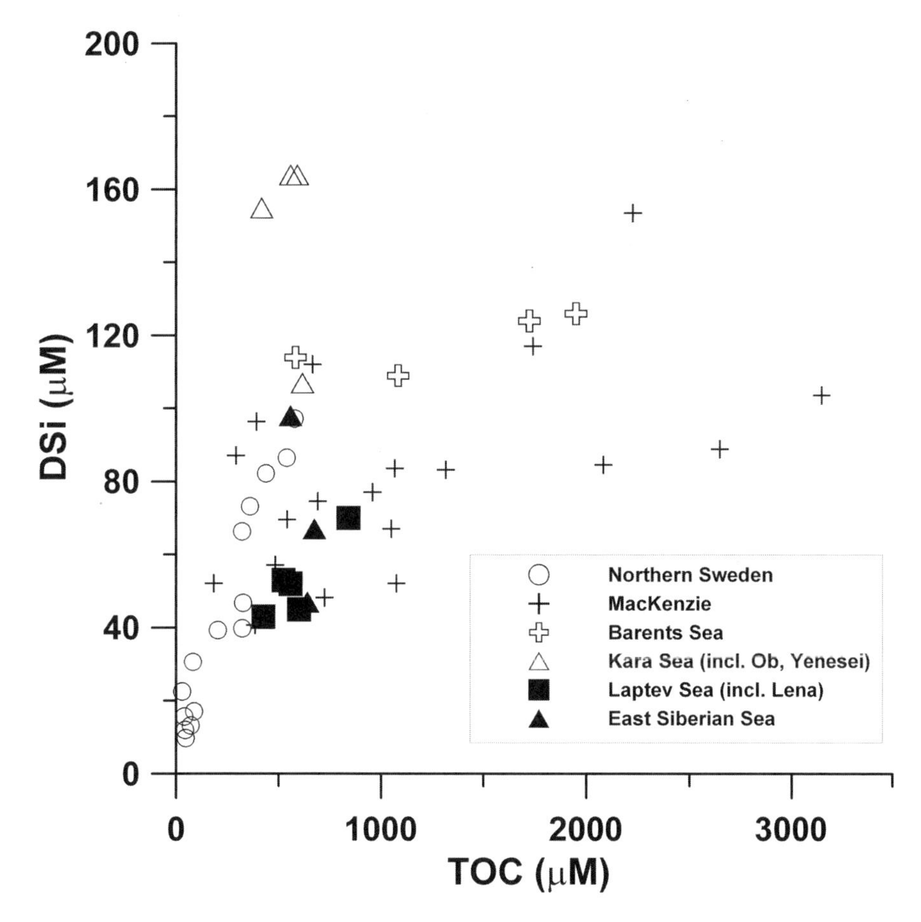

Figure 5.1. DSi versus total organic carbon (TOC) in major boreal and arctic watersheds of Eurasia and North America.

Vegetation Control of River Chemistry: A Case Study in Northern Sweden

Swedish Rivers as Model Systems for High-Latitude Rivers

The northern Swedish river systems of Torneälven, Kalixälven, Råneälven, Luleälven, Piteälven, Skellefteälven, and Umeälven provide ideal case studies with sufficient background data for investigating the role of vegetation (vascular plants) on nutrient land–sea fluxes of boreal and arctic rivers.

These rivers are midsize, and their water discharge ranges between 2 and 17 km^3/yr. They are located between N 64° and N 69° in northern Sweden, several originating above the Arctic Circle. The whole area is sparsely inhabited and, apart from the regulation of the rivers Luleälven, Skellefteälven, and Umeälven, might be characterized as unperturbed. Cold conditions favor boreal biomes (taiga, tundra), as is typical of the large Siberian and Canadian rivers also (Kohfeld and Harrison 2000). The spatial vegetation gradient found in watersheds of these rivers, from essentially unvegetated rocks at the top elevations, through alpine pasture, shrub, and brushwood, to coniferous and deciduous forests and mires in the river valleys, represents dominating biotopes of recently deglaciated environments. To cover finer spatial scales, allowing us to test whether an effect of vegetation cover on weathering rates can be found over various spatial catchment scales, we also investigated some subcatchments of the rivers Kalixälven (K1–K3) and Luleälven (L1–L9) so that in total nineteen river catchments and subcatchments, ranging in size from 30 to 40,000 km^2, were analyzed (Figure 5.2). A large effect of temperature on river chemistry can be ruled out because the temperature differences between the mountainous headwaters and the river mouths are quite small in these regions (about 3°C; the weathering rate increases by about 6 percent for each °C; Lasaga 1998).

Organic matter and inorganic nitrogen and phosphorus concentrations in boreal and arctic rivers are among the highest and lowest, respectively, reported worldwide (Dittmar and Kattner 2003). Thus, high-latitude rivers are generally oligotrophic, and autochthonous retention processes of DSi in their watersheds (as, for example, through diatom blooms in lakes) are naturally low. Mean global concentrations of TOC, DIN, dissolved inorganic phosphorus (DIP), and DSi for recent unperturbed arctic and boreal rivers are estimated as 750, 6, 0.1, and 110 µM, respectively (Meybeck 1979, 1982). The investigated Swedish rivers, and especially the unperturbed forest and wetland catchments (major catchments shown in Figure 5.2a and L7 and L9 in Figure 5.2b), represent these conditions reasonably well (Humborg et al. 2003). Most riverine nutrient inputs to the Arctic Ocean come from Russia, and an extensive data set has recently been intercalibrated (Holmes et al. 2001). DIN values reported from the three largest Siberian rivers (the Ob, Lena, and Yenesei; Holmes et al. 2000) show low concentrations comparable to those of the studied rivers of northern Sweden. Time series of DIP from the rivers Lena and Yenesei indicate concentrations between 0.1 and 0.3 µM, respectively (Holmes et al. 2000, 2001).

Concentrations of TOC and DSi in the major Siberian rivers likewise resemble our records.

Mean TOC and DSi concentrations of major Siberian rivers, including the Lena, Ob, Yenesei, and Onega, are 615 µM and 95 µM, respectively (Gordeev 2000). Thus,

Figure 5.2. (*opposite page*)
(a) Catchment areas and land cover characteristics of the investigated major river systems in northern Sweden. *(b)* Headwater area of the rivers Kalixälven and Luleälven, showing location of sampling sites, subcatchment areas, and land cover characteristics.

a

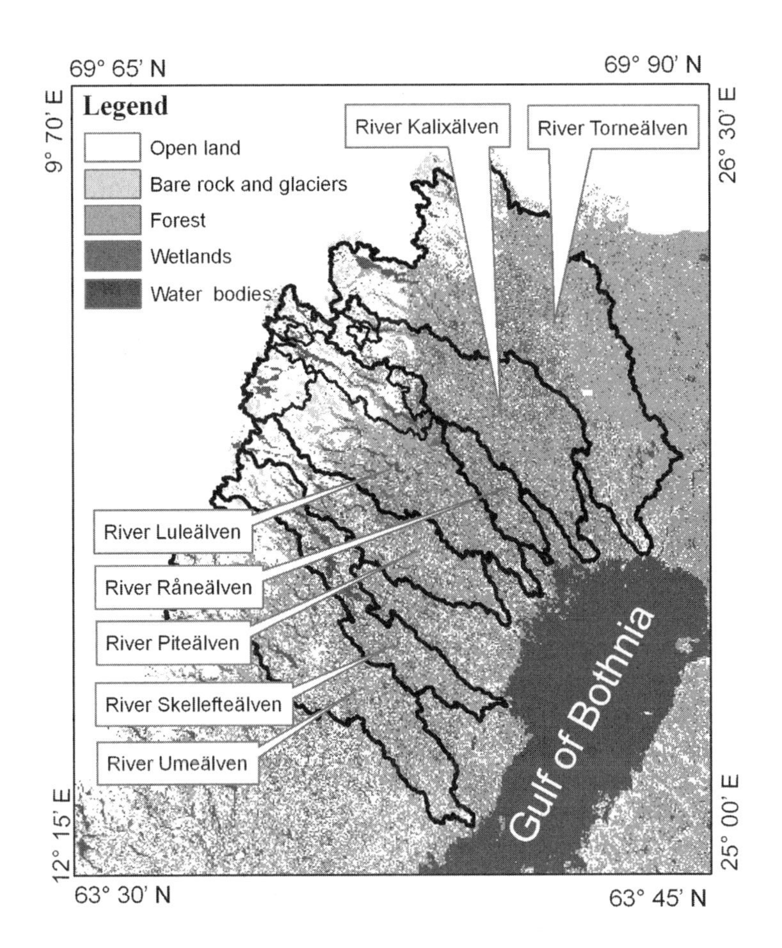
69° 65' N
69° 90' N
9° 70' E
26° 30' E
Legend
Open land
Bare rock and glaciers
Forest
Wetlands
Water bodies
River Kalixälven
River Torneälven
River Luleälven
River Råneälven
River Piteälven
River Skellefteälven
River Umeälven
Gulf of Bothnia
12° 15' E
25° 00' E
63° 30' N
63° 45' N

b

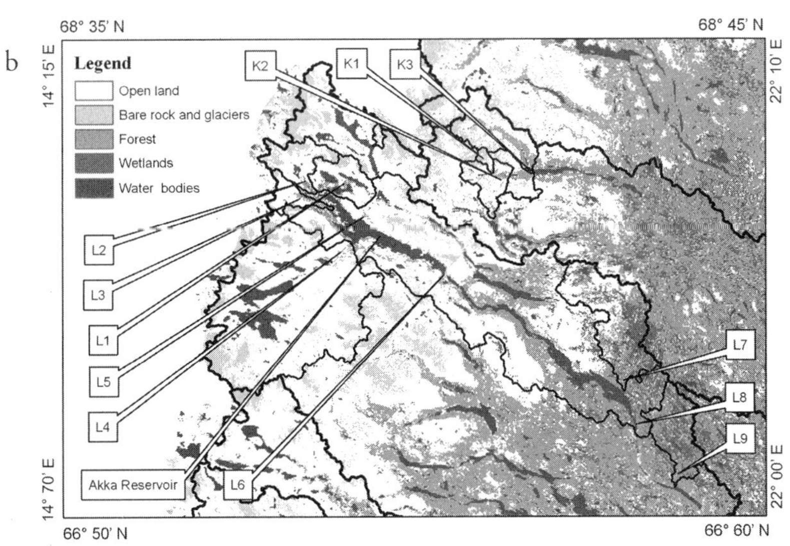
68° 35' N
68° 45' N
14° 15' E
22° 10' E
Legend
Open land
Bare rock and glaciers
Forest
Wetlands
Water bodies
K2
K1
K3
L2
L3
L1
L5
L4
L7
L8
L9
Akka Reservoir
L6
14° 70' E
22° 00' E
66° 50' N
66° 60' N

the fragmented data sets of major arctic Siberian rivers indicate that their nutrient concentrations and dynamics are rather similar to those presented here for the northern Swedish rivers.

Regression Analysis on River Chemistry Versus Landscape Variables

The highest concentrations of TOC and DSi were found in the forested catchments and wetlands, the regulated rivers had much lower concentrations, and the lowest concentrations were observed in all mountainous headwaters (K1–K3, L1–L6; Figure 5.2). Statistically, TOC and DSi concentrations are significantly lower in headwaters (TOC = 56.9 ± 22.8 μM, DSi = 26.3 ± 19.0 μM) than in the forested catchments (major river catchments in Figure 5.2a) and wetlands (L7 and L9) (TOC = 579.1 ± 125.5 μM, DSi = 67.7 ± 21.2 μM). Discharge-weighted DIN, DIP, and DSi concentrations in the various streams and rivers were statistically related to the discharge-weighted TOC concentration and to the fractions of the total catchment area covered by forest, wetlands, and lakes (Figure 5.3). It is striking that DSi and DIP (which originate from weathering) both increased by an order of magnitude with increasing concentration of TOC (range = 29–540 μM) and showed similar increases with percentage area of forest (range = 0–56 percent) and wetland (range = 0–39 percent). There was also a strong positive linear correlation between TOC concentration (not shown) and the percentage area of forest (R^2 = 0.90) and wetland (r^2 = 0.70).

The strongest correlation (Figure 5.3) appeared between DSi and TOC (R^2 = 0.92), followed by DIP and percentage forest area (r^2 = 0.78). The TOC concentration reflects soil organic matter and includes weathering agents such as organic acids. The concentrations of weathering products other than DSi, such as dissolved Ca, Mg, K, and Na, were also positively correlated to vegetation cover. Note that the high DSi and DIP concentrations reported in Figure 5.3 refer to small (L7, L9), midsized (Råneälven), and larger catchment (Torneälven, Kalixälven) areas. Finally, a negative but statistically nonsignificant correlation resulted from regression of DSi concentration on percentage area occupied by lakes and reservoirs (r^2 = 0.29). Previously, a significant correlation was reported for rivers in Sweden and Finland (Conley et al. 2000).

The DIN concentration pattern differed from that of DSi and DIP (Figure 5.3). DIN concentrations were generally low, about 2–4 μM, regardless of whether the lotic waters originated from catchments dominated by barren ground, forest, or wetland; consequently, coefficient values in the regression analyses were much lower.

Principal Component Analysis on Landscape Variables and River Chemistry

To analyze how typical landscape characteristics of these boreal and subarctic watersheds are related to river biogeochemistry, a multivariate technique, principal component

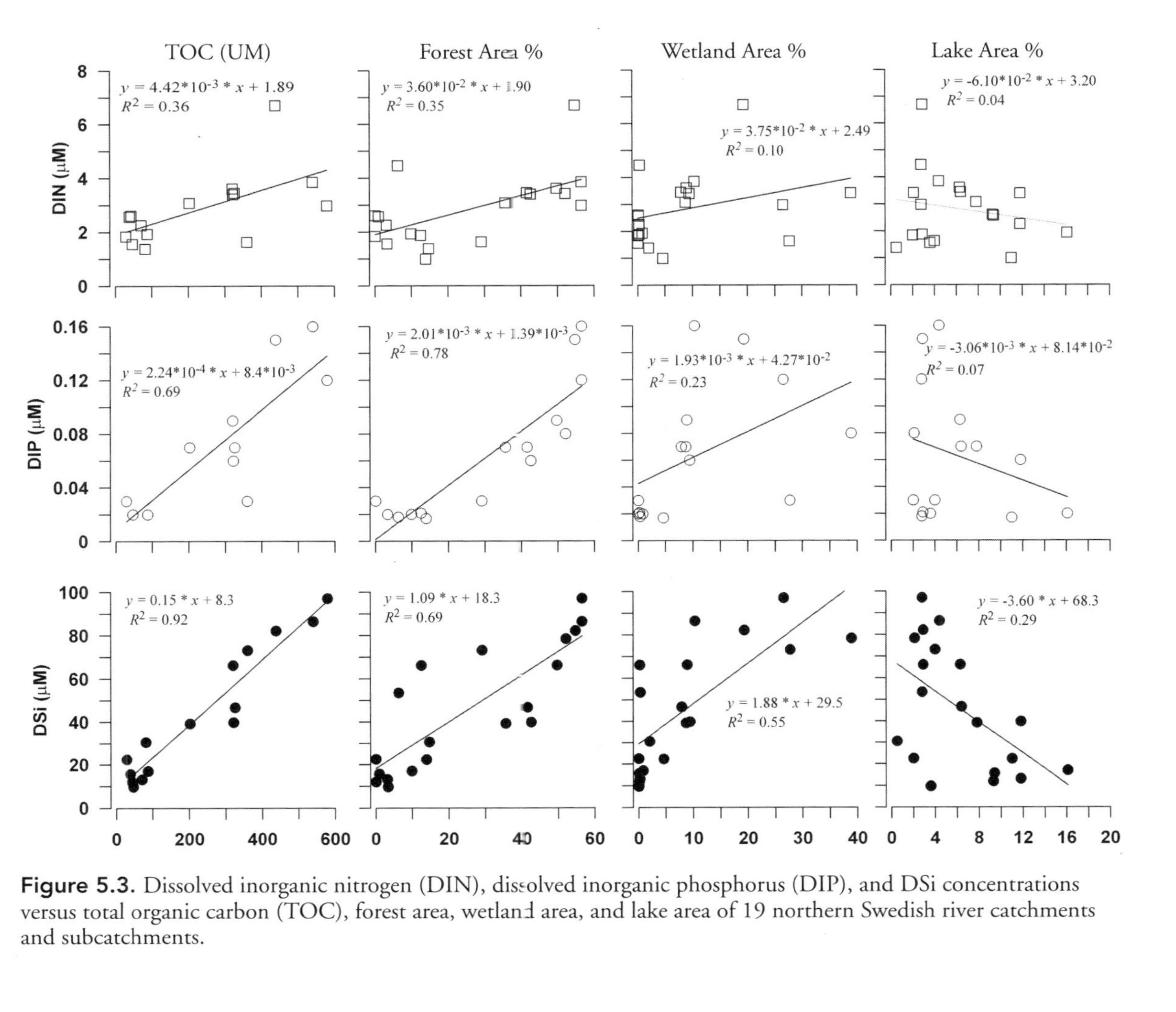

Figure 5.3. Dissolved inorganic nitrogen (DIN), dissolved inorganic phosphorus (DIP), and DSi concentrations versus total organic carbon (TOC), forest area, wetland area, and lake area of 19 northern Swedish river catchments and subcatchments.

analysis (PCA), was used. PCA ordinates a large number of variables along an equal number of orthogonal components. Each of these principal components is a linear combination of all variables, with the first component describing the largest part of the variation in the original data and so on. Landscape variables typical for subarctic and boreal watersheds analyzed in this study were land cover (wetland area, lake area, ice and snow area, unvegetated rock area, open land area, coniferous and deciduous forest areas), soil types (till above and below the tree line, peat, bare rock and thin soils, glaciofluvial sediment, sand and gravel, glacier, and silt and clay), and bedrock types (shale, carbonate-rich shale, sandstone, quartzite, gneiss, alkaline rock, and granite and acid volcanic rock). PCA of the data using these landscape variables and data on river chemistry explained 49 percent and 18 percent of the variance in data along the first and second axes, respectively (Figure 5.4).

The positive side of the first axis is related to land cover (e.g., coniferous forest and wetland areas), soil types (e.g., till below tree line, silt and clay, and peat), bedrock type (e.g., granite and acid volcanic rock), and also dissolved river constituents such as TOC as well as elements originating from weathering (i.e., DSi, DIP, Ca, Mg, K, and Na). Thus, all landscape variables connected with vegetation are grouped together with river chemistry variables that originate from weathering. On the negative side of the first axis, land cover types such as open land area and ice and snow area, as well as bare rock and thin soils as the soil type, were primarily highlighted. Notably, till above the tree line was positioned on the negative side of the first axis. Thus, environments with high physical weathering, thin soil coverage, and sparse vegetation were not connected to river chemistry variables. On the second axis, variables with high explanatory variance positioned on the positive side were lake area (as a land cover type) and shale as well as carbonate-rich shale (as bedrock types), whereas the negative side was related to the variables glacier as a (soil type) and alkaline rock as a (bedrock type).

To summarize, it can be concluded from this case study that the presence of vegetation (vascular plants), in combination with chemically unweathered soils and organic-rich deposits in wetlands (essentially mires), appear to be crucial for the TOC, DSi, and DIP concentrations in the boreal rivers studied. It appears that physical weathering, by glacial, ice, and wind erosion, of bedrock containing Si and P produces primary clay minerals but does not immediately lead to higher fluxes of DSi and DIP in high-latitude watersheds.

Contribution of High-Latitude Rivers to DSi Land–Sea Fluxes over Glacial Cycles

Boreal and arctic rivers might have had global significance for the DSi budget of the oceans because of the huge variations in river discharge that fluctuated from essentially nil during the last glacial maximum (about 21 kyr BP) to an average value of 4,500 km^3 during periods of active deglaciation in the Holocene, with a major contribution com-

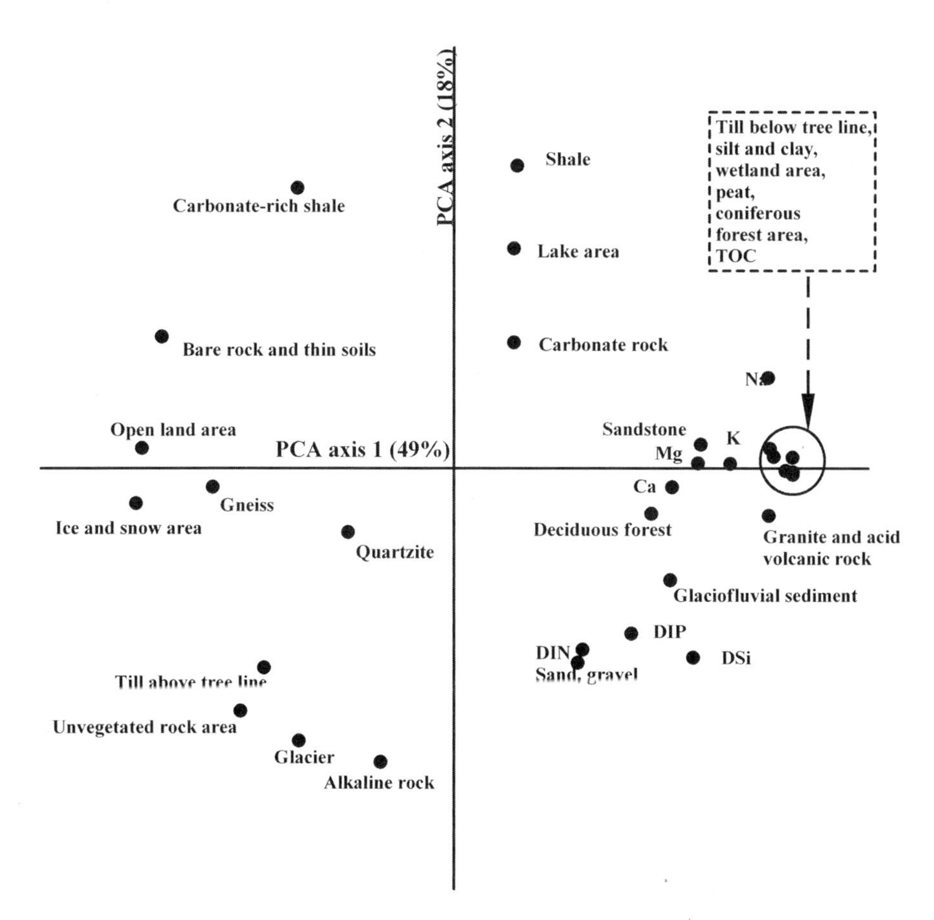

Figure 5.4. Principal component analysis (PCA) ordination of data on landscape characteristics (land cover, soil types, and bedrock types), and on river biogeochemistry of 17 river catchments of northern Sweden. DIN = dissolved inorganic nitrogen, DIP = dissolved inorganic phosphorus, TOC = total organic carbon.

ing from North American and Eurasian ice shields (Lambeck 1990; Lambeck et al. 2002). For periods of very rapid ice sheet decay, the water discharge of boreal and arctic rivers increased up to 14,500 km^3 (Bard et al. 1996). These water masses passed forested watersheds that appeared simultaneously during deglaciation events along the ice shields (Kohfeld and Harrison 2000), thus allowing the water to be charged with some 100 μM of DSi before reaching the oceans. During the mid-Holocene (about 9 kyr BP), the arctic forest limit was north of its present position in the Mackenzie Delta (Edwards et al. 2000), Europe (Prentice et al. 1996), and western and central Siberia (Texier et al. 1997). Simultaneously, forest biomes encroached on the present-day steppe in southeastern Europe and central Asia (Tarasov et al. 2000). Thus, such deglaciation events could have contributed some 1.6 Tmol/yr DSi input, which is three times more than today or 30 percent of the estimated input by all global rivers today (Tréguer et al. 1995).

Perturbations by Hydrological Alterations and River Eutrophication

River damming converted extended river valleys, where formerly the lentic waters percolated through wetlands and forests, into huge water bodies (reservoirs) that no longer have significant exchange with the riparian zone because the water level fluctuations in the reservoirs have eroded the marginal zone completely. These reservoirs often are some 50–100 km long, so major parts of the former riverbed are affected. In reaches between the reservoirs, underground channeling of water and a reduction of water level fluctuations result in further decreases in soil–water contact, consequently diminishing weathering rates. Apparent changes with damming are the much lower and year-round uniform DSi concentrations in the River Luleälven, which is the most heavily dammed river in Eurasia, compared with the high seasonal variations in the River Kalixälven, which forms the neighboring unregulated watershed of the Luleälven, as measured at the river mouths (Figure 5.5 a–b).

Lower concentrations in the River Luleälven are likewise observed for other major cations, nutrients, and TOC. The impact of damming can be demonstrated even on a larger scale by comparing the reservoir live storage and the DSi concentrations of twelve major Swedish rivers, all draining the Scandinavian mountain chain (i.e., rivers with similar climatic conditions and watershed properties such as bedrock types, soil types, and vegetation structure). The DSi concentration is clearly inversely correlated with the reservoir live storage.

Thus, the extensive damming of Swedish boreal rivers can be seen as a large-scale experiment that confirms the positive effect of vegetation cover on DSi land–sea fluxes because regulation of major Swedish rivers resulted mainly in less contact of surface waters with vegetated soils (Humborg et al. 2002).

Another drastic change in DSi land–sea fluxes can be assumed as an effect of eutrophication in major watersheds as a result of increasing fertilizer use and wastewater inputs, although the consequences of eutrophication for DSi fluxes are far less documented than for N and P (Humborg et al. 2000). A possible explanation for the lower DSi concentrations in eutrophic rivers is that ongoing eutrophication increases deposition of diatoms in lotic waters of a watershed (Schelske et al. 1983). Time series from the Mississippi River (Turner and Rabalais 2003) and also investigations in boreal and subarctic rivers such as various Norwegian catchments (Hessen 1999) have clearly shown that DSi runoff is negatively correlated with fertilizer use in the watersheds (i.e., greater P concentrations in watercourses, lakes, and reservoirs lead to higher DSi

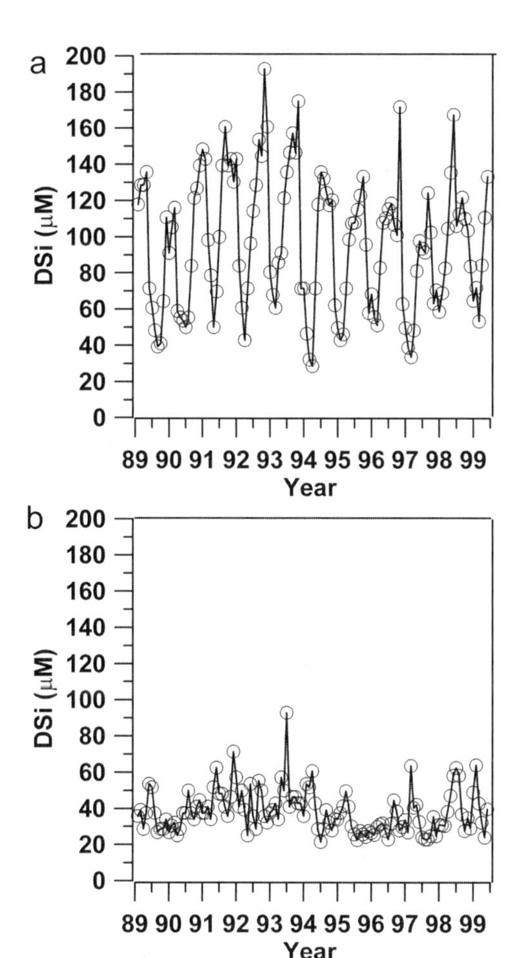

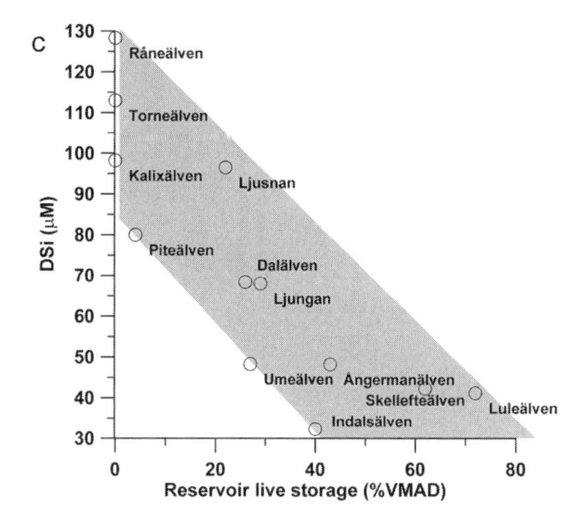

Figure 5.5. DSi concentrations *(a)* versus time at the mouth of the Kalixälven (nonregulated), *(b)* versus time at the mouth of the River Luleälven (regulated) (data from monthly measurements monitored at the river mouths by the Swedish University of Agricultural Sciences, Uppsala, Sweden), and *(c)* versus reservoir live storage of 12 Swedish rivers draining the Scandinavian mountain chain. The reservoir live storage is expressed as the percentage of the virgin mean annual discharge (%VMAD) of the river system that can be contained in reservoirs and thus is a measure of the degree of damming of a river; *live storage* refers to the volume that can be withheld in, and subsequently released from, the reservoir.

deposition in the watersheds because of diatom blooms). In Figure 5.6 the median DSi concentrations from all major rivers (n = 84) draining into the various subbasins of the Baltic Sea are shown. Highest concentrations were found in the Bothnian Bay and Bothnian Sea, which are sparsely populated, and eutrophication is much less than in the southeastern part of the Baltic Sea drainage basin, where most of the 90 million inhabitants of the Baltic Sea catchment live and cultural eutrophication, as in Poland, Germany, Denmark, and the Baltic states, is well reported (Stålnacke et al. 1999; Wulff et al. 1990). It would be expected that the DSi concentrations would behave in the opposite way, that is, much higher concentrations would be found in the rivers of the southeastern watershed of the Baltic Sea (Baltic proper, Gulf of Finland, and Gulf of Riga) because they are much less dammed, the annual mean temperature is higher, and vegetation cover much denser. Therefore, the low DSi concentrations may be attributed to river eutrophication (i.e., DSi sequestration in the lentic waters of the watersheds). In fact, the rivers with large percentages in both lake area and agricultural area, such as the Neva and the Götaälven, which discharge into the Baltic proper and Kategatt, respectively, showed DSi concentrations as low as less than 10 μM.

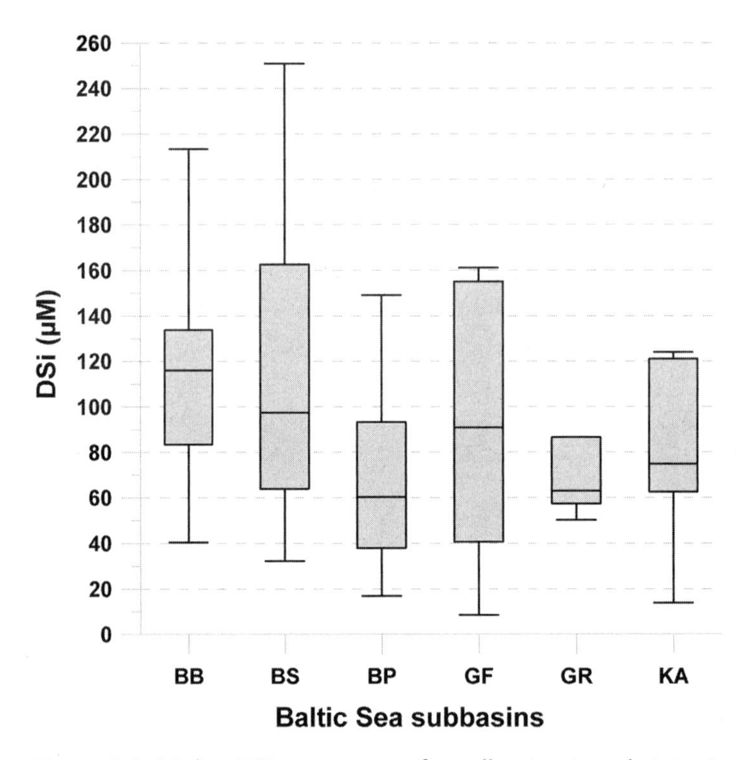

Figure 5.6. Median DSi concentration from all major rivers draining into the various subbasins of the Baltic Sea. BB = Bothnian Bay, BP = Baltic proper, BS = Bothnian Sea, GF = Gulf of Finland, GR = Gulf of Riga, KA = Kattegatt.

The effect of decreased DSi land–sea fluxes as a result of damming and cultural eutrophication in the river catchments is most obvious in enclosed seas with long water residence times. A drastic decrease of DSi concentration has been reported for the surface waters of the Black Sea after the erection of a major dam in the Danube (Humborg et al. 1997). A similar DSi decrease appears also to be the case in the Baltic Sea (Figure 5.7).

Significant decrease in DSi stocks of the Baltic Sea have been reported since the 1970s (Rahm et al. 1996; Sandén and Rahm 1993). This decrease occurred simultaneously with significant increases in N and P concentrations.

The potential effect of land use changes in high-latitude watersheds such as the conversion of peat land and forest to arable land is not well understood. Two effects are conceivable. The first is a decrease in DSi fluxes as a result of the buildup of Si in certain crop types. The second is an increase in DSi fluxes caused by the greater weathering of arable land compared with forests, as observed in the watershed of the Mississippi (Raymond and Cole 2003) or a mobilization of phytoliths, a significant Si source in soils (Conley 2002; see also Chapter 3, this volume). However, Conley et al. (2000) did not

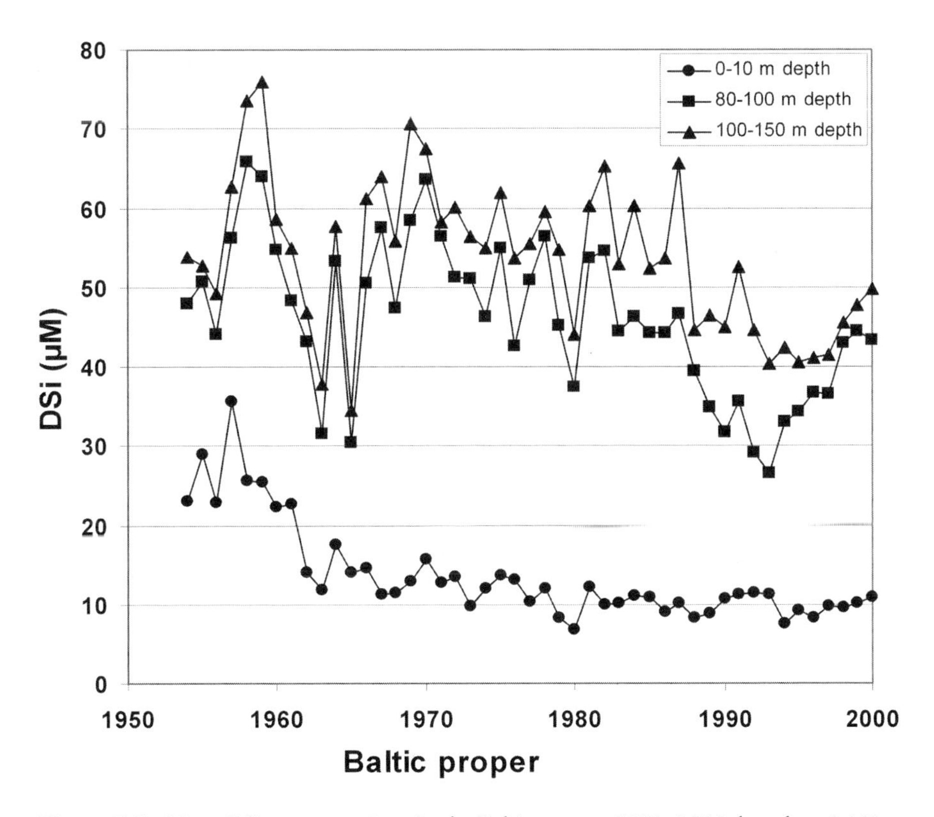

Figure 5.7. Mean DSi concentrations in the Baltic proper, 1950–2000, based on 9,193 stations (vertical casts, 0–10 m) and 4,919 stations (80–150 m) (data from the Baltic Environmental Database).

find any relationship between the proportion of arable land and DSi concentrations in Swedish and Finnish rivers.

Conclusion

The DSi concentrations of boreal and arctic rivers are 30–50 percent lower than the global average of 150 μM (Tréguer et al. 1995). Terrestrial vegetation cover in high-latitude watersheds that varied significantly over glacial cycles might be an additional if not the leading variable besides physical denudation (Gaillardet et al. 1999) and temperature (Meybeck 1979) that controls riverine DSi input to the northern oceans. Physical weathering of bedrock containing silicates have been greatest during the last glacial maximum, and it has been speculated that this led to higher land–sea fluxes of DSi (Froelich et al. 1992). However, our study on the mountainous headwaters of the northern Swedish rivers and other studies in glaciated terrain in Alaska (Anderson et al. 2000) show that physical weathering in high-latitude watersheds leads mostly to the formation of primary clay minerals but does not immediately lead to higher DSi fluxes, as observed for tropical rivers draining areas characterized by active tectonics and young geology (see Chapter 4, this volume). High DSi concentrations and fluxes occur only when the water has passed the tree line.

Boreal and arctic rivers are characterized by a weathering-limited regime (i.e., transport processes remove weathered material faster than weathering processes produce new material; Drever 1997) as a result of low temperature, thin soils or exposed bedrock outcrops with low vegetation cover. Thus, these rivers might be affected in two ways by human interventions: Hydrological alterations change the weathering regimes of entire watersheds, and river eutrophication increases DSi sequestration in both natural lakes and reservoirs. Finally, because global warming is believed to be especially pronounced at high latitudes in the Northern Hemisphere (IPCC 2001), a change in structure and cover of vegetation could alter quite rapidly the biogeochemistry of river catchments and land–ocean interactions along the coasts of the Arctic Ocean.

Literature Cited

Anderson, S. P., J. I. Drever, C. D. Frost, and P. Holden. 2000. Chemical weathering in the foreland of a retreating glacier. *Geochimica et Cosmochimica Acta* 64:1173–1189.

Bard, E., B. Hamelin, M. Arnold, L. Montaggioni, G. Cabioch, G. Faure, and F. Rougerie. 1996. Deglacial sea-level record from Tahiti corals and the timing of global meltwater discharge. *Nature* 382:241–244.

Bennett, P. C., J. R. Rogers, and W. J. Choi. 2001. Silicates, silicate weathering, and microbial ecology. *Geomicrobiology Journal* 18:3–19.

Brantley, S. L., and L. Stillings. 1996. Feldspar dissolution at 25 degrees C and low pH. *American Journal of Science* 296:101–127.

Broecker, W. S., J. P. Kennett, B. P. Flower, J. T. Teller, S. Trumbore, G. Bonani, and W.

Wolfli. 1989. Routing of meltwater from the Laurentide ice-sheet during the Younger Dryas cold episode. *Nature* 341:318–321.

Brzezinski, M. A. 1985. The Si:C:N ratio of marine diatoms: Interspecific variability and the effect of some environmental variables. *Journal of Phycology* 21:347–357.

Clark, P. U., N. G. Pisias, T. F. Stocker, and A. J. Weaver. 2002. The role of the thermohaline circulation in abrupt climate change. *Nature* 415:863–869.

Conley, D. J. 2002. Terrestrial ecosystems and the global biogeochemical silica cycle. *Global Biogeochemical Cycles* 16. doi: 10.1029/2002GB001894.

Conley, D. J., P. Stälnacke, H. Pitkänen, and A. Wilander. 2000. The transport and retention of dissolved silicate by rivers in Sweden and Finland. *Limnology and Oceanography* 45:1850–1853.

Dittmar, T., and G. Kattner. 2003. The biogeochemistry of the river and shelf ecosystem of the Arctic Ocean: A review. *Marine Chemistry* 83:103–120.

Drever, J. I. 1997. *The geochemistry of natural waters.* Englewood Cliffs, NJ: Prentice Hall.

Drever, J. I., and J. Zobrist. 1992. Chemical-weathering of silicate rocks as a function of elevation in the southern Swiss Alps. *Geochimica et Cosmochimica Acta* 56:3209–3216.

Edwards, M. E., P. M. Anderson, L. B. Brubaker, T. A. Ager, A. A. Andreev, N. H. Bigelow, L. C. Cwynar, W. R. Eisner, S. P. Harrison, F. S. Hu, D. Jolly, A. V. Lozhkin, G. M. MacDonald, C. J. Mock, J. C. Ritchie, A. V. Sher, R. W. Spear, J. W. Williams, and G. Yu. 2000. Pollen-based biomes for Beringia 18,000, 6000 and 0 C-14 yr BP. *Journal of Biogeography* 27:521–554.

Froelich, P. N., V. Blanc, R. A. Mortlock, S. N. Chillrud, W. Dunstan, A. Udomkit, and T.-H. Peng. 1992. River fluxes of dissolved silica to the oceans were higher during glacials: Ge/Si in diatoms, rivers and oceans. *Paleoceanography* 7:739–767.

Gaillardet, J., B. Dupre, P. Louvat, and C. J. Allegre. 1999. Global silicate weathering and CO_2 consumption rates deduced from the chemistry of large rivers. *Chemical Geology* 159:3–30.

Gordeev, V. V. 2000. River input of water, sediment, major ions, nutrients and trace metals from Siberian territory to the Arctic Ocean. Pp. 297–322 in E. L. Lewis (ed.), *The freshwater budget of the Arctic Ocean.* Dordrecht, The Netherlands: Kluwer.

Gordeev, V. V., J. M. Martin, I. S. Sidorov, and M. V. Sidorova. 1996. A reassessment of the Eurasian river input of water, sediment, major elements, and nutrients to the Arctic Ocean. *American Journal of Science* 296:664–691.

Grayston, S. J., D. Vaughan, and D. Jones. 1997. Rhizosphere carbon flow in trees, in comparison with annual plants: The importance of root exudation and its impact on microbial activity and nutrient availability. *Applied Soil Ecology* 5:29–56.

Hessen, D. O. 1999. Catchment properties and the transport of major elements to estuaries. *Advances in Ecological Research* 29:1–41.

Hinsinger, P., O. N. F. Barros, M. F. Benedetti, Y. Noack, and G. Callot. 2001. Plant-induced weathering of a basaltic rock: Experimental evidence. *Geochimica et Cosmochimica Acta* 65:137–152.

Holmes, R. M., B. J. Peterson, V. V. Gordeev, A. V. Zhulidov, M. Meybeck, R. B. Lammers, and C. J. Vorosmarty. 2000. Flux of nutrients from Russian rivers to the Arctic Ocean: Can we establish a baseline against which to judge future changes? *Water Resources Research* 36:2309–2320.

Holmes, R. M., B. J. Peterson, A. V. Zhulidov, V. V. Gordeev, P. N. Makkaveev, P. A. Stunzhas, L. S. Kosmenko, G. H. Köhler, and A. I. Shiklomanov. 2001. Nutrient chemistry of the Ob' and Yenisey rivers, Siberia: Results from June 2000 expedition and evaluation of long-term data sets. *Marine Chemistry* 75:219–227.

Humborg, C., S. Blomqvist, E. Avsan, Y. Bergensund, E. Smedberg, J. Brink, and C. M. Mörth. 2002. Hydrological alterations with river damming in northern Sweden: Implications for weathering and river biogeochemistry. *Global Biogeochemical Cycles* 16:1039.

Humborg, C., D. J. Conley, L. Rahm, F. Wulff, A. Cociasu, and V. Ittekkot. 2000. Silicon retention in river basins: Far-reaching effects on biogeochemistry and aquatic food webs in coastal marine environments. *Ambio* 29:45–50.

Humborg, C., Å. Danielsson, B. Sjøberg, and M. Green. 2003. Nutrient land–sea fluxes in oligotrophic and pristine estuaries of the Gulf of Bothnia, Baltic Sea. *Estuarine, Coastal and Shelf Science* 56:781–793.

Humborg, C., V. Ittekkot, A. Cociasu, and B. von Bodungen. 1997. Effect of Danube River dam on Black Sea biogeochemistry and ecosystem structure. *Nature* 386:385–388.

IPCC. 2001. *Climate change 2001: The scientific basis.* Cambridge: Cambridge University Press.

Jacobsen, M. C., R. J. Charlson, H. Rodhe, and G. H. Orians. 2000. *Earth system science.* San Diego, CA: Academic Press.

Kohfeld, K. E., and S. P. Harrison. 2000. How well can we simulate past climates? Evaluating the models using global palaeoenvironmental datasets. *Quaternary Science Reviews* 19:321–346.

Lambeck, K. 1990. Late Pleistocene, Holocene and present sea-levels: Constraints on future change. *Global and Planetary Change* 89:205–217.

Lambeck, K., T. M. Esat, and E. K. Potter. 2002. Links between climate and sea levels for the past three million years. *Nature* 419:199–206.

Landeweert, R., E. Hoffland, R. D. Finlay, T. W. Kuyper, and N. van Breemen. 2001. Linking plants to rocks: Ectomycorrhizal fungi mobilize nutrients from minerals. *Trends in Ecology and Evolution* 16:248–254.

Lasaga, A. 1998. *Kinetic theory in the earth science.* Princeton, NJ: Princeton University Press.

Meybeck, M. 1979. Pathways of major elements from land to ocean through rivers. Pp. 18–30 in *River inputs to ocean systems,* edited by J. M. Martin, J. D. Burton, and D. Eisma. Rome: UNEP IOC SCOR United Nations.

Meybeck, M. 1982. Carbon, nitrogen, and phosphorus transport by world rivers. *American Journal of Science* 282:401–450.

Millot, R., J. Gaillardet, B. Dupre, and C. J. Allegre. 2003. Northern latitude chemical weathering rates: Clues from the Mackenzie River Basin, Canada. *Geochimica et Cosmochimica Acta* 67:1305–1329.

Oliva, P., J. Viers, B. Dupre, J. P. Fortune, F. Martin, J. J. Braun, D. Nahon, and H. Robain. 1999. The effect of organic matter on chemical weathering: Study of a small tropical watershed: Nsimi-Zoetele site, Cameroon. *Geochimica et Cosmochimica Acta* 63:4013–4035.

Prentice, I. C., J. Guiot, B. Huntley, D. Jolly, and R. Cheddadi. 1996. Reconstructing biomes from palaeoecological data: A general method and its application to European pollen data at 0 and 6 ka. *Climate Dynamics* 12:185–194.

Rahm, L., D. Conley, P. Sandén, F. Wulff, and P. Stalnacke. 1996. Time series analysis of nutrient inputs to the Baltic Sea and changing DSi:DIN ratios. *Marine Ecology Progress Series* 130:221–228.

Raymond, P. A., and J. J. Cole. 2003. Increase in the export of alkalinity from North America's largest river. *Science* 301:88–91.

Sandén, P., and L. Rahm. 1993. Nutrient trends in the Baltic Sea. *Environmetrics* 4:75–103.
Schelske, C. L., E. F. Stoermer, D. J. Conley, J. A. Robbins, and R. M. Glover. 1983. Early eutrophication in the lower Great Lakes: New evidence from biogenic silica in sediments. *Science* 222:320–322.
Shiklomanov, I. A., R. B. Shiklomanov, B. J. Lammers, B. J. Peterson, and C. J. Vörösmarty. 2000. The dynamics of river water inflow to the Arctic Ocean. Pp. 281–296 in *The freshwater budget of the Arctic Ocean*, edited by E. L. Lewis. Dordrecht, The Netherlands: Kluwer.
Stålnacke, P., A. Grimvall, K. Sundblad, and A. Tonderski. 1999. Estimation of riverine loads of nitrogen and phosphorus to the Baltic Sea, 1970–1993. *Environmental Monitoring and Assessment* 58:173–200.
Tarasov, P. E., V. S. Volkova, T. Webb, J. Guiot, A. A. Andreev, L. G. Bezusko, T. V. Bezusko, G. V. Bykova, N. I. Dorofeyuk, E. V. Kvavadze, I. M. Osipova, N. K. Panova, and D. V. Sevastyanov. 2000. Last glacial maximum biomes reconstructed from pollen and plant macrofossil data from northern Eurasia. *Journal of Biogeography* 27:609–620.
Texier, D., N. de Noblet, S. P. Harrison, A. Haxeltine, D. Jolly, S. Joussaume, F. Laarif, I. C. Prentice, and P. Tarasov. 1997. Quantifying the role of biosphere–atmosphere feedbacks in climate change: Coupled model simulations for 6000 years BP and comparison with palaeodata for northern Eurasia and northern Africa. *Climate Dynamics* 13:865–882.
Tréguer, P., D. M. Nelson, A. J. van Bennekom, D. J. Demaster, A. Leynaert, and B. Queguiner. 1995. The silica balance in the world ocean: A reestimate. *Science* 268:375–379.
Turner, R. E., and N. N. Rabalais. 2003. Linking landscape and water quality in the Mississippi River Basin for 200 years. *BioScience* 53:563–572.
Wetzel, R. G. 2001. *Limnology, lake and river ecosystems*, 3rd ed. San Diego: Academic Press.
Wulff, F., A. Stigebrandt, and L. Rahm. 1990. Nutrient dynamics of the Baltic Sea. *Ambio* 19:126–133.

6

Dissolved Silica in the Changjiang (Yangtze River) and Adjacent Coastal Waters of the East China Sea

Jing Zhang, Su Mei Liu, Ying Wu, Xiao Hong Qi, Guo Sen Zhang, and Rui Xiang Li

Biogeochemistry of dissolved silica (DSi) in coastal environments has received great attention recently because of its possible role in the global carbon cycle through the alteration of marine food web structures (cf. Ittekkot et al. 2000). Scarcity of macronutrients, such as DSi, favors species other than diatoms that determine the efficiency of the biological carbon pump in the sea and thus the sequestration of atmospheric CO_2 (Brzezinski et al. 2003). In the Danube–Black Sea system, the retention of Si in the upstream reservoir has been found to cause an overall drop in annual load of 60–70 percent since the 1970s (Humborg et al. 1997). In the Huanghe (Yellow River)–Bohai system, land source freshwater input decreased by 65 percent in the 1990s relative to the 1950s, and the DSi:N molar ratio dropped from about 10 to 1.5–2. Simultaneously, the diatom to dinoflagellate biomass ratio fell from 57 to 6.4 (Zhang et al. 2004).

We examine data on DSi in the Changjiang (Yangtze River) and adjacent coastal waters of the East China Sea and compare them with those reported in the literature. The field observations were made in 2001–2002 and included analyses of water samples and mesocosm experiments.

Hydrographic Alteration and Riverine Regime for Dissolved Silica

The Changjiang originates in the Qinghai–Tibet Plateau at an elevation of 5,000–6,000 m. It drains an area of about 1.8×10^6 km^2 and empties into the East China Sea. In the upper reaches, the river flows through mountainous terrain with steep slopes, where the rock outcrops are composed of clastic and metamorphic rocks.

Carbonates are abundant in the middle reaches, and magmatic rock outcrops are not uncommon in the southeastern part of the drainage basin; in the lower reaches, the Changjiang drains the fluvial and lacustrine plain of East China. The middle and lower reaches of the Changjiang drainage area are home to about one third of the population in China, and the region contributes about 40 percent of the national crop production. According to historical data sets, the Changjiang carries 958×10^9 m^3/yr of freshwater and 0.4×10^9 tons/yr of terrestrial sediments to the East China Sea shelf (Yang et al. 2002).

Figure 6.1 shows the concentrations of DSi in the lower reaches of the Changjiang between 1959 and 2001. They range from 78.2 to 155 μM, with a mean of 115 ± 20 μM (Table 6.1).

During almost the same period (1955–2001) the annual water discharge observed at Datong Hydrographic Station (DHS), located 640 km inland from the river mouth, was between 21,400 m^3/s and 39,400 m^3/s. There was a slight increase at the end of the twentieth century; the annual discharge in the 1990s was 15 percent higher than in the 1950s. According to Liu et al. (2003) and Ding et al. (2004), DSi in the Changjiang decreases gradually over a distance of 4,000 km from upstream areas to the river mouth by 10–20 percent. The DSi concentration in the mid-1960s was about 30 percent higher than in the 1970s. After that, until the 1990s, DSi remained rather stable at a concentration of about 110 μM (Figure 6.1). Then DSi concentrations increased between 1999 and 2001 (Figure 6.1). High DSi concentrations at DHS in the early 1960s may have been related to an increase in soil erosion over the drainage basin that induced the mobilization of DSi. This is corroborated by the higher total suspended matter concentrations measured at DHS in the 1960s (average 0.635 g/L), which were about 25 percent higher than in the 1970s (0.502 g/L). Furthermore, in 1957 the area of erosion accounted for 20 percent of the whole Changjiang drainage basin (0.36×10^6 km^2), and 22 percent of the drainage basin was covered by forests. By 1986, the erosion area had increased to 0.74×10^6 km^2 (i.e., 41.0 percent of the drainage basin), and the forest cover had decreased to 10 percent (Yin and Li 2001). An estimate for the tributaries in the northern part of the drainage basin indicates that the yield of DSi is 40.7 ± 14.8 mmol/m^2/yr, whereas in the southern tributaries it is 72.4 ± 30.9 mmol/m^2/yr (Liu et al. 2003). Isotopic measurements of DSi in the Changjiang yield values of δ^{30}Si

Table 6.1. Concentration of DSi and silicon isotopes in the Changjiang in comparison with the Amazon and Congo.

River	SiO_3^{2-}(μM)	δ^{30}Si (‰)	References
Changjiang	78.2–155	+0.7 to +3.4	This study, Ding et al. (2004)
Amazon	115–120	+0.6 to +0.9	Gaillardet et al. (1997), De La Rocha et al. (2000)
Congo	165–210	+0.4 to +0.8	Dupre et al. (1996), De La Rocha et al. (2000)

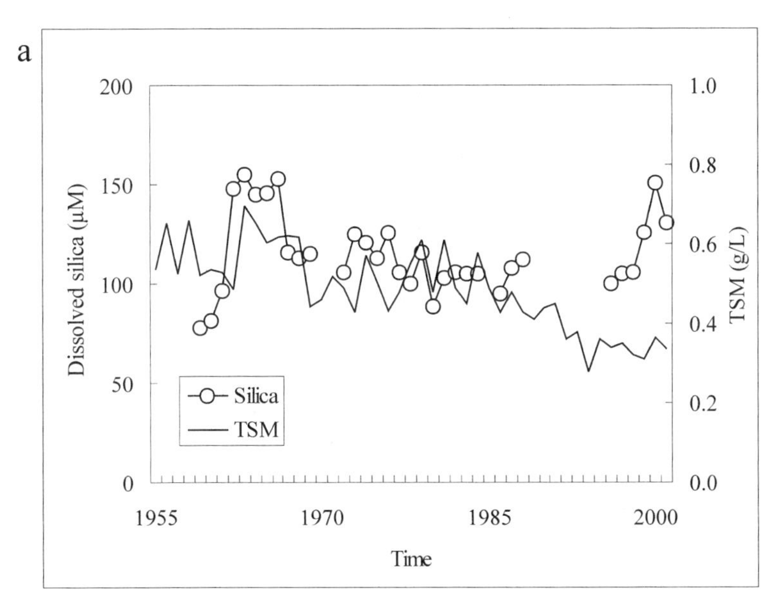

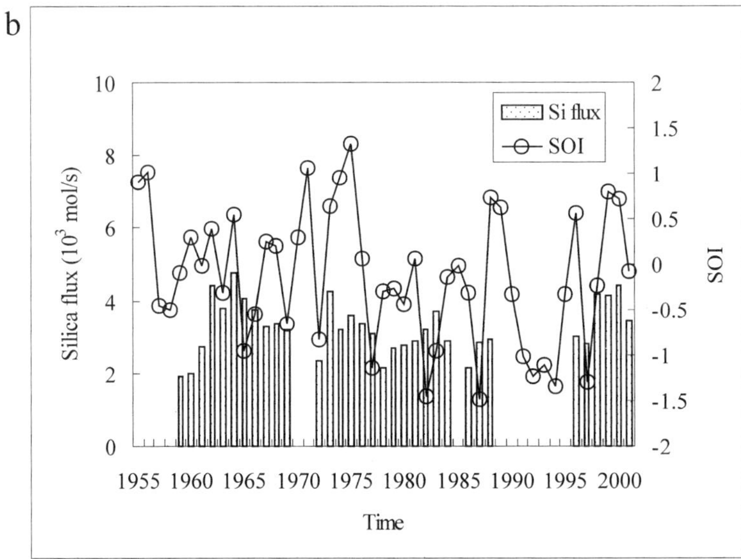

Figure 6.1. *(a)* Distribution of DSi and total suspended matter (TSM) in the Changjiang, volume-weighted annual average in 1959–2001. TSM data are from the Datong Hydrographic Station, and DSi concentrations in 1959–1984 are from the Changjiang Hydrographic Commission and adopted by Li and Cheng (2001). *(b)* Annual average DSi flux and Southern Oscillation Index (SOI) are shown for comparison.

= +0.7‰ to +3.4‰ (Table 6.1), increasing gradually from upstream to the river mouth and regulated by the weathering regime (e.g., precipitation and temperature) and the type of drained ecosystem (e.g., wetland and paddy field), particularly in the middle and lower reaches of the river (Ding et al. 2004). Moreover, the annual average transport of DSi in the Changjiang shows variations that could be related to climate-induced changes in the hydrological cycle at a decadal time scale (Figure 6.1). Total suspended matter decreased after the 1970s in the lower reaches and fell by 40 percent between 1995 and 2000, presumably in part because of sediment trapping behind the several dams in the drainage basin. However, there is no straightforward relationship with the observed variation of DSi.

Dynamics of Dissolved Silica in Coastal Waters

DSi behavior can be either conservative or nonconservative in coastal waters, depending on the hydrodynamic and biogeochemical processes involved. In the Changjiang Estuary and adjacent coastal environment of the East China Sea, DSi concentration is dominated by the simple dilution of freshwater end member by marine waters in summer, when the Changjiang discharge is highest (Figure 6.2). This implies that response of coastal ecosystems to the land source delivery of DSi can be approximated by taking into account the freshwater input. Whereas in the estuarine mixing zone, surface, and near-bottom concentrations are almost identical for DSi, in the coastal waters further offshore surface water is depleted of DSi, and near-bottom samples may have concentrations up to ten times higher.

The influence of Changjiang effluent plumes on the dispersal of DSi derived from land can be seen in Figure 6.3. In surface waters, the high DSi plume can be identified off the Changjiang with a strong eastward gradient. At a distance of 150–200 km, the concentration can still be 5–10 μM. In near-bottom waters, the horizontal gradient is weak. At the river mouth, near-bottom concentrations are lower than at the surface; in offshore regions, they are comparable to or even higher than in surface waters.

Silicate and Coastal Ecosystems

Here we compare field observations of diatom blooms with results from mesocosm experiments conducted in coastal waters off the Changjiang in May 2002 (Table 6.2). The mesocosm was designed as a funnel with a diameter of 1 m and height of 1 m, containing about 700 L of seawater. Nutrients (i.e., NO_3^- and PO_4^{3-}) were added in one of the mesocosms, and samples for particulate biogenic silica (BSi), chlorophyll-*a* (Chl-*a*), and species identification were taken on the morning of six consecutive days (Li et al. 2003).

In field observations, *P. dentatum* and *S. costatum* are the two dominant species in the water column, with biomasses of up to 2.0 × 10^6 cell/L and 0.97 × 10^6 cell/L,

a

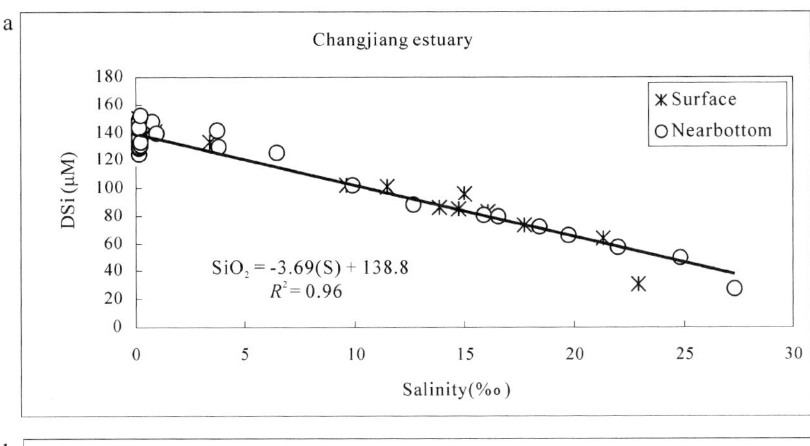

b

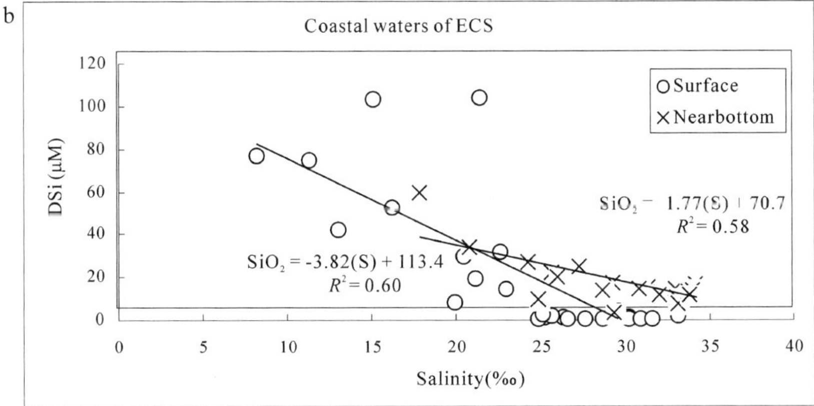

Figure 6.2. Distribution of DSi in *(a)* the Changjiang estuary (July 2001) and *(b)* the adjacent coastal waters of the East China Sea (ECS, August 2002). Note that surface and near-bottom concentrations are almost identical in the estuary, but in the coastal waters the near-bottom samples have higher silicate than at the surface.

respectively (Li et al. 2003). In the mesocosms, cell numbers of *P. dentatum* and *S. costatum* reached peak abundances after 1–2 days for batch control. In the case of nutrient addition, the cell abundances of both species increased more slowly but reached higher biomasses: The cell numbers were twice as high for *P. dentatum*; for *S. costatum*, they increased up to tenfold.

This is also shown by a dramatic change in Chl-*a* in nutrient addition culture, which is twice as high as in the control experiment. Changes in cell abundance in nutrient amendment culture probably result from an increase in species growth rate, which was higher than in the control by 150 percent for *S. costatum* and 50 percent for *P. dentatum* (Li et al. 2003).

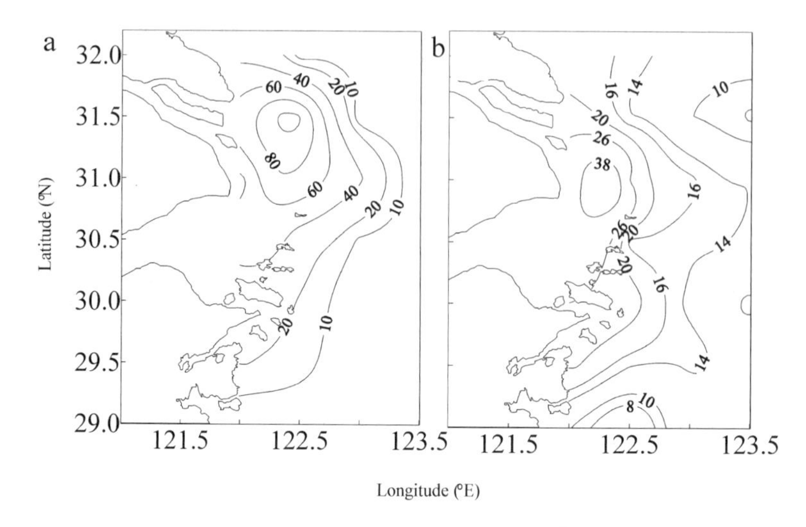

Figure 6.3. Dispersal of DSi (μM) from the Changjiang to the coastal area of the East China Sea in August 2002, *(a)* at the surface and *(b)* in near-bottom waters. The field observation is limited in the region of 29.0°–32.0°N and 122.0°–123.5°E because of logistical problems.

Table 6.2. Mesocosm experiments, which show the initial concentration (μM) of nutrient species for control and amendment series. The samples used for incubation were collected from coastal waters off the Changjiang at 30°29.00′N and 122°37.50′E at a water depth of about 20 m, with salinity of 22.3 and temperature of 19.75°C.

Mesocosm	SiO_3^{2-}	PO_4^{3-}	NO_3^-	NO_2^-	NH_4^+	DIN/P	DIN/Si
Control	33.8	0.284	37.1	0.56	1.81	139	1.17
N and P addition	33.5	4.90	80.4	0.67	1.10	16.8	2.45

DIN = dissolved inorganic nitrogen.

These changes are also seen in BSi. Figure 6.4 shows that BSi in experiments with nutrient addition can be two to three times higher than in the batch control and showed a rapid increase after 2 days of incubation. As Changjiang effluent plumes are characterized by higher dissolved inorganic nitrogen (DIN = NO_3^- + NO_2^- + NH_4^+) and silicate relative to phosphate, an increase in river flow in spring may promote the growth rate of diatoms relative to other species. Given a DIN/PO_4^{3-} similar to the Redfield ratio (i.e., N:P = 16), an increase in DIN/DSi apparently induces more efficient uptake of silicate (Table 6.2 and Figure 6.4).

a

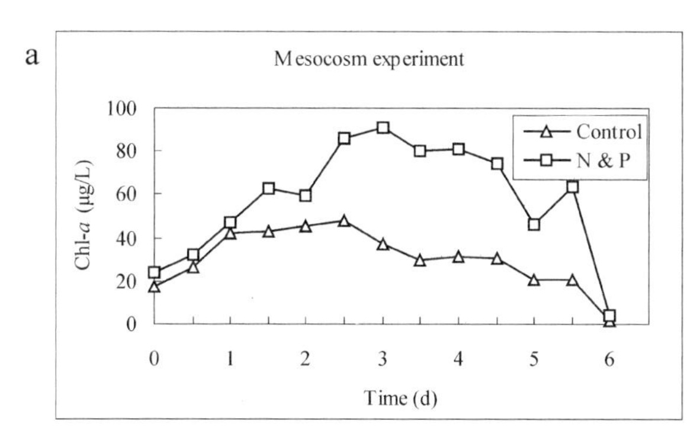

b

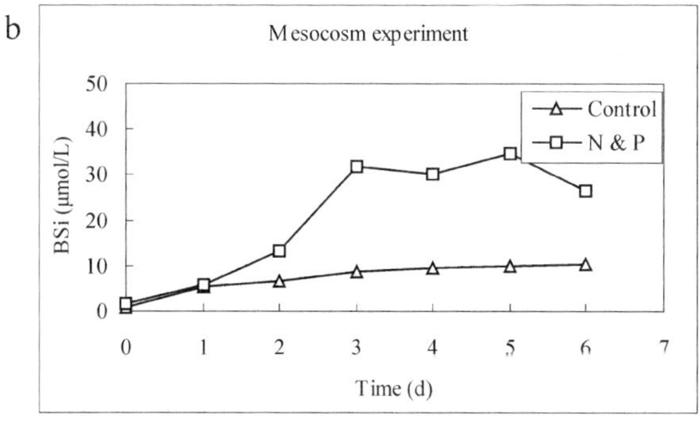

c

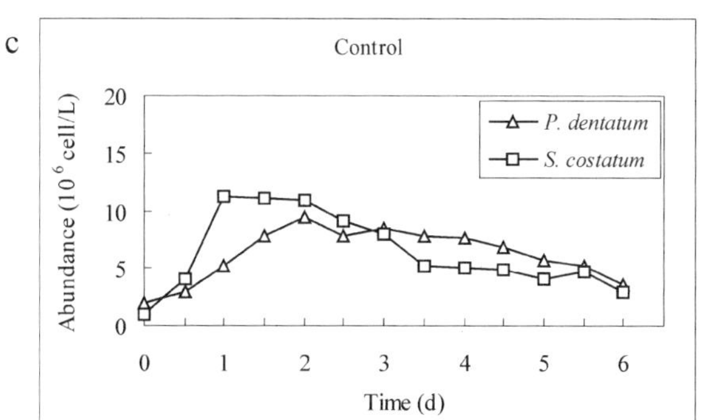

d

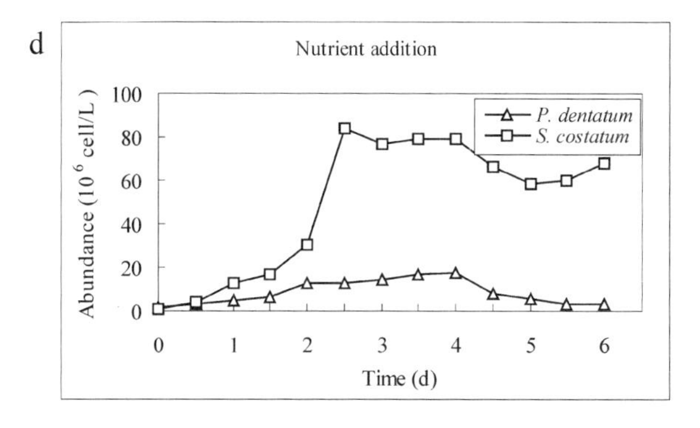

Figure 6.4. Data of mesocosm experiments, which show 2 phytoplankton species, *P. dentatum* and *S. costatum*, responding to nutrient amendments, with *(a)* chlorophyll-*a* (Chl-*a*), *(b)* BSi, *(c)* cell abundance for control, and *(d)* nutrient addition series. Initially, *P. dentatum* is blooming in the coastal waters; later *S. costatum* becomes dominant in the incubation system after addition of nitrate and phosphate.

The sedimentary records of BSi in core samples from the inner shelf areas can be related to the Chl-*a* or primary production and effluent flux of nutrients from the Changjiang. Measurements reveal temporal variations in these processes. BSi contents ranged from 0.3 percent to 0.7 percent over a period of 20 years, as ascertained from ^{210}Pb dating. The BSi in settling particles is 0.4–0.5 percent, as shown by sediment trap data (Iseki et al. 2003). High BSi levels were found in core samples corresponding to the years 1982–1984 and 1996–1999, periods with elevated phytoplankton biomass (e.g., Chl-*a*) (Figure 6.5). BSi and Chl-*a* in the early 1980s and late 1990s thus were twice as high as in 1985–1990.

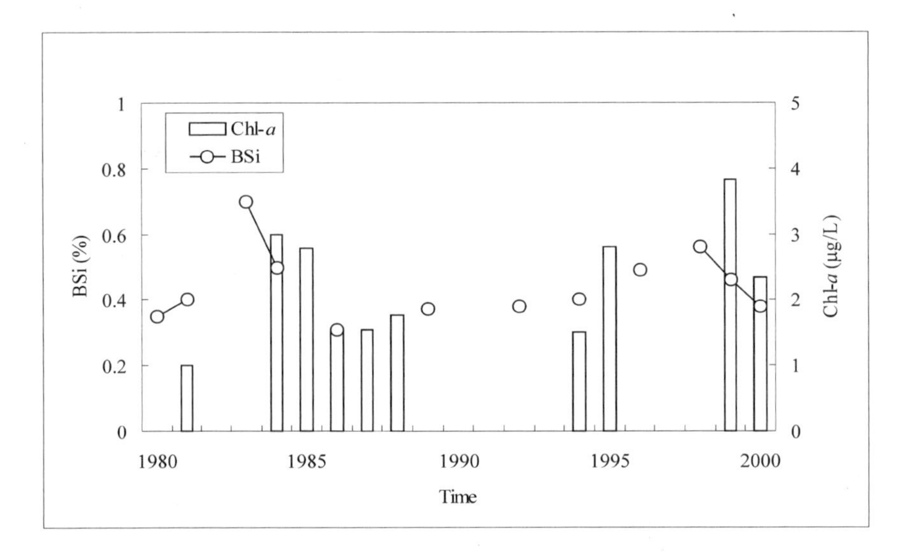

Figure 6.5. Comparison of BSi (%) in core sediment samples at water depth of about 25 m in coastal environment (Station E4, 31°00.00′N and 122°37.00′E) and summer chlorophyll-*a* (Chl-*a*) concentration (μg/L) off the Changjiang estuary. High concentration of BSi in core sediment occurs in 1983 and 1998, corresponding to peak annual flow of 35.2×10^3 m^3/s and 39.4×10^3 m^3/s, respectively, in the Changjiang together with exceptionally high nutrient fluxes.

Conclusion

Over the last 40–50 years, the concentration of DSi in the lower reaches of the Changjiang has exhibited great interannual variability. This variability is related to extensive soil erosion and mobilization of silicate in the 1960s and the possible influence of changing hydrological regimes in the drainage basins caused by climate change as seen from the observed correlation, though weak, between DSi fluxes and the Southern Oscillation Index. In the estuaries and adjacent coastal waters the behavior is conservative. Effective uptake of silicate by diatoms under an N:P ratio that is similar to the Redfield value causes blooms of diatoms and an increase in BSi in coastal waters. Records of BSi in coastal sediments mirror the temporal changes in silicate fluxes from the Changjiang and the associated diatom blooms. Information currently available does not allow us to distinguish between the impacts of climate variability and hydrological alterations (e.g., damming).

Acknowledgments

This study was funded by the Ministry of Science and Technology of China (No. 2006CB400601 and 2001CB409703) and by the Natural Science Foundation of China (No. 40476036). The field observations in the Changjiang Estuary in July 2001 were funded by the Science and Technology Committee of Shanghai. The manuscript was prepared under the Priority Academic Discipline framework provided by the Shanghai Municipal Government. We express our gratitude to colleagues from the Marine Biogeochemistry Group (Ocean University of China and State Key Laboratory of Estuarine and Coastal Research, East China Normal University) for their assistance in field observations and laboratory analysis.

Literature Cited

Brzezinski, M. A., M. L. Dickson, D. M. Nelson, and R. Sambrotto. 2003. Ratios of Si, C and N uptake by microplankton in the Southern Ocean. *Deep-Sea Research II* 50:619–633.

De La Rocha, C., M. A. Brzezinski, and M. J. DeNiro. 2000. A first look at the distribution of the isotopes of silicon in natural waters. *Geochimica et Cosmochimica Acta* 64:2467–2477.

Ding, T., D. Wan, C. Wang, and F. Zhang. 2004. Silicon isotope compositions of dissolved silicon and suspended matter in the Yangtze River, China. *Geochimica et Cosmochimica Acta* 68:205–216.

Dupre, B., D. Gaillardet, D. Rousseau, and C. J. Allegre. 1996. Major and trace elements of river-borne materials: The Congo Basin. *Geochimica et Cosmochimica Acta* 60:1301–1321.

Gaillardet, J., B. Dupre, C. J. Allegre, and P. Negrel. 1997. Chemical and physical denudation in the Amazon River Basin. *Chemical Geology* 142:141–173.

Humborg, C., V. Ittekkot, A. Cociasu, and B.von Bodungen. 1997. Effect of Danube River dam on Black Sea biogeochemistry and ecosystem structure. *Nature* 386:385–388.

Iseki, K., K. Okamura, and K. Kiyomoto. 2003. Seasonality and composition of downward particulate fluxes at the continental shelf and Okinawa Trough in the East China Sea. *Deep-Sea Research II* 50:457–473.

Ittekkot, V., C. Humborg, and P. Schäfer. 2000. Hydrological alterations on land and marine biogeochemistry: A silicate issue? *BioScience* 50:776–782.

Li, M. T., and H. Q. Cheng. 2001. Changes of dissolved silicate fluxes from the Changjiang River into the sea and its influences since the last 50 years. *China Environmental Sciences* 21:193–197 (in Chinese).

Li, R. X., M. Y. Zhu, Z. L. Wang, X. Y. Shi, and B. Z. Chen. 2003. Mesocosm experiment on competition between two HAB species in the East China Sea. *Journal of Applied Ecology* 14:1049–1054 (in Chinese).

Liu, S. M., J. Zhang, H. T. Chen, Y. Wu, H. Xiong, and Z. F. Zhang. 2003. Nutrients in the Changjiang and its tributaries. *Biogeochemistry* 62:1–18.

Yang, S. L., Q. Y. Zhao, and I. M. Belkin. 2002. Temporal variation in the sediment load of the Yangtze River and the influences of human activities. *Journal of Hydrology* 263:56–71.

Yin, H. F., and C. A. Li. 2001. Human impact on floods and flood disasters on the Yangtze River. *Geomorphology* 41:105–109.

Zhang, J., Z. G. Yu, T. Raabe, S. M. Liu, A. Starke, L. Zou, H. W. Gao, and U. Brockmann. 2004. Dynamics of inorganic nutrient species in the Bohai seawaters. *Journal of Marine Systems* 44:189–212.

7

Atmospheric Transport of Silicon

Ina Tegen and Karen E. Kohfeld

Silicon (Si) can be transported through the atmosphere associated with airborne particles. Fly ash produced by industrial burning can contain Si, and those particles are abundant in heavily industrialized areas. However, global estimates of industrial Si emissions are not available. On the other hand, abundant amounts of Si are transported with airborne soil particles. Soil-derived mineral dust contributes significantly to the global aerosol load; estimates of global dust emissions range from 1,000 to 3,000 Mt/yr (Houghton et al. 2001). Dust aerosol consists of micrometer-size windblown soil particles that can be carried over thousands of kilometers through the atmosphere, and dust plumes are predominant features in satellite retrievals of global aerosol patterns (Husar et al. 1997; Herman et al. 1997). Silicon is one of the largest constituents of soil-derived mineral aerosol. Estimates of crustal Si content range between 26 and 44 wt% (Orlov 1992), where most Si is present in the form of quartz (SiO_2). The mean value of crustal Si has been estimated at 26.9 percent (Wedepohl 1978). Tréguer et al. (1995) estimate that about 10 Tmol (280 Tg) particulate Si reaches the ocean surface by eolian dust transport, of which they assume 5–10 percent to be dissolved in seawater. Thus, the transport of soil-derived dust in the atmosphere is an effective means of redistributing Si in the environment.

Global Cycle of Soil Dust Aerosol

Dust aerosol has a high spatial and temporal variability, and large uncertainties exist in quantitative estimates of large-scale dust loads. Modern global dust distribution and properties can be characterized by satellite retrievals and by concentration measurements at surface stations. So far, quantitative estimates of dust optical thickness by satellites are possible only over ocean surfaces, although qualitative patterns of dust distribution can be observed over land, for example, by the Total Ozone Mapping Spectrometer satellite instrument (Herman et al. 1997). In contrast, surface station measurements of dust concentrations (e.g., Prospero and Lamb 2003) and deposition flux measurements are too sparse to obtain regional averages through extrapolation of the local observations.

Therefore, assessments of dust deposition and of the effect of dust on climate rely on distributions computed with global dust cycle models, which predict emissions, transport within the atmosphere, and deposition. These models must be constrained and validated with available in situ observations or remotely sensed aerosol products.

Mineral aerosol is generated in arid and semiarid continental regions. Major dust sources include the Saharan and Sahelian region, the Arabian Peninsula, the Gobi and Takla Makan deserts in Asia, and the Australian desert (Duce 1995). The majority of atmospheric dust is emitted from North African (50–70 percent) and Asian (10–25 percent) deserts. Dust emission occurs when the surface wind shear velocity in the source area exceeds a threshold value, which is a function of surface properties such as the presence of roughness elements, the size of soil grains, and soil moisture. With the parameterization given by Marticorena and Bergametti (1995), the threshold wind shear velocities to lift grains from the ground are approximately 60 cm/s, 30 cm/s, 40 cm/s, and 150 cm/s for particles with diameters of 1,000, 100, 20, and 2 μm, respectively. Theoretical considerations and wind tunnel experiments show a dependence of dust emissions to the third or fourth power of the surface wind shear velocity (Gillette and Passi 1988; Shao et al. 1993). Fine soil particles (less than 5 μm) that can be transported over large distances are released when larger windblown sand grains impact on the soil surface, mobilizing the smaller particles through the process of saltation. Dust emissions occur preferentially in areas that contain fine and loose sediment. Prospero et al. (2002) find a good agreement of satellite-observed dust maxima on land and the location of topographic depressions, where fine sediment could accumulate either because such depressions were lakes during wetter climate periods or through riverine sediment input into those areas. Rough surfaces containing structural elements such as rocks and vegetation increase the threshold wind velocity needed for dust emission (Marticorena and Bergametti 1995) because the obstacles absorb wind energy. Most of these relevant processes have been incorporated into source parameterizations of models of the global dust cycle. Recently developed models include the use of an offline vegetation model to prescribe vegetation cover and therefore the extent of dust sources (Mahowald et al. 1999) and the occurrence of topographic depressions in dry, unvegetated areas as indicators of preferential sources for fine, loose sediment that can be easily deflated under strong wind conditions (Ginoux et al. 2001; Zender et al. 2003) and explicit simulation of a seasonal vegetation distribution to determine potential dust source regions (Tegen et al. 2002; Werner et al. 2002). Crusting of soil surfaces can reduce dust release from a source region. In addition, when fine soil material is lost by deflation, dust emission from a soil can decrease over time. Because of the lack of appropriate input data, those processes have not yet been implemented into global dust emission models. Surface disturbance as a consequence of cultivation in dry regions can increase dust emissions (e.g., during the Dust Bowl events in the United States in the 1930s and 1950s). Currently available estimates suggest that up to 50 percent of the global dust emissions could originate from such anthropogenically disturbed surfaces (Tegen and Fung 1995; Sokolik and Toon 1996). However, newer research suggests that such sources probably contribute much less to the global dust burden than the percentage given in those estimates.

Dust injected at altitudes of several kilometers can be transported over thousands of kilometers by strong wind systems. Most prominently, dust emitted in the Sahara desert (where at least 50 percent of the global dust aerosol is produced) is transported across the North Atlantic to North and Central America and is found as far downwind as the Amazon Basin. Also, large amounts of Asian dust (Asian deserts produce about 10–25 percent of the global dust aerosol) are transported over the North Pacific toward the mid-Pacific islands and North America. State-of-the-art global dust models reproduce the large-scale regional and seasonal patterns (e.g., the Saharan dust plume over the North Atlantic, the transport of Asian dust to the Pacific, and the Arabian dust plume in the Northern Hemisphere summer) of the atmospheric dust loads reasonably well. However, none of them are able to realistically reproduce the magnitude of both Saharan and Asian dust export at the same time. These shortcomings reflect oversimplifications in the treatment of dust emissions in the current generation of models.

Dust is removed from the atmosphere by wet deposition (i.e., scavenging through precipitation in the water or ice phase) or through dry deposition, which is the gravitational settling and turbulent mixing toward the surface. Close to source regions, gravitational sedimentation is responsible for the major part of dust deposition, whereas wet deposition dominates the removal of far-traveled dust particles over remote ocean areas (Figure 7.1).

Although global models simulate the process of scavenging by dry sedimentation and subcloud removal, in-cloud removal of dust usually is neglected. Mostly, the process of subcloud removal is not simulated explicitly. Instead, a so-called scavenging ratio (i.e., the ratio of the dust concentration in rainwater and the dust concentration in air) is used to determine the efficiency of aerosol removal by rain.

Duce et al. (1991) report observed scavenging ratios between 200 and 1,000 for different regions of the world. Global models generally use a single scavenging ratio from within this range. Uncertainties in the choice of an appropriate scavenging ratio and failure to take into account changes in dust hygroscopicity increase the uncertainty of dust distributions in remote regions. Whether dust is removed from the atmosphere by dry or wet processes probably will affect the solubility of any trace elements transported with the dust. Unless they are removed by precipitation, the length of time that dust aerosol particles remain in the atmosphere depends on the deposition velocities of the different particle sizes (Figure 7.2). Atmospheric lifetimes of dust range from a few hours for particles larger than 10 μm to 10–15 days for submicron particles (Ginoux et al. 2001).

Table 7.1 summarizes estimates of dust deposition into different ocean basins, derived from an extrapolation of dust concentration measurements combined with a simple model of dust deposition (Duce et al. 1991; Prospero 1996) and from results of recent global models of the dust cycle (Ginoux et al. 2001; Werner et al. 2002; Zender et al. 2003). For the global dust models, the total annual dust emission is given as well.

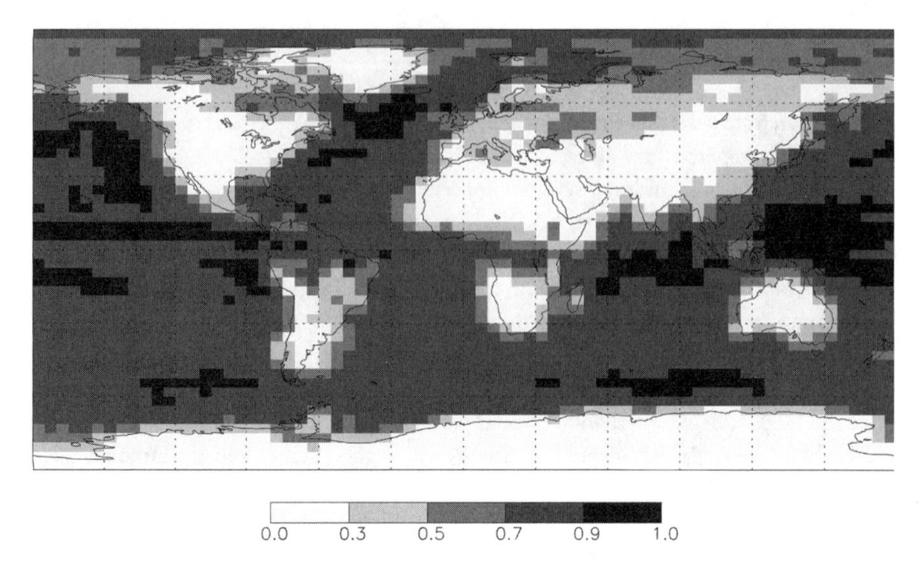

Figure 7.1. Ratio of wet to total deposition of dust aerosol, from the results by Tegen et al. (2002). Whereas over source regions more than 90% of dust is removed from the atmosphere by dry deposition, the main removal mechanism over remote ocean regions is wet deposition.

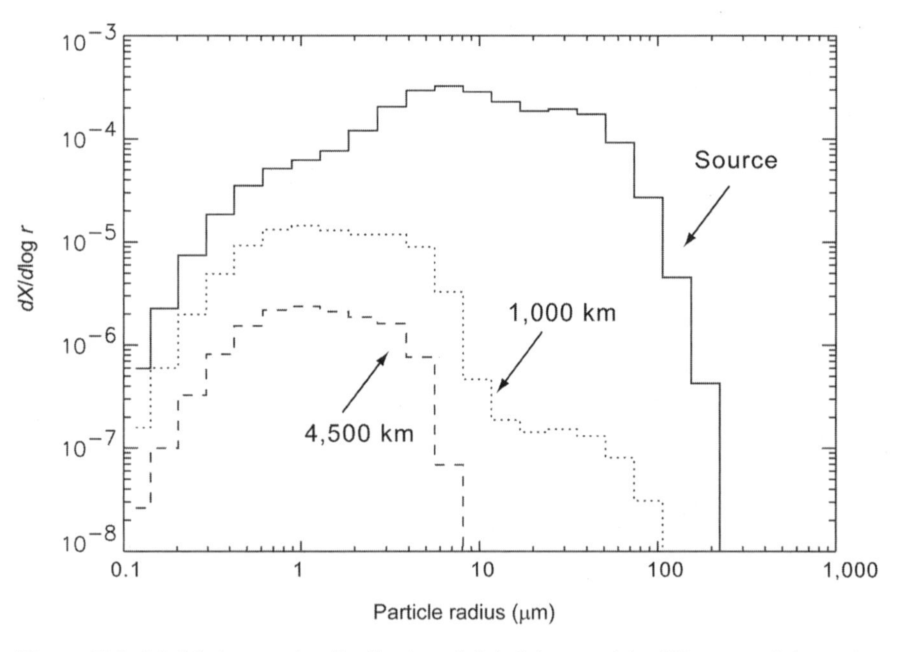

Figure 7.2. Modeled mass–size distribution of global dust particles (X = mass mixing ratio, in g dust/kg air) near the source *(solid line)*, about 1,000 km downwind *(dotted line)*, and about 4,500 km downwind of a source area *(dashed line)*. Model results are from Tegen et al. (2002). Particles larger than 10μm are quickly removed from the atmosphere.

Table 7.1. Deposition of dust and Si into ocean basins extrapolated from observations or derived from global dust cycle models. Estimates of Si deposition and emission fluxes were computed assuming that dust contains 26.9 wt% Si.

	Extrapolated from Observations		Global Dust Cycle Models		
	Duce et al. 1991 (dust/Si, Mt/yr)	Prospero 1996 (dust/Si, Mt/yr)	Ginoux et al. 2001 (dust/ Si, Mt/yr)	Werner et al. 2002 (dust/ Si, Mt/yr)	Zender et al. 2003 (dust/ Si, Mt/yr)
North Pacific	480/129	96/26	92/25	24/6	31/8
South Pacific	39/10	8/2	28/8	17/5	8/2
North Atlantic	220/59	220/59	184/49	201/54	178/48
South Atlantic	24/4	5/1	20/5	18/5	29/8
Indian Ocean	144/39	29/8	154/41	70/19	48/13
Global emissions	—	—	1,814/489	1,035/278	1,490/401

The estimates from Duce et al. (1991) and Prospero (1996) are based on extrapolation of the same measurements; however, Prospero computed his results using lower, more realistic scavenging ratios than Duce et al., except in the North Atlantic. The high scavenging ratio of 1,000 used by Duce et al. is the reason for their unusually high estimate of dust deposition in the North Pacific. The estimates of dust deposition in the North Atlantic agree well because a reasonable amount of data exists to constrain estimates of dust export from the Sahara.

Changes in Soil Aerosol in Past Climates

The atmospheric cycle of soil-derived dust aerosol is expected to respond to changes in global climate conditions, such as the overall intensity of winds that entrain dust into the air, and changes in the hydrological cycle that can affect the extent and dryness of source areas and the length of time that dust remains in the atmosphere. At the same time, dust can affect global and regional climate conditions in several ways, by affecting the radiative balance of the atmosphere or by providing micronutrient fertilizers to marine and terrestrial plants, thereby influencing the uptake of CO_2 in the biosphere and atmospheric CO_2 concentrations.

Dust deposition data from ice cores, marine sediments, and terrestrial sites provide information on dust in past climate periods (Kohfeld and Harrison 2001). The longest land records suggest that continuous atmospheric dust deposition has occurred on the Chinese loess plateau since about 7 million years ago (Ding et al. 1999). Significant efforts have been made to quantify changes in dust deposition through time. From these records we know that during glacial periods, dust deposition was about 2–10 times higher than in interglacial periods (Figure 7.3).

The reason for the greater dust deposition probably is a combination of different factors, such as higher aridity, lower vegetation, greater availability of fine sediment, and higher surface wind speeds in glacial climates. Higher surface wind speeds could lift greater amounts of dust aerosols from surfaces than modern conditions, greater aridity would increase the areas with drier soils and lower vegetation, where soil deflation by wind is possible, and increase availability of fine sediments that could easily be picked up by surface winds. Atmospheric loadings of dust particles could also be higher under cold climate conditions as a consequence of the weakened hydrological cycle. Lower atmospheric water content leads to less precipitation, reducing the washout of atmospheric dust particles and thus leading to longer atmospheric lifetimes.

The best-documented period of high dust deposition in the Quaternary is the last glacial period (approximately 17,000–23,000 years ago). The last glacial maximum (LGM) has been a major focus of dust cycle modeling (e.g., Andersen et al. 1998; Mahowald et al. 1999; Lunt and Valdes 2002; Werner et al. 2002), partly because dust deposition rates were high and well documented for comparison (e.g., Kohfeld and Harrison 2001) and partly because the changes in boundary conditions (e.g., ice sheet extent, atmospheric composition) are well known. An LGM simulation using a full seasonal cycle of vegetation cover as a mask for dust emissions, as well as dry lake beds as

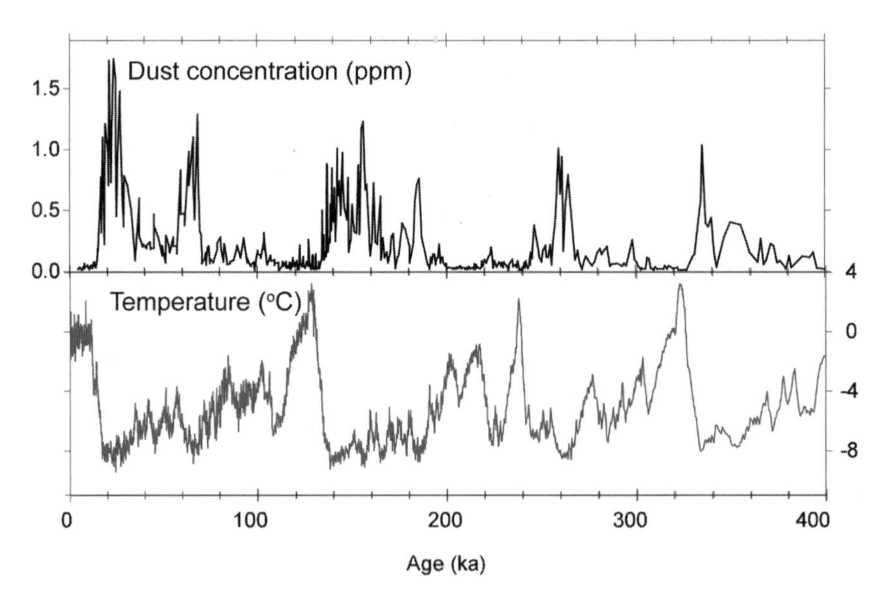

Figure 7.3. Records of dust concentrations and temperature, taken from the Vostok ice core, Vostok, Antarctica (Petit et al. 1999). Dust concentrations are much higher during glacial periods.

preferential dust emission regions, shows a more moderate increase in dust emissions by factor of two to three during the LGM, in agreement with the available marine sediment and ice core data (Werner et al. 2002). The potentially large effect of dust supplied by glacial outwash and the increase in dry lake beds at the LGM has not yet been evaluated by such global dust cycle models.

Properties of Soil Dust

The sizes of soil dust aerosol particles determine their atmospheric lifetimes and thus the distance that the particles can travel. Close to source regions the median dust particle radii are around 30–50 μm because particles in this size range can be lifted most effectively by surface winds. Because large dust particles fall out quickly, dust that has been transported away from the source has a median radius of about 1 μm (Duce 1995; see also Figure 7.2).

The mineral composition of dust particles reflects the mineralogy of the rocks at the earth's surface; some main constituents are quartz, feldspars, carbonates, and clays. Dust particles from different source regions have specific mineralogical composition and, as a consequence, different chemical composition. Claquin et al. (1999) summarized available information to compile global maps of mineral composition of soils in dry areas by relating soil mineralogy to soil types, for which global maps were available (Zobler 1986). They find that quartz makes up about 70 percent of the mass in the silt fraction of the soil (particles 2–50 μm diameter), and illite is the most abundant clay mineral (which contributes up to 50 percent regionally to the clay fraction). However, many uncertainties remain in that investigation. Additionally, it can be expected that during atmospheric transport the mineral composition of dust would change when the size distribution shifts toward smaller particles.

The Si content of different minerals can vary by a factor of approximately two. The average crustal Si content has been estimated to be 26.9 wt%. Whereas magmatites can contain 21–35 wt% Si and sediments 28–33 wt% (Wedepohl 1978), the Si content of clays is lower, on the order of 19–28 wt% (Newmann 1987).

Similar to the method used to estimate iron deposition by dust for investigations of the role of iron fertilization of the oceanic biosphere (e.g., Archer et al. 2000), we can obtain an estimate of the amount of Si deposited by dust aerosol by multiplying the deposited dust mass by the crustal abundance of 26.9 wt% Si. With this approximation we disregard the regional variability of the Si content in dust. However, as long as no more detailed information on the global mineral composition of atmospheric dust is available, this is a useful first-order estimate. With this assumption, global Si emissions of 250–500 Mt/yr associated with soil dust particles less than 10 μm (which can be transported over large distances) are computed with global dust cycle models (Table 7.1). Estimates of Si deposition into the ocean show that the highest amounts of dust Si are deposited into the tropical North Atlantic and the western North Pacific (Figure 7.4, Table 7.1).

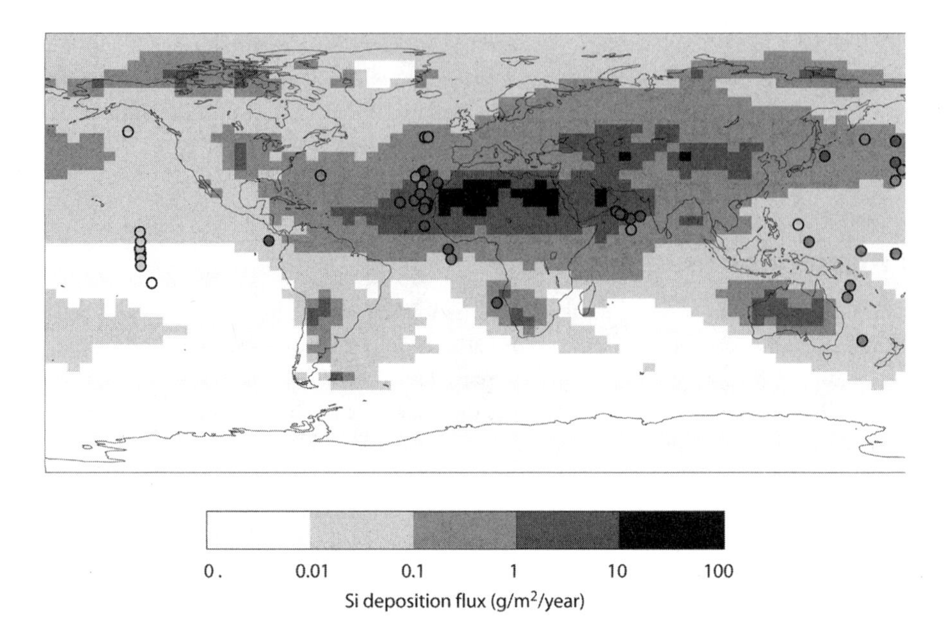

Figure 7.4. Deposition pattern of atmospheric Si, from global model results of dust deposition (*shaded areas*, Werner et al. 2002) and compiled sediment trap deposition fluxes (*symbols*, Kohfeld and Harrison 2001). Si content of 26.9% in the dust particles was applied globally.

Si deposition into the Southern Hemisphere ocean basins is low, with a total of probably less than 20 Mt/yr. If we assume that far-traveled dust consists mainly of clay particles, which have lower Si contents than the average crustal abundance, the deposition fluxes of Si may actually be lower than the estimates given in Table 7.1. However, this assumption would be based only on the median particle size of far-traveled dust, which may not reflect the mineral composition. The results in Table 7.1 show that the uncertainties associated with the particulate Si deposition into the different ocean regions are large; the estimates differ up to an order of magnitude for the North Pacific. This is a consequence of the large uncertainties in the methods that are used to estimate the global patterns of dust deposition.

Only a small fraction of the deposited Si dissolves in seawater. The percentage of particulate Si that dissolves in the water column depends on the mineral composition and size of the dust particles and is also uncertain. Whereas Tréguer et al. (1995) assume that 5–10 percent of the Si transported to the ocean surface by aeolian dust dissolves in the surface water, Guerzoni et al. (1999) report 1 percent solubility for Si in Saharan dust samples collected at Sardinia. Gehlen et al. (2003) assume a 3 percent solubility of aluminosilicates in seawater in their model investigation. As for the total Si content in dust particles, the solubility is likely to change during transport, with the removal of large, insoluble quartz particles and possible chemical transformation during the cycling of dust particles through clouds.

Conclusion

The deposition of atmospheric Si into the global oceans can be estimated from oceanic dust deposition fluxes. According to the estimates given in Table 7.1, the annual eolian deposition of dust into the oceans ranges between approximately 300 and 900 Tg per year (i.e., between 80 and 240 Mt particulate Si). This flux is smaller than the estimate of the particulate riverine transport into the oceans. However, whereas the riverine flux of Si influences mostly the coastal regions, eolian dust can be deposited on remote ocean surfaces. The eolian oceanic Si input is equivalent to about 3–9 Tmol Si per year. Assuming solubilities of particulate Si between 1 and 10 percent, 0.03–1 Tmol of dissolved Si is added by deposited soil dust to the oceans each year. During glacial times, this value would have been two to three times higher. This is a small fraction of the total content of dissolved Si of approximately 10^5 Tmol (Tréguer et al. 1995), but regional variations may be of importance. For example, in the North Atlantic, up to 50 $g/m^2/year$ is deposited at some locations (Tegen et al. 2002), which would result in a regional flux of up to 0.05 mol Si at the locations of maximum dust deposition.

Given those uncertainties, it is unlikely that these estimates could be much improved with the implementation of a more detailed mineralogy to better represent the regional variability in Si content in the dust before the uncertainties in the estimates of dust deposition themselves can be reduced.

Literature Cited

Andersen, K. K., A. Armengaud, and C. Genthon. 1998. Atmospheric dust under glacial and interglacial conditions. *Geophysical Research Letters* 25:2281–2284.

Archer, D., A. Winguth, D. Lea, and N. Mahowald. 2000. What caused the glacial/interglacial atmospheric pCO_2 cycles? *Reviews of Geophysics* 38:159–189.

Claquin, T., M. Schulz, and Y. Balkanski. 1999. Modelling the mineralogy of atmospheric dust sources. *Journal of Geophysical Research* 104:22,243–22,256.

Ding, Z. L., S. F. Xiong, J. M. Sun, S. L. Yang, Z. Y. Gu, and T. S. Liu. 1999. Pedostratigraphy and paleomagnetism of a similar to 7.0 Ma eolian loess–red clay sequence at Lingtai, Loess Plateau, north-central China and the implications for paleomonsoon evolution. *Palaeogeography, Palaeoclimatology, Palaeoecology* 152:49–66.

Duce, R. A. 1995. Sources, distributions, and fluxes of mineral aerosols and their relationship to climate. Pp. 43–72 in *Aerosol forcing of climate: Report of the Dahlem Workshop on Aerosol Forcing of Climate Berlin 1994, April 24–29,* edited by R. J. Charlson and J. Heintzenberg. Chichester, UK: Wiley.

Duce, R. A., P. Liss, J. T. Merrill, E. L. Atlas, P. Buat-Menard, B. B. Hicks, J. M. Miller, J. M. Prospero, T. Arimoto, T. M. Church, W. Ellis, J. N. Galloway, L. Hansen, T. D. Jickells,

A. H. Knap, K. H. Reinhardt, B. Schneider, A. Soudine, J. J. Tokos, S. Tsunogai, R. Wollast, and M. Zhou. 1991. The atmospheric input of trace species into the world ocean. *Global Biogeochemical Cycles* 5:193–259.

Gehlen, M., C. Heinze, E. Maier-Reimer, and C. I. Measures. 2003. Coupled Al–Si geochemistry in an ocean general circulation model: A tool for the validation of oceanic dust deposition fields? *Global Biogeochemical Cycles* 17:1028.

Gillette, D. A., and R. Passi. 1988. Modeling dust emission caused by wind erosion. *Journal of Geophysical Research* 93:14,233–14,242.

Ginoux, P., M. Chin, I. Tegen, J. M. Prospero, B. Holben, O. Dubovik, and S.-J. Lin. 2001. Sources and distributions of dust aerosols simulated with the GOCART model. *Journal of Geophysical Research* 106:20,255–20,273.

Guerzoni, S., E. Molinaroli, P. Rossini, G. Rampazzo, G. Quarantotto, G. De Falco, and S. Cristini. 1999. Role of desert aerosol in metal fluxes in the Mediterranean area. *Chemosphere* 39:229–246.

Herman, J. R., P. K. Bhartia, O. Torres, C. Hsu, C. Seftor, and E. Celarier. 1997. Global distribution of UV-absorbing aerosols from Nimbus-7 TOMS data. *Journal of Geophysical Research* 102:16,911–16,922.

Houghton, J. T., Y. Ding, D. J. Griggs, M. Noguer, P. J. van der Linden, X. Dai, K. Maskell, and C. A. Johnson (eds.). 2001. *Climate change.* New York: Cambridge University Press.

Husar, R. B., J. M. Prospero, and L. Stowe. 1997. Characterization of tropospheric aerosols over the oceans with the NOAA advanced very high resolution radiometer optical thickness operational product. *Journal of Geophysical Research* 102:16,889–16,909.

Kohfeld, K. E., and S. P. Harrison. 2001. DIRTMAP: The geological record of dust. *Earth-Science Review* 54:81–114.

Lunt, D. J., and P. J. Valdes. 2002. Dust deposition and provenance at the last glacial maximum and present day. *Geophysical Research Letters* 29:2085.

Mahowald, N., K. E. Kohfeld, M. Hansson, Y. Balkanski, S. P. Harrison, I. C. Prentice, M. Schulz, and H. Rodhe. 1999. Dust sources and deposition during the last glacial maximum and current climate: A comparison of model results with palaeodata from ice cores and marine sediments. *Journal of Geophysical Research* 104:15,895–15,916.

Marticorena, B., and G. Bergametti. 1995. Modeling the atmospheric dust cycle: 1. Design of a soil-derived dust emission scheme. *Journal of Geophysical Research* D8 100:16,415–16,430.

Newmann, A. C. D. 1987. *Chemistry of clays and clay minerals.* London: Mineralogical Society.

Orlov, D. S. 1992. *Soil chemistry.* Russian Translation Series 92. Rotterdam: A.A. Balkema.

Petit, J. R., J. Jouzel, D. Raynaud, N. I. Barkov, J. M. Barnola, I. Basile, M. Bender, J. Chappellaz, M. Davis, G. Delaygue, M. Delmotte, V. M. Kotlyakov, M. Legrand, V. Y. Lipenkov, C. Lorius, L. Pepin, C. Ritz, E. Saltzman, and M. Stievenard. 1999. Climate and atmospheric history of the past 420,000 years from the Vostok ice core, Antarctica. *Nature* 399:429–436.

Prospero, J. M. 1996. The atmospheric transport of particles to the ocean. Pp. 19–52 in *Particle flux to the ocean,* edited by V. Ittekkot, P. Schäfer, and P. J. Depetris. SCOPE 57. Chichester, UK: Wiley.

Prospero, J. M., P. Ginoux, and O. Torres. 2002. Environmental characterization of global sources of atmospheric soil dust identified with the Nimbus-7 TOMS absorbing aerosol product. *Reviews of Geophysics* 40(1): doi: 10.1029/2000RG000095.

Prospero, J. M., and P. J. Lamb. 2003. African droughts and dust transport to the Caribbean: Climate change implications. *Science* 302:1024–1027.

Shao, Y., M. R. Raupach, and P. A. Findlater. 1993. Effect of saltation bombardment on the entrainment of dust by wind. *Journal of Geophysical Research* 98:12,719–12,726.

Sokolik, I., and O. B. Toon. 1996. Dust radiative forcing by anthropogenic airborne mineral aerosols. *Nature* 381:681–683.

Tegen, I., and I. Fung. 1995. Contribution to the mineral aerosol load from land surface modification. *Journal of Geophysical Research* 100:18,707–18,726.

Tegen, I., S. P. Harrison, K. E. Kohfeld, I. C. Prentice, M. C. Coe, and M. Heimann. 2002. The impact of vegetation and preferential source areas on global dust aerosol: Results from a model study. *Journal of Geophysical Research* 107: doi: 10.1029/2001JD000963.

Tréguer, P., D. M. Nelson, A. J. Vanbennekom, D. J. DeMaster, A. Leynaert, and B. Queguiner. 1995. The silica balance in the world ocean: A reestimate. *Science* 268:375–379.

Wedepohl, K. H. 1978. *Handbook of geochemistry*. Berlin: Springer.

Werner, M., I. Tegen, S. P. Harrison, K. E. Kohfeld, I. C. Prentice, Y. Balkanski, H. Rodhe, and C. Roelandt. 2002. Seasonal and interannual variability of the mineral dust cycle under present and glacial climate conditions. *Journal of Geophysical Research* 107: doi: 101029/2002JD002365.

Zender, C. S., H. Bian, and D. Newman. 2003. Mineral Dust Entrainment and Deposition (DEAD) model: Description and 1990s dust climatology. *Journal of Geophysical Research* 108:4416.

Zobler, L. 1986. *A world soil file for global climate modeling*. NASA Technical Memorandum 87802. New York: NASA Goddard Institute for Space Studies.

8

Estuarine Silicon Dynamics

Lei Chou and Roland Wollast†

Silicon is a major dissolved constituent and represents about 10 percent of the total dissolved load in average river water. The mean concentration of dissolved silicate (DSi) in river water is around 200 μM (Meybeck 1979), and that in oceanic water is around 100 μM (Wilson 1975). In contrast to other major elements, DSi is the first by order of importance in that it is more concentrated in rivers than in the ocean.

In estuaries where freshwater mixes with seawater, one observes a rapid decrease in DSi concentration toward the sea. This observation has led to several investigations concerning the possible mechanisms responsible for the removal of this element during estuarine mixing. Two types of processes, biotic and abiotic, have been suggested and studied. It is known that DSi is an essential nutrient for aquatic siliceous organisms such as diatoms, radiolarians, and sponges. The removal of this element from estuarine waters by phytoplankton has been recognized for a long time (Atkins 1926). Abiotic uptake of DSi in estuaries resulting from physicochemical processes occurring during mixing of freshwater and saltwater was proposed by Bien et al. (1958) for the Mississippi. It was postulated in the early 1960s that reactions between DSi and aluminosilicate minerals might play an important role in controlling the composition of seawater (Sillén 1961; Mackenzie et al. 1967). This hypothesis has generated several studies on the abiotic uptake of DSi that may occur during estuarine mixing of freshwater and seawater.

In order to evaluate the DSi fluxes delivered to the ocean by rivers, it is essential to identify and quantify the various processes responsible for DSi removal, retention, and regeneration in the estuary. In addition, the uncertainty involved in the assessment of Si fluxes in its biogeochemical cycle is largely related to our poor understanding of the distribution and behavior of biogenic silica (BSi) produced not only by aquatic organisms but also by terrestrial plants as phytoliths. Unlike that of N and P, the regeneration of DSi by dissolution of biogenic opal is slow. In fact, the export

†Roland Wollast died on July 28, 2004.

flux of particulate BSi by estuaries is essentially unknown. Furthermore, the budget of Si in estuarine sediments, including BSi deposition and redissolution, DSi consumption during diagenesis, and diffusion toward the overlying water column, has been little investigated.

In this chapter we discuss the various abiotic and biotic DSi removal processes occurring in the estuarine zone that may affect Si fluxes to the open ocean. The discussion includes the behavior of opal in estuarine waters and sediments.

Definition of Estuaries and Mixing Properties of Seawater and Freshwater

The most widely accepted definition of estuaries is that given by Pritchard (1967:3): "An estuary is a semi-enclosed coastal body of water which has a free connection with the open sea and within which sea water is measurably diluted with fresh water derived from land drainage." However, it should be recognized that the region where a river enters the sea and mixes with ocean water can represent a large stretch of the river that is tidally mixed and that cannot be ignored in estuarine studies, especially if sediment transport is considered.

From a biogeochemical point of view, the characteristics of the water circulation induced by the mixing of freshwater and saltwater play an important role in silica cycling. In a salt wedge estuary, freshwater tends to spread out over the denser saltwater that flows landward along the bottom. Under these stratified conditions, which occur mainly when the river discharge is large and the tidal amplitude low, vertical mixing is limited, and the residence time of freshwater in the estuarine zone is very short. Nearly all freshwater is discharged into the adjacent coastal ocean in the surface water layer, giving rise to a well-established river plume and an external estuarine circulation induced by density gradients.

When the tidal effect becomes more important and the freshwater discharge diminishes, salt intrusion increases. Friction with the bottom and at the pycnocline interface tends to mix vertically the water masses, giving rise to a vertically well-mixed estuary. Under this condition, the volume of saltwater entering the estuary per tide is much larger than the volume of freshwater discharged by the river during the same period of time. Because of this large dilution by seawater, the residence time of freshwater and riverborne material in the estuarine zone may reach several months. This usually results in potential changes in the composition and properties of the riverine components.

When seawater is mixed with freshwater in the estuary, it is possible to define a mixing line by plotting the concentration of a component as a function of the concentration of a conservative property such as salinity (Figure 8.1).

If the estuarine system is at steady state, which requires that the material fluxes remain constant at both boundaries of the system, the mixing line is represented by a

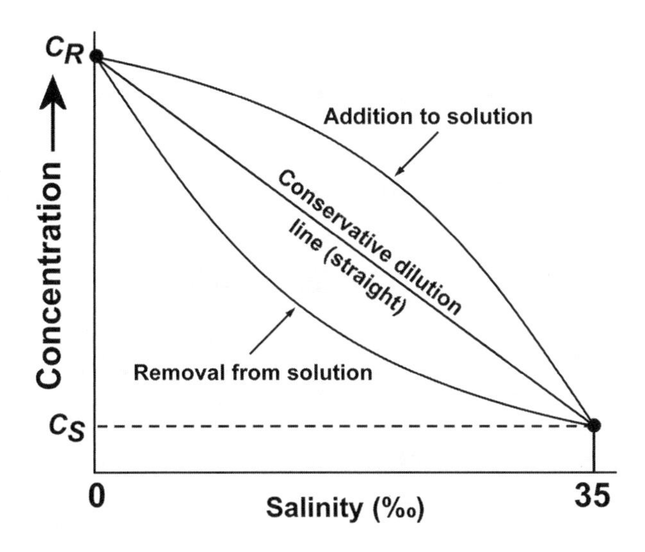

Figure 8.1. Idealized plot of the concentration of dissolved components and salinity during estuarine mixing. C_R and C_S are the concentrations of the dissolved species in river water and in seawater, respectively.

straight line for a conservative component. This straight line joins the two end member components of freshwater and seawater. If the component is added to the system during mixing, the point representative of the mixture is above the straight line, and it is below the straight line if the component is consumed.

This two–end member mixing approach has been used extensively to evaluate the conservative or nonconservative behavior of DSi and other dissolved components, mainly in strong tidal estuaries. It must be used with caution because estuaries often deviate substantially from steady-state conditions at the time scale of the residence time of freshwater in the estuary (Officer and Lynch 1981). This is especially the case for the river end member because of fluctuations in water discharge and composition. The situation often is more complex because several sources of freshwater are present in the estuarine system. Nevertheless, a straight line can be obtained only if the dissolved component behaves conservatively. However, departure from the straight line does not necessarily imply nonconservative behavior. This may simply result from the fact that the system has not been maintained at steady state during the residence time of the freshwater in the estuary. In this case, it is still possible to evaluate the uptake or release of a component by using a simplified one-dimensional, non–steady-state hydrodynamic model coupled with a biogeochemical model (e.g., Regnier et al. 1998).

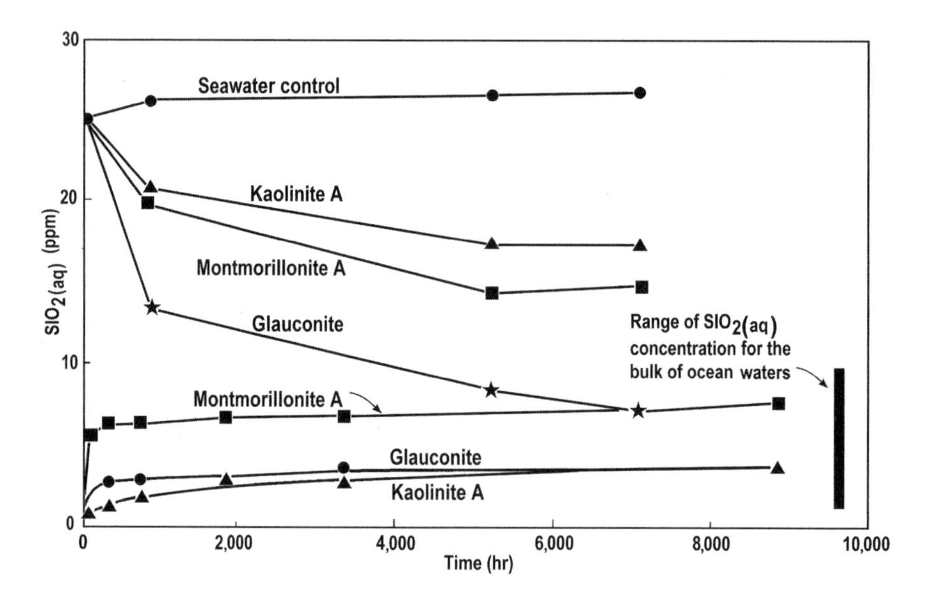

Figure 8.2. Concentration of DSi as a function of time for seawater–clay suspension interactions (after Mackenzie et al. 1967).

Abiotic Removal Processes of Dissolved Silica

Role of Abiotic Processes in the Global Silica Cycle

Unable to explain the mean concentration of DSi observed in the ocean simply by biological uptake, Sillén (1961) proposed a chemical model involving reactions of silicate minerals with dissolved species. Mackenzie and Garrels (1966) suggested reactions between DSi and suspended clays of continental origin to support the model of Sillén. These reactions require high concentrations of alkali or alkaline earth ions to produce marine clay minerals. A typical example of the synthesis of new clays favored in the marine environment is the transformation of kaolinite to chlorite according to the following simplified reaction:

$$\underset{\text{kaolinite}}{Al_2Si_2O_5(OH)_4} + H_4SiO_4 + 5Mg^{2+} + 10HCO_3^- =$$

$$\underset{\text{chlorite}}{Mg_5Al_2Si_3O_{10}(OH)_8} + 10CO_2 + 5H_2O$$

Mackenzie and Garrels (1966) showed that this concept of reverse weathering was consistent with a mass balance for the oceans based on removal of riverborne cations. Furthermore, these reactions are important because they release CO_2 to the atmosphere and modify the buffering capacity of seawater.

Mackenzie et al. (1967) conducted laboratory experiments on the uptake and release of DSi by clay minerals in seawater. Their results showed rapid release of Si to seawater initially depleted in DSi. The reverse reaction occurred in DSi-enriched seawater but at a much slower rate (Figure 8.2). Siever (1968) recognized the usefulness of this model but noted that these reactions probably are postdepositional or postburial.

Abiotic Processes in the Water Column

Krauskopf (1956) was among the first to suggest that the increase of ionic strength during the mixing of river water with seawater could lead to the formation of colloidal silica, removing DSi from the water column. This idea was widely adopted at that time as an explanation of the lower DSi concentration found in seawater compared with freshwater. However, this hypothesis was refuted by Burton et al. (1970a) and Siever (1971), who demonstrated that DSi was present only in its monomeric form in natural waters. It was observed that artificial polymers of silicic acid undergo rapid depolymerization when added to surface waters. The removal of DSi by polymerization in estuaries therefore is unlikely.

The longitudinal profile of DSi in the Mississippi observed by Bien et al. (1958) exhibited nonconservative behavior, suggesting a significant removal of this element during estuarine mixing, which could not be attributed to diatoms. In addition, these authors conducted various laboratory experiments in which they demonstrated that this uptake was caused by adsorption of DSi on suspended matter from the river when in contact with electrolytes from seawater. Later studies of the Mississippi estuary by Fanning and Pilson (1973) indicated a much lower removal of only about 7 percent DSi.

Burton et al. (1970b) investigated the behavior of DSi during estuarine mixing in several small estuaries, which exhibited a moderate uptake of DSi. Because this removal can occur under conditions of low biological activity, many authors have generally ascribed it to nonbiological processes. Fanning and Pilson (1973) had difficulty in determining whether the removal of DSi occurred in estuaries of the Orinoco and Savannah rivers because of uncertainty in the composition of seawater entering the estuary. Liss and Spencer (1970) carried out laboratory experiments using mixtures of river water and seawater from the Conway estuary to which additional DSi and a natural suspended solid had been added. They found abiological removal of DSi at low salinities, with uptake increasing as suspended solid concentration increased. Sholkovitz (1976) conducted similar experiments in which various proportions of the filtered (0.45 μm) freshwater end member of Scottish rivers were mixed with seawater. He investigated the transfer with time of dissolved elements (including silica) to the particulate phase through flocculation, coagulation, and precipitation. Rapid and important removal was observed upon mixing of the two end members for most elements investigated, except for silica, which showed a lack of reactivity (3–6 percent removal). Another possible uptake mechanism of DSi is that of the adsorption of silica on ferrihydrites (Anderson and Benjamin 1985); this process has not yet been studied in estuarine systems.

Other field and laboratory studies on DSi uptake in estuaries have shown that abiotic removal in the absence of diatoms is moderate (10–30 percent) or even insignificant (for review, see Liss 1976). In conclusion, the aforementioned studies have shown that in most cases the abiotic removal of DSi is not very important and often insignificant in the water column of estuarine systems. Optimal conditions for abiotic uptake correspond to low-salinity water (0–5‰), high concentrations of suspended matter and DSi, and rapid increase in salinity. The reverse weathering hypothesis is attractive because it allows one to balance the Si budget at a global geological scale together with those of major cations and CO_2. However, it is important to point out that experimental DSi concentrations are always much higher than those encountered in seawater and in coastal or estuarine waters rich in DSi. The uptake of DSi by clay minerals therefore is not likely to occur in these water column environments but may occur in the underlying sediment in porewater systems.

Abiotic Processes in Sediments

The reaction between DSi and clay minerals should be highly favored in sediments because of the high concentrations of DSi present in porewaters. These high concentrations result from dissolution of siliceous organisms after deposition. The increase in DSi concentration in porewaters of marine sediments is a well-established general trend (Sayles 1981). The DSi concentration at depth often approaches a constant value, but in some sediments the gradient reverses itself, suggesting neoformation of a solid silica phase. However, the variable asymptotic porewater DSi levels in the core cannot be explained solely by differences in opal solubility (Van Cappellen and Qiu 1997). In sediments, the diagenetic reactions discussed earlier prevent the porewater from reaching saturation with respect to opaline silica. Furthermore, the near-surface concentration gradient across the sediment–water interface produces an upward flux of DSi toward the water column.

In estuaries where the rate of deposition of detrital material usually is very high, only a small fraction of the BSi is remineralized and diffuses into the overlying water column. The opal may be retained and preserved in the sedimentary column, where it tends to react slowly with increasing burial depth to cristobalite and quartz (Calvert 1983). The DSi released to porewaters from the dissolution of BSi could be consumed by the formation of authigenic aluminosilicate minerals. The importance of this reverse weathering reaction in estuarine and coastal environments has been recognized in recent field and laboratory studies (Dixit et al. 2001; Michalopoulos and Aller 1995, 2004; Michalopoulos et al. 2000), confirming the observational results obtained in earlier investigations (Ristvet 1978; Mackenzie et al. 1981; Mackin and Aller 1984, 1986, 1989).

Biological Removal of Dissolved Silica

Estuaries usually are characterized by a high biological activity caused by the large input of nutrients derived from land. However, they also transport an important amount of suspended matter that limits light penetration. Thus, primary production is possible only when turbidity is sufficiently reduced. In estuaries, the removal of suspended particles is favored by flocculation and coagulation induced by the increase in salt concentration seaward in the estuary. This phenomenon usually is observed when the salinity reaches 5‰ and is often associated with a turbidity maximum. Primary production generally increases toward the mouth of the estuary (Joint and Pomroy 1981), indicating that the decrease in nutrients is more than compensated for by the increased water transparency.

The euphotic depth in estuaries often is much shallower than the mixing depth, which in well-mixed estuaries covers the entire water column. Thus, in nonstratified estuaries the phytoplankton spend a large part of their life cycle in the dark, where they continue to respire. Under these conditions, the development of phytoplankton is limited. Vertical turbulent mixing, which is elevated in estuaries, is thus an important factor controlling primary production. The phytoplankton community in estuaries often is dominated by diatoms that perform well under low light conditions (Richardson et al. 1983) and have high salinity tolerance (Heip et al. 1995). Because of the abundance of intertidal flats in many estuaries, benthic diatom production may be an important part of the primary production (Heip et al. 1995).

Anthropogenic activities have significantly increased the riverine flux of N and P to the coastal zones. In contrast, the major land source of dissolved Si is through natural chemical weathering of silicate minerals; the global silica flux associated with this process has not changed significantly in past centuries. However, large-scale land use activities, such as river damming and river diversion, can result in important silica retention on land, causing severe reductions of DSi input to the sea (Humborg et al. 1997, 2000; Ittekkot et al. 2000). As a consequence, there can be a change in nutrient ratios (Si:N, Si:P) in river discharge, affecting significantly the development of spring diatom blooms in estuaries and in the adjacent seas. This excess delivery of N and P, compared with Si, has led to modification of phytoplankton species composition and thus of ecosystem functioning in many coastal and other aquatic environments (Conley et al. 1993; Humborg et al. 1997, 2000; Ittekkot et al. 2000; Lancelot et al. 1987; Rousseau et al. 2002). One adverse consequence is the alteration of the marine food web by the reduced importance of the diatom blooms. Another effect is the frequent occurrence of toxic algae that are harmful to the environment.

The major input of nutrients (N, P, Si) to the open ocean is through river and groundwater flow, estuaries, and the coastal zone. This continuum of aquatic systems acts as an efficient, selective, and successive filter for nutrients before they reach the land–ocean boundary. The retention of Si within the continuum can be expected to be greater than that of other nutrients because this element is recycled through dissolution of diatom frustules, which is a slow process compared with the bacterial mineralization of N and P. Silica retention and its magnitude are poorly known at present because our knowledge of the processes affecting the Si biogeochemical cycling in the continuum is insufficient. This flux deserves more attention and quantification so that we can evaluate its impact on coastal eutrophication and on the modification of the marine food web.

Case Studies

Amazon

The Amazon is the world's largest river, with an annual discharge of 6.5×10^3 km^3/yr and a drainage basin area of 6.1×10^6 km^2 (GEMSWATER, www.gemswater.org). The maximum river discharge (spring) is only three times the minimum discharge (fall). The concentration of DSi in freshwater before mixing with seawater is around 115 μM (DeMaster et al. 1983), and the suspended matter concentration is about 186 mg/L. This estuary typically is an outer estuary in which the freshwater plume from the river interacts with the north Brazilian coastal current and is transported northwest (Milliman et al. 1975) on the shelf (Figure 8.3).

Observations of DSi concentrations and salinity in the surface water of the Amazon shelf region (Milliman and Boyle 1975; Edmond et al. 1981) indicated the removal of DSi by diatoms when the salinity reached 3–7‰. However, little BSi was found accumulating in the sediments and was rather transported northwest or dissolved at the sediment–water interface (Milliman and Boyle 1975). DeMaster et al. (1983) studied the uptake and accumulation of silica on the Amazon continental shelf (Figure 8.3) during a period of low river discharge. Even under these conditions, the 10‰ isohaline was observed up to 100 km seaward of the river mouth.

Interestingly, cross-shelf profiles of salinity (Figure 8.4) show strong horizontal stratification in front of the North Channel, and vertical mixing is more pronounced in the South Channel, which is rather well mixed.

The depth of the photic zone ranges from 2 to 4 m at concentrations of suspended matter equal to 10 mg/L and decreases to 0.1–0.3 m when the turbidity is equal to 100 mg/L. A plot of DSi versus salinity shows a perfect dilution line up to a salinity of 8‰ for the southern stations and 20‰ for the northern stations. At higher salinity values, DSi concentrations in surface water drop below the dilution line. In bottom waters DSi concentrations are greater than those predicted by the ideal dilution line, suggesting the release of DSi from the bottom. Microscopic analysis of suspended solids shows that diatoms are abundant where DSi deviates from the dilution line. The most abundant genera include *Coscinodiscus*, *Skeletonema*, *Synedra*, and *Thalassiosira*. A map of the distribution of BSi in surface water is shown in Figure 8.5.

DeMaster et al. (1983) found no evidence of abiotic uptake of silica. Using the silica versus salinity plots, these authors estimated that 33 percent of the riverine DSi flux was removed in the northern stations but only 17 percent in the southern stations. The northern stations are characterized by low turbidity and a shallower mixing depth because of the horizontal stratification of the water column. Therefore, the diatom populations remain in the photic zone for longer periods of time. In the southern stations,

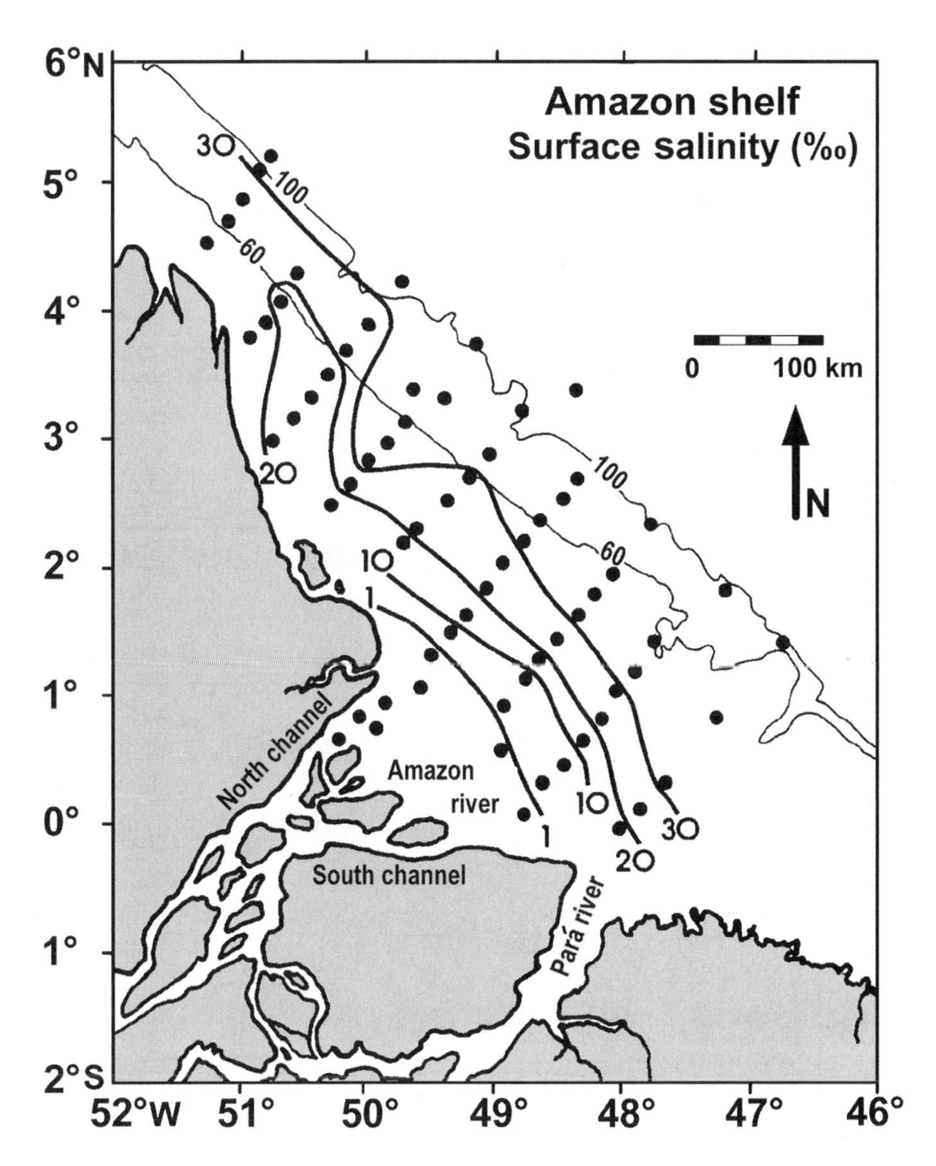

Figure 8.3. Surface salinity distribution on the Amazon shelf (after DeMaster et al. 1983).

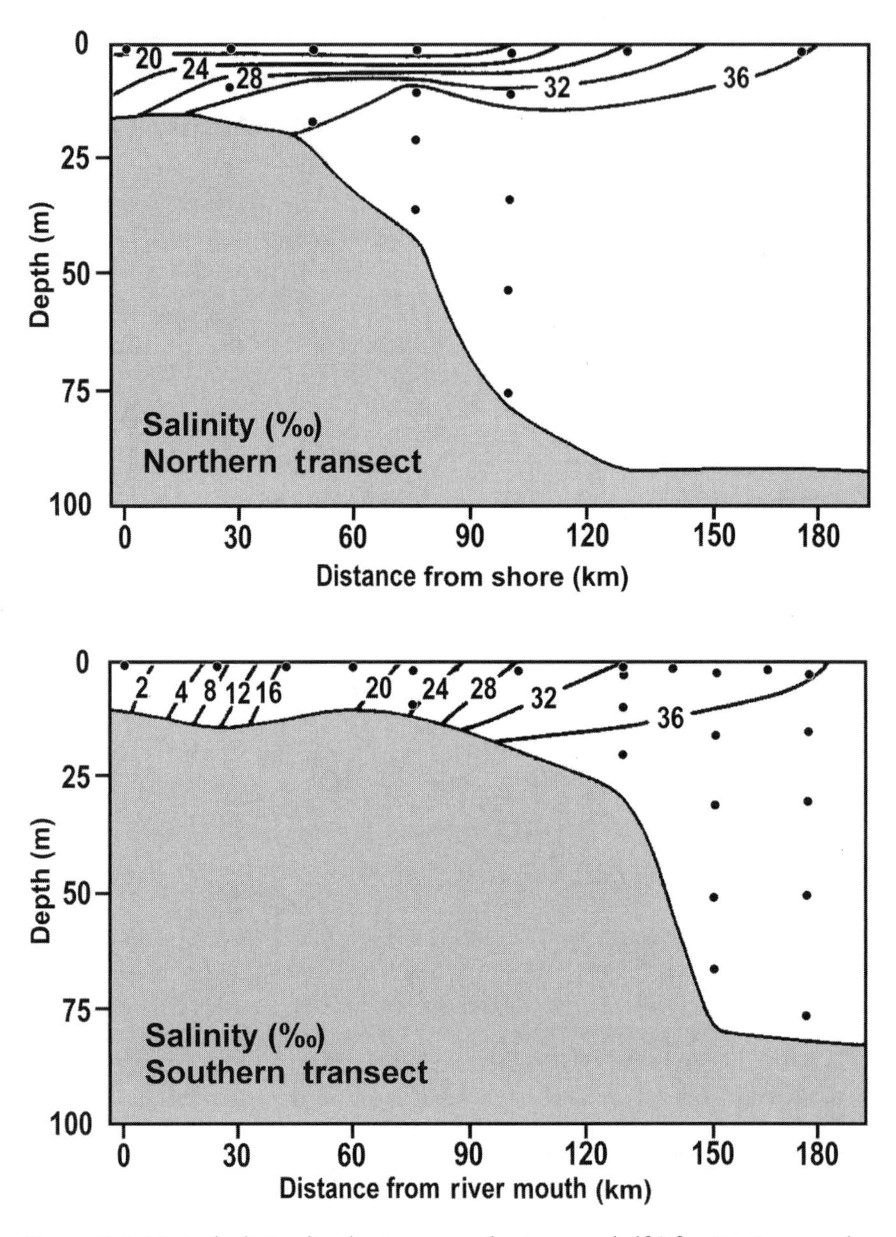

Figure 8.4. Vertical salinity distribution across the Amazon shelf (after DeMaster et al. 1983).

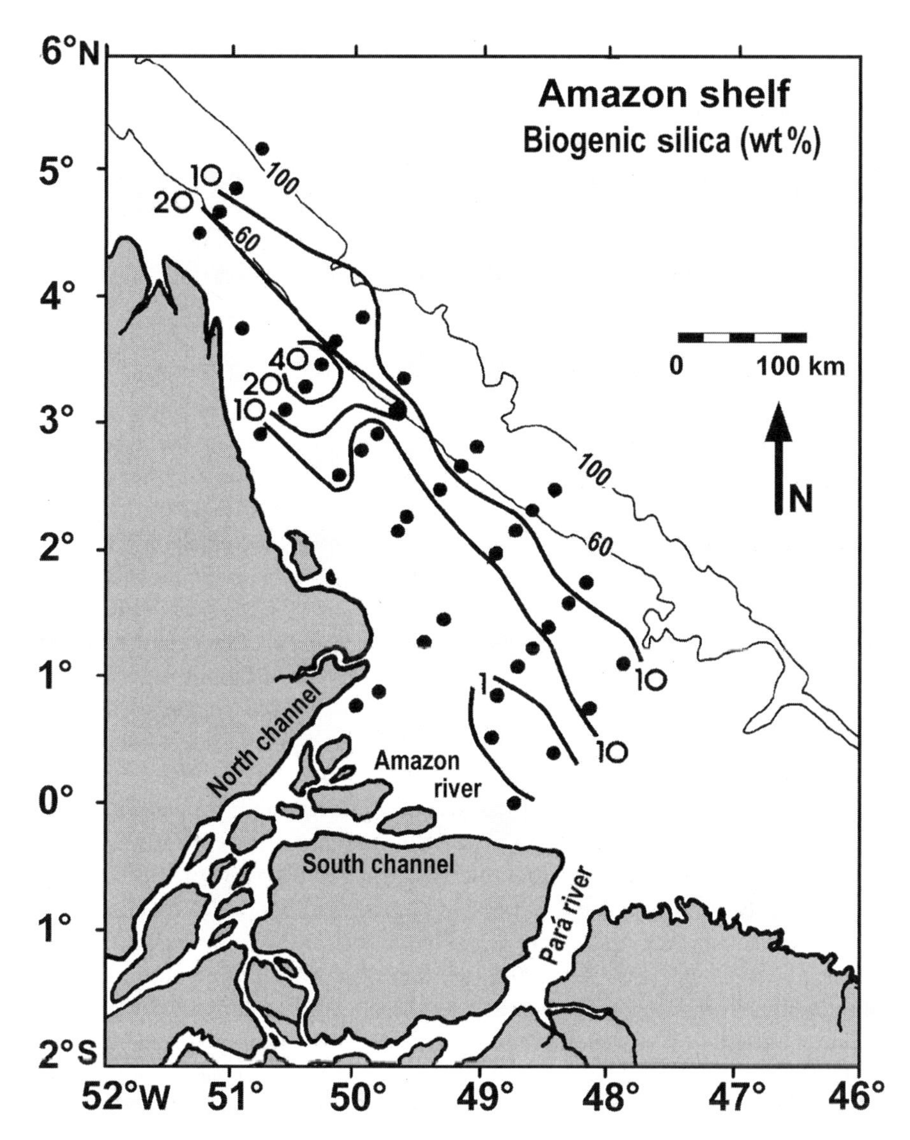

Figure 8.5. BSi content in surface suspended solids on the Amazon shelf (after DeMaster et al. 1983).

the sediments are resuspended because of the high vertical turbulence, and the diatom population spends a long period of time below the euphotic zone. Under these conditions, the mixing depth exceeds the critical depth where integrated respiration equals integrated production.

Besides turbidity, nutrients also control the productivity of the Amazon shelf. According to Ryther et al. (1967) and Gibbs (1970), the concentrations of nitrate and phosphate are low in the river and are not sufficient to sustain the primary production needed to consume the high concentration of DSi. According to the authors, the primary production on the Amazon shelf region is maintained by upwelling of seawater rich in nitrate and phosphate.

DeMaster et al. (1983) also attempted to estimate a budget of silica for the Amazon plume. The river supply of DSi to the plume is about 49×10^{12} g SiO_2/yr. However, 17–33 percent of DSi entering the Amazon shelf is removed by diatoms, representing a flux of $8–16 \times 10^{12}$ g SiO_2/yr. The maximum value is comparable to the opal production given by Milliman and Boyle (1975), which is around 15×10^{12} g SiO_2/yr. From the rate of sedimentation on the shelf and a mean weight percentage of BSi in sediments (0.25 percent), DeMaster et al. (1983) estimated that a maximum of 2×10^{12} g SiO_2/yr is permanently removed from the water column. This low rate of BSi accumulation in sediments compared with production indicates that most of the BSi dissolves in the water column or at the sediment–water interface. From the 49×10^{12} g SiO_2/yr of DSi delivered by the Amazon River, 47×10^{12} g SiO_2/yr reaches the open ocean. Thus the role of the Amazon continental shelf in trapping the riverine DSi flux is considered minimal (about 4 percent).

During the AmasSeds project, the biogeochemical processes and nutrient dynamics in the Amazon shelf waters were investigated seasonally (DeMaster and Pope 1996; DeMaster et al. 1996). Mass balance calculations of DSi removed from the water column and the standing crop of BSi indicated massive loss of particulate BSi, which could be attributed to zooplankton grazing and particle sedimentation. In the framework of the same project, Michalopoulos and Aller (2004) reexamined the storage of BSi in the Amazon continental shelf sediments and evaluated more thoroughly the role of transformation of opaline Si into neoformed minerals during early diagenesis. These authors modified the procedure for the determination of BSi by including an initial mild leach step with 0.1 N HCl to remove the metal oxide coatings and to activate poorly crystalline authigenic phases for alkaline dissolution in 1 percent Na_2CO_3. Thus their measurements of BSi include the early diagenetic alteration products such as cation-rich authigenic aluminosilicates. The two-step procedure indicates that about 90 percent of the BSi originally deposited is converted to clay or otherwise altered, leading to a higher BSi content of about 1.8 percent SiO_2, compared with 0.25 percent SiO_2 as originally determined by DeMaster et al. (1983) for the Amazon shelf sediments. Michalopoulos and Aller revised the estimates of reactive Si burial and concluded that the deltaic storage of riverine Si reached about 22 percent of the Amazon River input, higher than previously thought.

Zaire

The behavior of silica in the Zaire, the world's second largest river, with a discharge of 1.4×10^3 km^3/yr, has been less studied. The concentration of DSi (186 μM) is similar to that observed in the Amazon, but the turbidity is significantly lower (32 mg/L) (GEMSWATER, www.gemswater.org). During its dilution with seawater in the plume, DSi appears to behave conservatively based on observations to date, except at high salinities, where some removal can be detected (van Bennekom et al. 1978). This can be explained by the low concentration of N and P nutrients in the river water. As in the Amazon, primary production is related to the upwelling of nutrient-rich oceanic waters. Thus, only a very small fraction of DSi carried by the river may be trapped during estuarine mixing, which occurs in the coastal zone and adjacent open ocean. Accumulation rates and relative abundances of siliceous microfossils in sediments in the Congo fan area have been determined by Uliana et al. (2001). The authors found that the major contributors to productivity were marine diatoms.

Small Stratified Estuaries

In smaller stratified estuaries such as the Rhône (Chou and Wollast 1997) and the Rhine (unpublished data), the mixing of freshwater and seawater in the river plume is fast, and the residence time of freshwater in the outer estuary is reduced to a few days. This time is not sufficient to observe any significant removal of DSi in the thin freshwater surface layer (Figure 8.6).

The low concentrations of DSi in the riverine end member of the Rhône could be attributed to the consumption of this nutrient by freshwater diatoms farther upstream. Thus, in rivers with high freshwater discharge or in systems where the tidal amplitude is negligible, it is likely that riverine DSi is almost quantitatively transferred to the oceanic system, and little is removed in the estuary. However, in the case of mountainous rivers of Hawaii, DSi is removed rapidly during estuarine mixing (F. T. Mackenzie, personal communication, 2005).

Tidal Estuaries

Rivers with low freshwater discharge that are under the influence of large tidal oscillations are characterized by a strong input of saltwater and a deep propagation of the tide in the river system. An important consequence of this situation is the long residence time of freshwater in the estuarine system, which may reach several months. The long residence time and the fluctuations of the water composition within the estuary are very favorable for the occurrence of physical, chemical, and biological transformations, which may modify significantly the fluxes of components in the system. Because of their limited size, these systems are also very sensitive to anthropogenic perturbations. Silica is an example of a component whose behavior may be significantly affected during its journey in the estuaries.

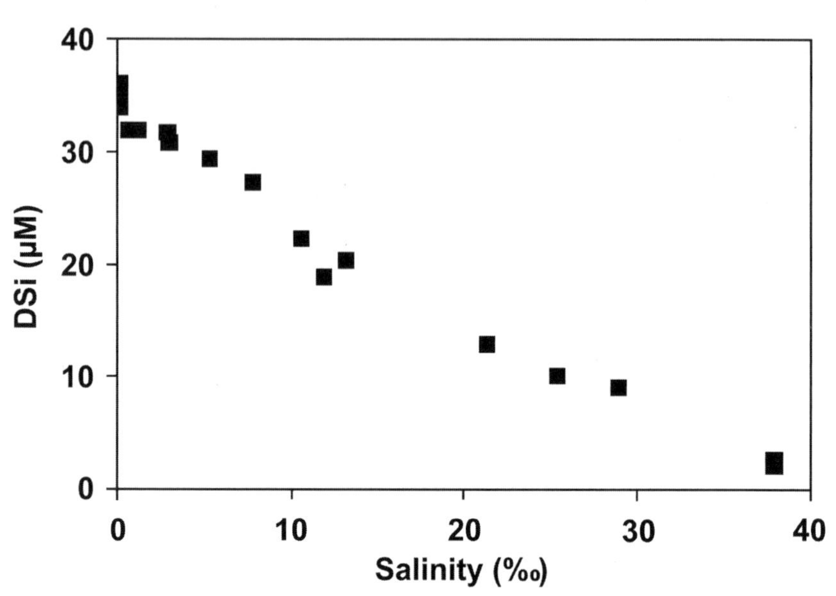

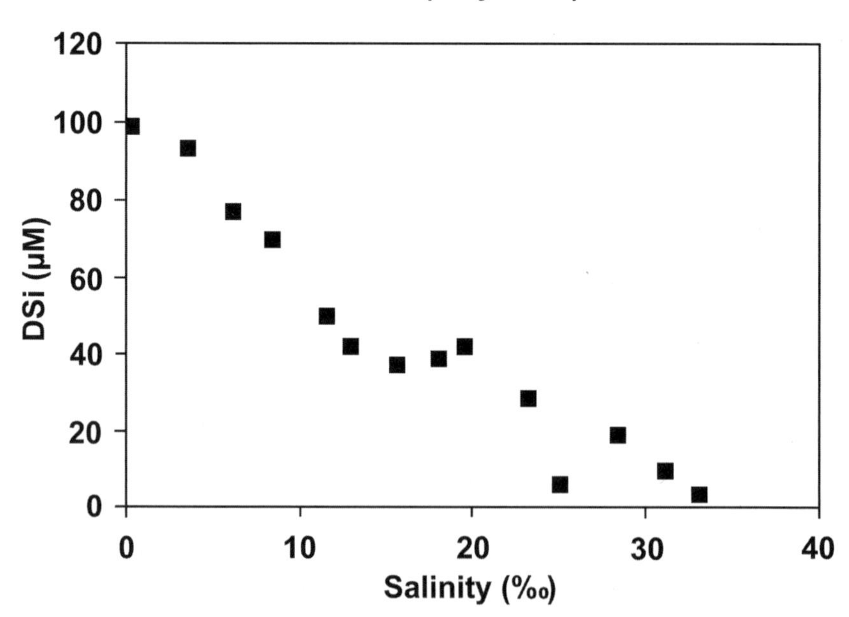

Figure 8.6. Longitudinal profile of DSi as a function of salinity in the Rhône (after Chou and Wollast 1997) and Rhine (unpublished data) estuaries.

We will describe here, as an example, the behavior of silica in the Scheldt estuary (Figure 8.7), where intensive field studies concerning this element and others were performed.

It is a strong tidal estuary with a mean annual river discharge of 3.8 km^3/yr and a surface area of the hydrographic basin equal to 21.6 $\times$ 10^3 km^2 (Wollast 1988). There are very large seasonal fluctuations of freshwater discharge, with mean minimal values of 20 m^3/s to maximal values exceeding 350 m^3/s. The tidal amplitude at the mouth is about 4 m, allowing an input of 1 $\times$ 10^9 m^3 per tide. During the same tidal period, the volume of freshwater leaving the estuary is only 6 $\times$ 10^3 m^3 per tide. The salt intrusion covers a distance of about 100 km from the mouth, and the tide is still close to 2 m at 160 km from the mouth, where tidal changes are blocked by a sluice. High water velocities and bottom friction are sufficient to mix efficiently the water column, and little or no vertical stratification can be observed.

In the case of the Scheldt, the residence time of freshwater in the mixing zone may reach 3 months in summer and is still about 1 month during the high flood period in winter and early spring. The turbidity is elevated in the freshwater present in the upper tidal estuary, with a marked cycle of deposition and resuspension of particulate matter

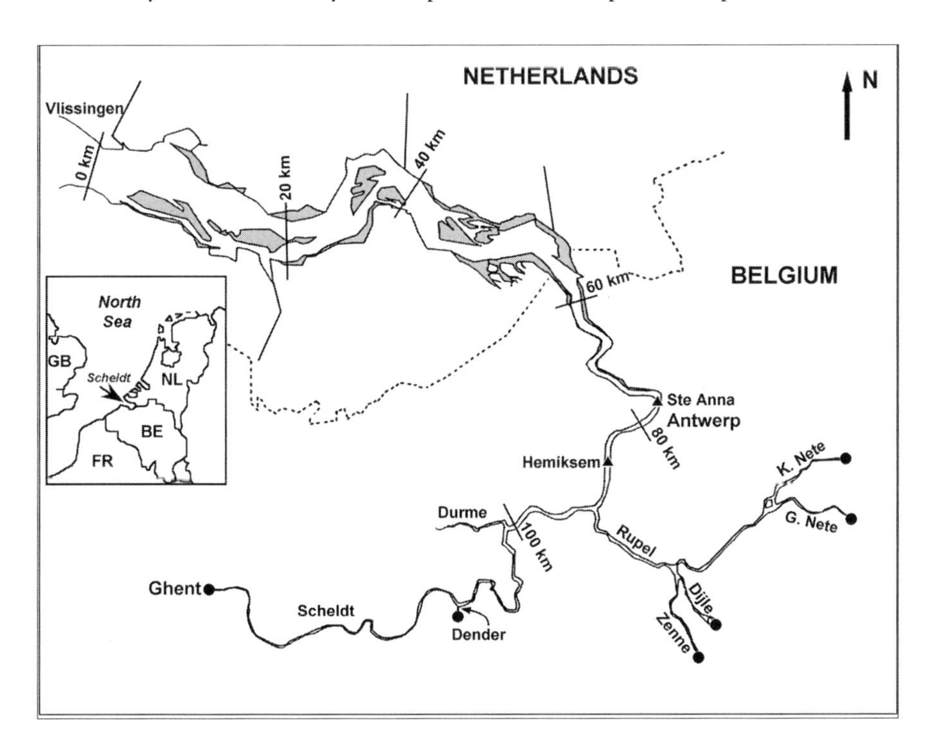

Figure 8.7. Map of the Scheldt estuary. The limits of the tidal influence are indicated by solid circles.

linked to tides. Beyond the turbidity maximum in the vicinity of salinity 5‰, the suspended matter content decreases progressively with increasing salinity. The Scheldt is also situated in a heavily populated and highly industrialized region. The concentration of N and P nutrients reflects the strong anthropogenic perturbations, with present-day mean freshwater values up to 600 μM of total dissolved N (NO_3 + NO_2 + NH_4) and 8 μM of PO_4. These values are more than one order of magnitude larger than the natural concentrations in unpolluted rivers. In contrast, the concentrations of DSi in freshwater in winter (250 μM) are comparable to those observed in similar rivers of the temperate region.

DSi behaves markedly as a nonconservative element in the Scheldt estuary. Wollast and De Broeu (1971) were the first to demonstrate that the silica uptake, which removed all DSi from the water column as salinities reached 20‰, was due to consumption by diatoms (Figure 8.8).

The authors also showed by laboratory experiments that the concentration of DSi was too low and the residence time too short to observe abiotic uptake of DSi in the

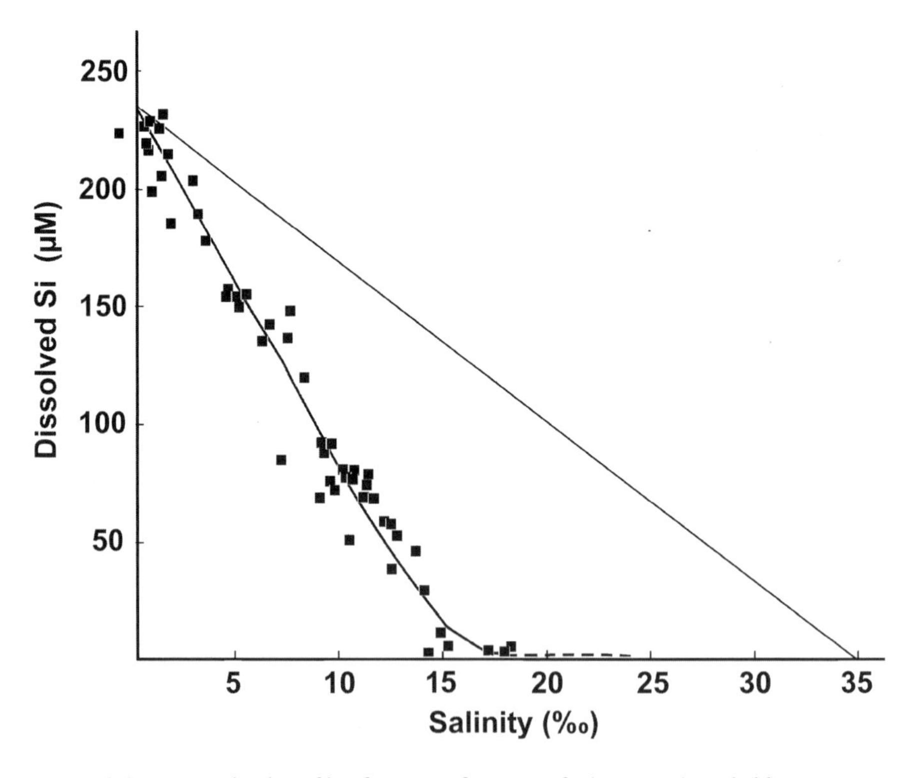

Figure 8.8. Longitudinal profile of DSi as a function of salinity in the Scheldt estuary in 1967 (after Wollast and De Broeu 1971).

Scheldt estuary. In the late 1960s, when this study was performed, the turbidity of the freshwater part of the upper estuary was too high to allow the development of diatoms. Phytoplankton productivity began to be active only after the turbidity maximum in the area of mixing of freshwater and seawater. Analysis of historical data shows that there were only limited fluctuations of DSi in space and time for the upper freshwater part of the estuary, suggesting insignificant diatom activity upstream.

The water quality of the Scheldt and its tributaries has improved since then because of the recent implementation of numerous wastewater treatment plants. Although the turbidity of the upper Scheldt estuary is still high (suspended particulate matter = 100–150 mg/L), the occurrence of diatoms in the upper reaches of the estuary is now well established (Muylaert et al. 2005). This is also demonstrated by the evolution of the concentration of DSi as a function of the distance to the sea. Figure 8.9 shows that the uptake of DSi occurs both in the freshwater part (km 100–km 165) and in the brackish zone (km 0–km 100).

In the freshwater part, above km 100, the consumption of DSi starts in March and becomes important in July and August, leading to its depletion in the upper reaches. This is further demonstrated in Figure 8.10, which shows the temporal evolution of DSi, BSi, and chlorophyll-*a* as measured weekly at Hemiksem (about km 85).

There is a slight and progressive uptake of DSi in spring followed by a bloom of diatoms covering the months of July and August. Delstanche (2004) showed that BSi concentration covaried with that of chlorophyll-*a* in the freshwater tidal reaches, suggesting that diatoms are the dominant phytoplankton species (Figure 8.11). The intensity of the diatom bloom decreases downstream, with the maximum observed during summer and early fall.

The uptake of DSi in the marine part of the estuary can be further established by the evolution of its concentration as a function of salinity in various months of the year (Figure 8.12).

The concentration of DSi follows strictly the ideal dilution line in January (not shown). A slight departure from this line is already present at low salinities in April, but the bloom of diatoms starts only in May and persists until September. In summer, DSi is completely depleted in the brackish zone, as is the case for the freshwater part of the estuary (Figures 8.10 and 8.12). However, one observes a small continuous input of DSi at around km 90 from the Rupel tributary (Figure 8.9).

The most significant change between the situation in July 1967 (Figure 8.8) and that in the 1990s (Figure 8.12) is the substantial decrease of the concentration of DSi in river water entering the mixing zone. In 1967, the concentration of DSi was still around 200 μM in the riverine end member in July. It is less than 10 μM at present because of the uptake of DSi by freshwater diatoms in the upper estuary. In aquatic systems, environmental factors influencing primary production are physical (e.g., temperature, light intensity), chemical (e.g., nutrient supply), and biological (e.g., grazing, salinity stress). In most temperate estuaries, the essential nutrients (N, P, Si) are abundant, and because

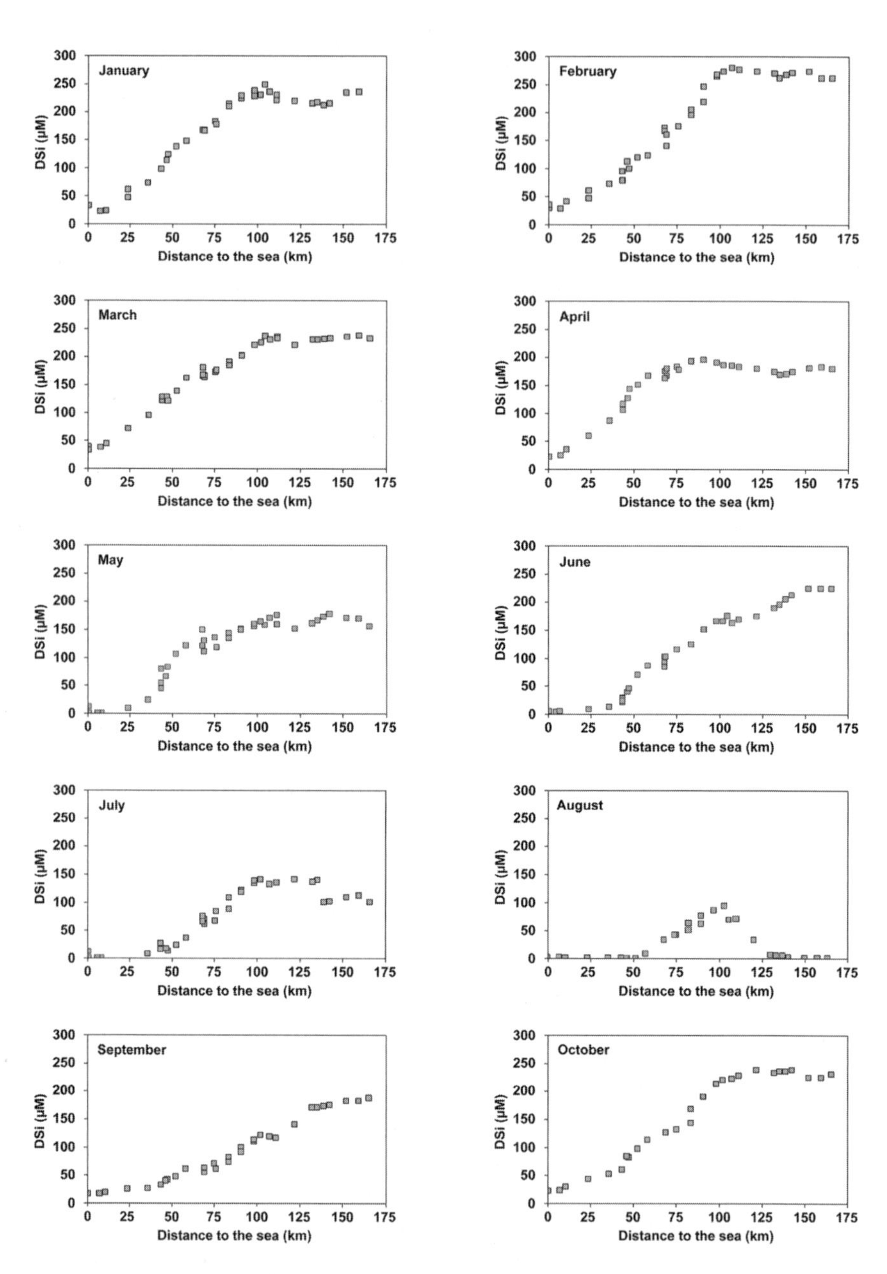

Figure 8.9. Longitudinal profile of DSi as a function of distance to the sea in the Scheldt estuary for 1998.

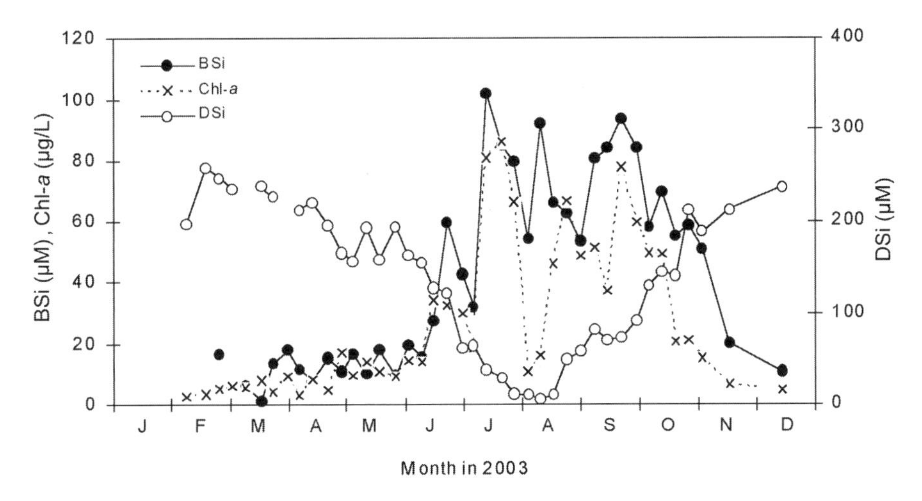

Figure 8.10. Temporal evolution of DSi, BSi, and chlorophyll-*a* (Chl-*a*) at Hemiksem (about km 85) in the Scheldt estuary for 2003. DSi and BSi data are from V. Carbonnel.

of the high turbidity, primary production usually is limited by light availability (Cloern 1987, Cole et al. 1992, Kromkamp and Peene 1995, Kocum et al. 2002, Monbet 1992). This is the case for the Scheldt estuary, where the longitudinal distribution of dissolved orthophosphate and of total inorganic nitrogen fluctuates seasonally, but their concentrations remain well above that needed for phytoplankton growth (Figure 8.12).

Billen et al. (2005) evaluated the nutrient fluxes and water quality in the drainage network of the Scheldt basin over the past five decades. Analysis of data on DSi concentrations available for a station situated at about km 60 from the sea showed a distinct decreasing trend, at least in the last 30 years. The authors interpreted this trend as the result of higher Si retention caused by higher diatom production in the cleaner drainage network upstream.

Because of the high turbidity prevailing in estuaries, the euphotic depth is limited to a very thin layer of water, which is typically less than 1 m. Because the water column is well mixed vertically, phytoplankton spend most of their time in the dark, where no photosynthesis is possible and where they continue to respire. The net primary production thus depends greatly on the ratio of euphotic depth to the mixing depth, which is the water depth in a well-mixed estuary. The water depth increases continuously downstream from the river and upstream to the mouth of the estuary. The turbidity usually decreases with increasing salinity, and thus the euphotic depth increases seaward. The improving light conditions toward the sea do not necessarily warrant higher net production in the entire water column. If the water depth increases more rapidly than the euphotic depth, the net primary production should decrease seaward. This situation occurs in the Scheldt, where the highly turbid upper freshwater zone corresponds also to the shallow

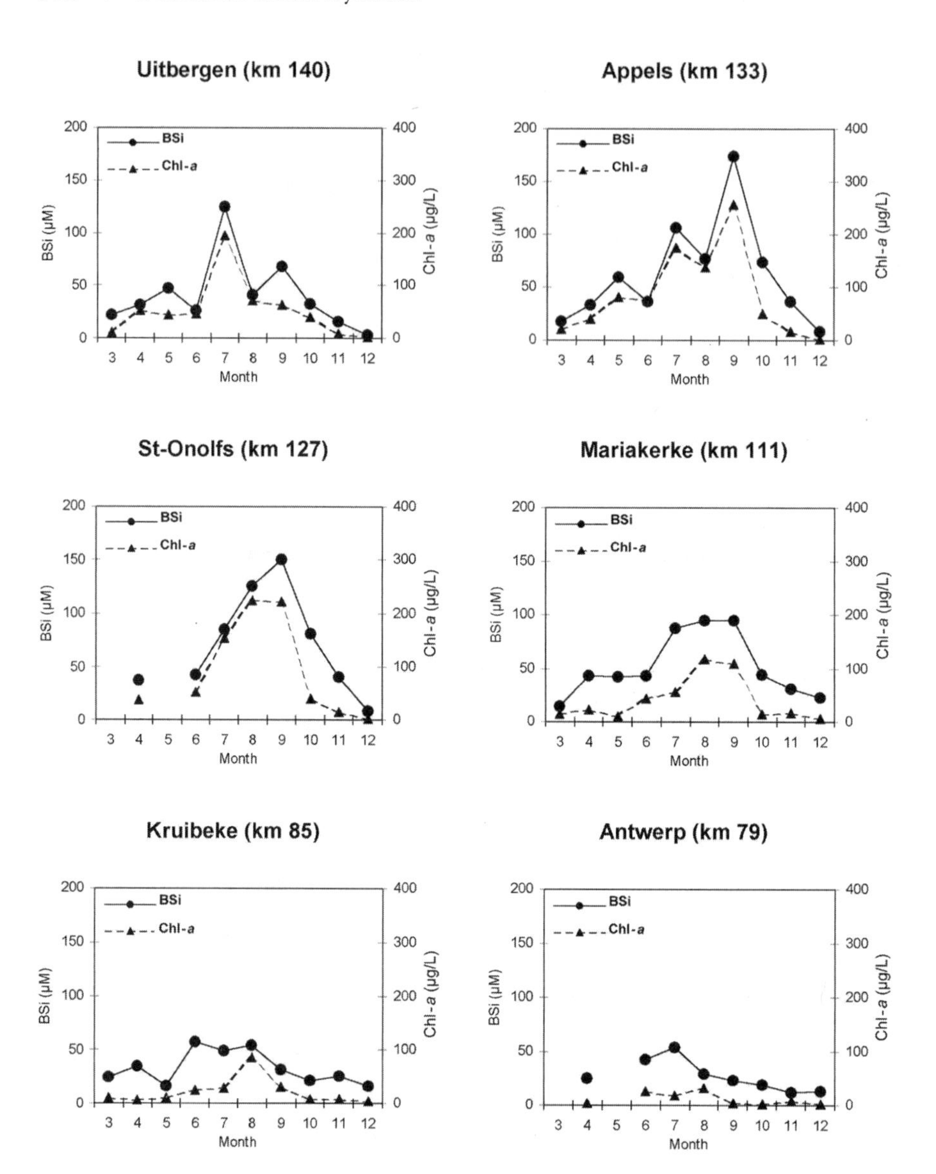

Figure 8.11. Seasonal evolution of BSi and chlorophyll-*a* (Chl-*a*) in the freshwater tidal reaches of the Scheldt estuary for 2002 (after Delstanche 2004). Distance to the sea for each station is indicated in parentheses.

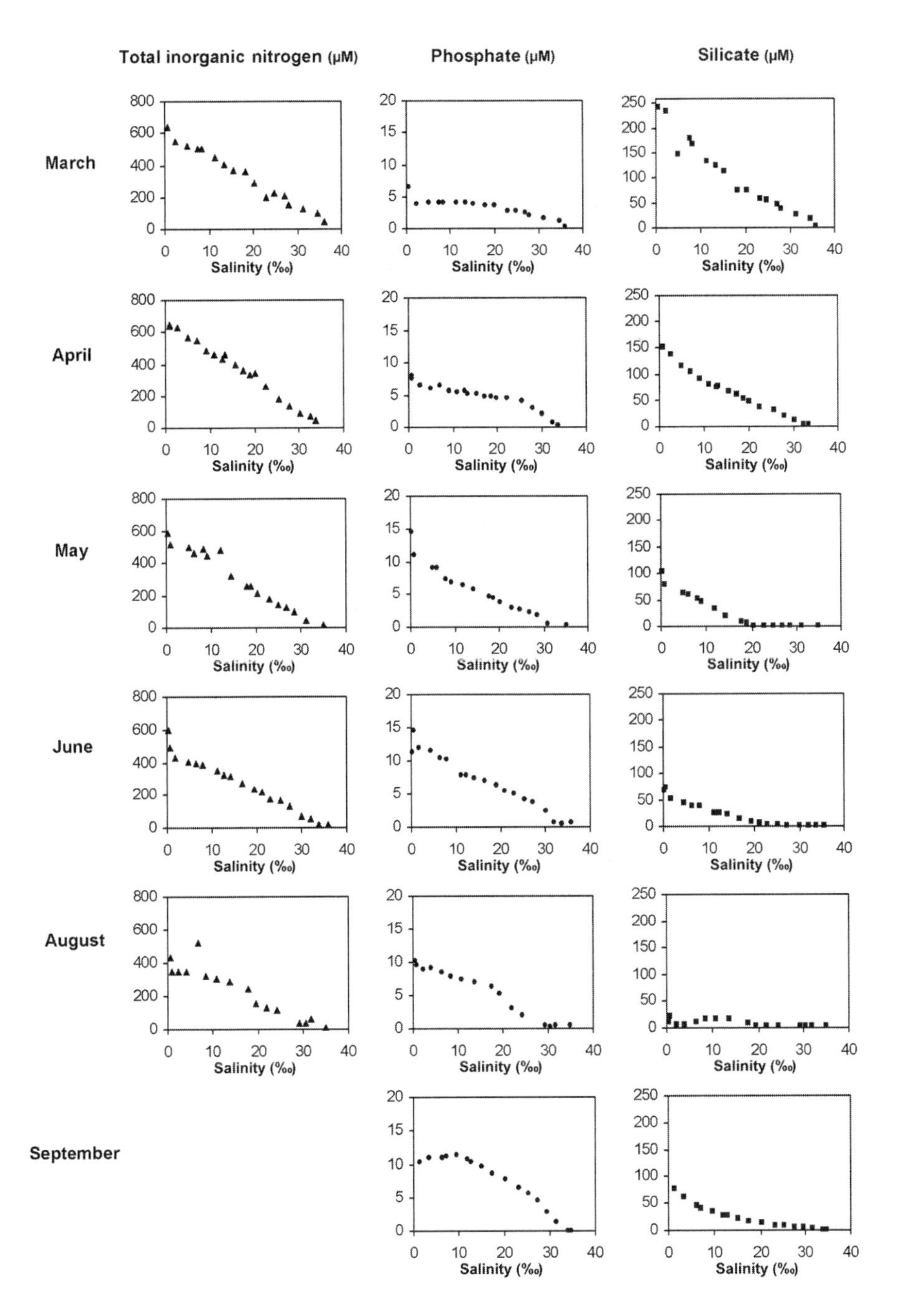

Figure 8.12. Seasonal evolution of dissolved nutrients (total inorganic nitrogen, phosphate, silicate) as a function of salinity in the Scheldt estuary in the 1990s.

area of the estuary. As a result, net primary production per unit surface area is the highest in the freshwater part, where higher turbidity is also encountered (Muylaert et al. 2005; Desmit et al. 2005). The rapid decrease in chlorophyll-*a* concentration observed in the Scheldt with increasing salinity above 5‰ may be explained largely by the increase of water depth in this area instead of osmotic stress (Muylaert et al. 2005).

As in many other freshwater tidal estuaries, the zooplankton community is dominated by rotifers in the freshwater reaches, which are inefficient in controlling phytoplankton biomass (Heinbokel et al. 1988; Pace et al. 1992; Holst et al. 1998; Muylaert et al. 2000; Park and Marshall 2000). In the brackish part, the zooplankton is dominated by marine calanoid copepods (Soetaert and Van Rijswijk 1993) that can contribute significantly to the transfer of opal to sediments.

Based on the one-dimensional model of Wollast (1978) for the Scheldt and using the DSi concentrations of the time series at km 100 as a boundary condition and concentrations along the estuary, a budget can be established for Si. The intensive diatom activity in the Scheldt estuary is responsible for a removal of DSi that can be roughly estimated to be 20 percent of the annual river input (Wollast 1978). A large fraction of the opal produced by the phytoplankton is trapped in the estuarine sediments, especially in the area of the turbidity maximum, where the sedimentation rate of suspended matter is maximal. An unknown fraction of opal nevertheless may be exported to the North Sea and further dissolved. A more detailed budget for the silicon biogeochemical cycling in the Scheldt river–estuary–coastal zone continuum is being established in the framework of the Belgian Silica Retention in the Scheldt Continuum and Its Impact on Coastal Eutrophication (SISCO) project. BSi content in the surface sediments (top 1 cm) of the subtidal zone of the Scheldt estuary could vary from less than 0.05 percent SiO_2 to about 1.5 percent SiO_2 (L. Rebreanu, personal communication, 2005). Results obtained from dissolution studies performed on surface sediment of the intertidal zone, containing 5 percent SiO_2, suggest that little BSi is remobilized in the sedimentary column (De Bodt 2005). Mass balance calculations based on DSi and BSi concentrations in the water column indicate that as much as 70 percent of the BSi produced in the freshwater tidal reaches of the Scheldt estuary could be retained in the sediments (V. Carbonnel, personal communication, 2005).

The increase of N and P inputs by the various estuaries bordering the North Sea and the decrease of DSi caused by eutrophication of the river system have significantly modified phytoplankton succession in the adjacent coastal zone (Lancelot et al. 1987; Cadée 1990; Billen et al. 1991; Rousseau et al. 2002). According to Redfield et al. (1963), diatom growth requires Si:N and Si:P molar ratios of 1 and 16, respectively. Analysis of long-term trends of the ratio of nutrients, delivered annually by the Scheldt Basin at the entrance of the estuarine zone, indicates Si:N ratios below 1 since the mid-1960s (Billen et al. 2005). The Si:P ratio also declined rapidly since the 1950s, reaching values below the Redfield ratio by the late 1960s, but began to increase again since the 1990s due to the decreased P loading (Billen et al. 2005). Seasonal cycling of nutrients in the Belgian coastal zone shows that the Si:N ratio could reach values below 0.1 in spring (van der Zee and Chou 2005), suggesting Si limitation. The early spring diatom bloom in the Southern Bight of the North Sea has been drastically reduced and replaced by the excessive development of flagellates (*Phaeocystis* sp.). As a consequence, the food web is reduced, and the *Phaeocystis* residues form anesthetic foams on the beaches, causing environmental concern (Lancelot 1995).

Conclusion

The distribution and behavior of DSi in estuaries depend greatly on physical and hydrodynamic properties of the system. In large stratified estuaries, DSi is almost quantitatively transferred to the coastal zone because of the high turbidity occurring in the freshwater system and the low concentrations of N and P nutrients (Table 8.1).

DSi may be consumed rapidly by diatoms in the outer estuary or in the adjacent coastal zone if the turbidity is sufficiently decreased by dilution or settling of the suspended matter. However, complete consumption of DSi in undisturbed systems requires additional inputs of N and P nutrients from the open ocean by upwelling or by vertical mixing of deep, nutrient-rich oceanic water at the shelf break. As shown in Table 8.1, the Si:N ratio of riverine nutrients is much larger in pristine rivers than is needed for the development of diatoms, and the same is true for P.

The situation is different in smaller estuaries that are tidally well mixed and that are

Table 8.1. Mean concentration of nutrients in selected rivers.

	Discharge (km^3/yr)	PO_4 (μM)	Total Dissolved Inorganic Nitrogen (μM)	SiO_2 (μM)	Si:N
Amazon	6,450	0.40	2.85	120	42
Zaire	1,350	0.80	6.40	190	30
Changjiang	873	0.47	55	108	2
Mississippi	580	6.7	75	126	1.7
Rhine	74.5	13	277	86	0.31
Rhône	49.5	4.5	68	67	1
Scheldt	3.8	10	558	200	0.37
Mean river water composition		0.33	8.5	180	21

Data from GEMSWATER (www.gemswater.org).

often strongly disturbed by anthropogenic activities. The concentrations of dissolved N and P species are very high and do not limit photosynthesis. During storms, inorganic nitrogen also tends to be leached from the soils and enters the Scheldt, contributing to higher N loading. However, the euphotic depth is very shallow and much less than the vertical mixing depth. As a consequence, phytoplankton spends a large part of the daytime in the dark, and respiration in the water column exceeds primary production. The ratio of euphotic depth to mixing depth therefore is the main factor controlling the development and growth of phytoplankton in tidal estuaries. In most cases, and as long as DSi is available, the phytoplankton population is dominated by diatoms, probably because of their ability to live under low light conditions. The diatoms can totally consume DSi in the estuarine system in summer, when the incident light is maximum, which then strongly affects the food web in the adjacent coastal zone.

Acknowledgments

This study was supported by the Belgian Federal Science Policy Office (SISCO project, contract no. EV/11/17A) and by the European Union (EUROTROPH project, contract no. EVK3-CT-2000-00040). This is also a contribution to the European Integrated Project CarboOcean (contract no. 511176-2). P. Meire and S. Van Damme kindly let us use 1998 DSi data for the Scheldt estuary from the Onderzoek Milieu-Effecten Sigmaplan database. V. Carbonnel provided water column DSi and BSi data for the Hemiksem station. We thank N. Canu, C. de Marneffe, N. Roevros, and M. Tsagaris for their assistance in field sampling and in various laboratory analyses. We would like to thank Fred Mackenzie for his constructive comments and suggestions, which improved the clarity of this manuscript. Finally, L.C. would like to dedicate this chapter to her coauthor and dear colleague, the late Roland Wollast, for his great passion for the silica cycle at regional and global scales.

Literature Cited

Anderson, P. R., and M. M. Benjamin. 1985. Effects of silicon on the crystallization and adsorption properties of ferric oxides. *Environmental Science and Technology* 19:1048–1053.

Atkins, W. R. G. 1926. Seasonal changes in the silica content of natural waters in relation to the phytoplankton. *Journal of the Marine Biological Association of the United Kingdom* 14:89–99.

Bien, G. S., D. E. Contois, and W. H. Thomas. 1958. The removal of soluble silica from fresh water entering the sea. *Geochimica et Cosmochimica Acta* 14:35–54.

Billen, G., J. Garnier, and V. Rousseau. 2005. Nutrient fluxes and water quality in the drainage network of the Scheldt Basin over the last 50 years. *Hydrobiologia* 540:47–67.

Billen, G., C. Lancelot, and M. Maybeck. 1991. N, P, and Si retention along the aquatic continuum from land to ocean. Pp. 19–44 in *Ocean margin processes in global change*, edited by R. F. C. Mantoura, J. M. Martin, and R. Wollast. Chichester, UK: Wiley.

Burton, J. D., T. M. Leatherland, and P. S. Liss. 1970a. The reactivity of dissolved silicon in some natural waters. *Limnology and Oceanography* 15:473–476.

Burton, J. D., P. S. Liss, and V. K. Venugopalan. 1970b. The behaviour of dissolved silicon during estuarine mixing. I. Investigations in Southampton water. *Journal du Conseil. Permanent International pour l'Exploration de la Mer* 33:134–140.

Cadée, G. C. 1990. Increase of *Phyaeocystis* blooms in the westernmost inlet of the Wadden Sea, the Mesdiep, since 1973. *Water Pollution Research Report* 12:105–112.

Calvert, S. E. 1983. Biological uptake and accumulation of silica on the Amazon continental shelf. *Geochimica et Cosmochimica Acta* 47:1713–1723.

Chou, L., and R. Wollast. 1997. Biogeochemical behavior and mass balance of dissolved aluminum in the western Mediterranean Sea. *Deep-Sea Research II* 44:741–768.

Cloern, J. E. 1987. Turbidity as a control on phytoplankton biomass and productivity in estuaries. *Continental Shelf Research* 5:1367–1381.

Cole, J. J., N. F. Caraco, and B. L. Peierls. 1992. Can phytoplankton maintain a positive carbon balance in a turbid, freshwater, tidal estuary? *Limnology and Oceanography* 37:1608–1617.

Conley, D. L., C. Schelske, and E. F. Stoermer. 1993. Modification of the biogeochemical cycle of silica with eutrophication. *Marine Ecology Progress Series* 101:179–192.

De Bodt, C. 2005. *Etude de la dissolution de la silice biogénique dans les sédiments de l'estuaire de l'Escaut.* Travail de fin d'études, Bioingénieur en chimie et bioindustries, Université Libre de Bruxelles.

Delstanche, S. 2004. *Contribution à l'étude du cycle biogéochimique de la silice dans le continuum aquatique de l'Escaut.* Mémoire, diplôme d'etudes spécialisées interuniversitaire en hydrologie, Université Catholique de Louvain.

DeMaster, D. J., G. B. Knapp, and C. A. Nittrouer. 1983. Biological uptake and accumulation of silica on the Amazon continental shelf. *Geochimica et Cosmochimica Acta* 47:1713–1723.

DeMaster, D. J., and R. H. Pope. 1996. Nutrient dynamics in Amazon shelf waters: Results from AmasSeds. *Continental Shelf Research* 16:263–289.

DeMaster, D. J., W. O. Smith Jr., D. M. Nelson, and J. Y. Aller. 1996. Biogeochemical processes in Amazon shelf waters: Chemical distributions and uptake rates of silicon, carbon and nitrogen. *Continental Shelf Research* 16:617–643.

Desmit, X., J. P. Vanderborght, P. Regnier, and R. Wollast. 2005. Control of phytoplankton production by physical forcing in a strongly tidal, well-mixed estuary. *Biogeosciences* 2:205–218.

Dixit, S., P. Van Cappellen, and A. J. Van Bennekom. 2001. Processes controlling solubility of biogenic silica and porewater build-up of silicic acid in marine sediments. *Marine Chemistry* 73:333–352.

Edmond, J. M., E. A. Boyle, B. Grant, and R. F. Stallard. 1981. The chemical mass balance in the Amazon plume: The nutrients. *Deep-Sea Research* 28:1339–1374.

Fanning, K. A., and M. E. Q. Pilson. 1973. The lack of inorganic removal of dissolved silica during river–ocean mixing. *Geochimica et Cosmochimica Acta* 37:2405–2415.

Gibbs, R. J. 1970. Circulation in the Amazon River estuary and adjacent Atlantic Ocean. *Journal of Marine Research* 28:113–123.

Heinbokel, J. F., D. W. Coats, K. W. Henderson, and M. A. Tyler. 1988. Reproduction rates and secondary production of three species of the rotifer genus *Synchaeta* in the estuarine Potomac River. *Journal of Plankton Research* 10:659–674.

Heip, C. H. R., N. K. Goosen, P. M. J. Herman, J. Kromkamp, J. J. Middelburg, and K. Soetaert. 1995. Production and consumption of biological particles in temperate tidal estuaries. *Oceanography and Marine Biology Annual Review* 33:1–149.

Holst, H., H. Zimmermann, H. Kausch, and W. Koste. 1998. Temporal and spatial dynamics of planktonic rotifers in the Elbe Estuary during spring. *Estuarine Coastal Shelf Science* 47:261–273.

Humborg, C., D. J. Conley, L. Rahm, F. Wulff, A. Cociasu, and V. Ittekkot. 2000. Silicon retention in river basins: Far-reaching effects on biogeochemistry and aquatic food webs "in coastal Marine Environments." *Ambio* 29:45–50.

Humborg, C., V. Ittekkot, A. Cociasu, and B. von Bodungen. 1997. Effect of Danube River dam on Black Sea biogeochemistry and ecosystem structure. *Nature* 386:385–388.

Ittekkot, V., C. Humborg, and P. Schäfer. 2000. Hydrological alterations on land and marine biogeochemistry: A silicate issue? *BioScience* 50(9):776–782.

Joint, I. R., and A. J. Pomroy. 1981. Primary production in a turbid estuary. *Estuarine Coastal Shelf Science* 13:303–316.

Kocum, E., G. J. C. Underwood, and D. B. Nedwell. 2002. Simultaneous measurement of phytoplanktonic primary production, nutrient and light availability along a turbid, eutrophic UK east coast estuary (the Colne Estuary). *Marine Ecology Progress Series* 231:1–12.

Krauskopf, K. B. 1956. Dissolution and precipitation of silica at low temperatures. *Geochimica et Cosmochimica Acta* 10:1–26.

Kromkamp, J., and J. Peene. 1995. On the possibility of net primary production in the turbid Scheldt estuary (SW Netherlands). *Marine Ecology Progress Series* 121:249–259.

Lancelot, C. 1995. The mucilage phenomenon in the continental coastal waters of the North Sea. *Science of the Total Environment* 165:83–102.

Lancelot, C., G. Billen, A. Sournia, T. Weisse, F. Colijn, M. J. W. Veldhuis, A. Davies, and P. Wassman. 1987. *Phaeocystis* blooms and nutrient enrichment in the continental coastal zone of the North Sea. *Ambio* 16:38–46.

Liss, P. S. 1976. Conservative and non-conservative behavior of dissolved constituents during estuarine mixing. Pp. 93–127 in *Estuarine chemistry*, edited by J. D. Burton and P. S. Liss. London: Academic Press.

Liss, P. S., and C. P. Spencer. 1970. Abiological processes in the removal of silicate from seawater. *Geochimica et Cosmochimica Acta* 34:1073–1088.

Mackenzie, F. T., and R. M. Garrels. 1966. Chemical mass balance between rivers and oceans. *American Journal of Science* 264:507–525.

Mackenzie, F. T., R. M. Garrels, O. P. Bricker, and F. Bickley. 1967. Silica in sea water: Control by silicate minerals. *Science* 155:1404–1405.

Mackenzie, F. T., B. L. Ristvet, D. C. Thorstenson, A. Lerman, and R. H. Leeper. 1981. Reverse weathering and chemical mass balance in a coastal environment. Pp. 152–187 in *River inputs to the ocean*, edited by J.-M. Martin, J. D. Burton, and D. Eisma. Geneva: United Nations Environment Programme, United Nations Educational, Scientific, and Cultural Organization.

Mackin, J. E., and R. C. Aller. 1984. Diagenesis of dissolved aluminium in organic-rich estuarine sediments. *Geochimica et Cosmochimica Acta* 48:299–313.

Mackin, J. E., and R. C. Aller. 1986. The effects of clay mineral reactions on the dissolved Al distributions in sediments and waters of the Amazon continental shelf. *Continental Shelf Research* 6:245–262.

Mackin, J. E., and R. C. Aller. 1989. The nearshore marine and estuarine chemistry of dissolved aluminum and rapid authigenic mineral precipitation. *Reviews in Aquatic Sciences* 1:537–554.

Meybeck, M. 1979. Concentration des eaux fluviales en éléments majeurs et apports en solutions aux océans. *Revue de Géologie Dynamique et de Géographie Physique* 21(3):215–246.
Michalopoulos, P., and R. C. Aller. 1995. Rapid clay mineral formation in Amazon delta sediments: Reverse weathering and oceanic elemental cycles. *Science* 270:614–617.
Michalopoulos, P., and R. C. Aller. 2004. Early diagenesis of biogenic silica in the Amazon delta: Alteration, authigenic clay formation, and storage. *Geochimica et Cosmochimica Acta* 68:1061–1085.
Michalopoulos, P., R. C. Aller, and R. J. Reeder. 2000. Conversion of diatoms to clays during early diagenesis in tropical, continental shelf muds. *Geology* 28:1095–1098.
Milliman, J. D., and E. Boyle. 1975. Biological uptake of dissolved silica in the Amazon River estuary. *Science* 189:995–997.
Milliman, J. D., C. P. Summerhayes, and H. T. Barretto. 1975. Oceanography and suspended matter off the Amazon River. February–March 1973. *Journal of Sedimentary Petrology* 45:189–206.
Monbet, Y. 1992. Control of phytoplankton biomass in estuaries: A comparative analysis of microtidal and macrotidal estuaries. *Estuaries* 15:563–571.
Muylaert, K., M. Tackx, and W. Vyverman. 2005. Phytoplankton growth rates in the freshwater tidal reaches of the Scheldt estuary (Belgium) estimated using a simple light-limited primary production model. *Hydrobiologia* 540:127–140.
Muylaert, K., R. Van Mieghem, K. Sabbe, M. Tackx, and W. Vyverman. 2000. Dynamics and trophic roles of heterotrophic protists in the plankton of a freshwater tidal estuary. *Hydrobiologia* 432:25–36.
Officer, C. B., and D. R. Lynch. 1981. Dynamics of mixing in estuaries. *Estuarine, Coastal and Shelf Science* 12:525–533.
Pace, M. L., E. G. S. Findlay, and D. Lints. 1992. Zooplankton in advective environments: The Hudson River community and a comparative analysis. *Canadian Journal of Fisheries and Aquatic Science* 49:1060–1069.
Park, G. S., and H. G. Marshall. 2000. The trophic contributions of rotifers in tidal freshwater and estuarine habitats. *Estuarine, Coastal and Shelf Science* 51:729–742.
Pritchard, D. W. 1967. What is an estuary? Physical point of view. Pp. 3–5 in *Estuaries*, edited by G. H. Lauff. Washington, DC: American Association for the Advancement of Science 83.
Redfield, A. C., B. H. Ketchum, and F. A. Richards. 1963. The influence of organisms on the composition of sea-water. Pp. 12–37 in *The sea*, edited by M. N. Hill. New York: Wiley.
Regnier, P., A. Mouchet, R. Wollast, and F. Ronday. 1998. A discussion of methods for estimating residual fluxes in strong tidal estuaries. *Continental Shelf Research* 18:1543–1571.
Richardson, K., J. Beardall, and J. A. Raven. 1983. Adaptation of unicellular algae to irradiance: An analysis of strategies. *New Phytologist* 93:157–191.
Ristvet, B. L. 1978. *Reverse weathering reactions within recent nearshore marine sediments, Kaneohe Bay, Oahu.* Kirtland AFB, NM: Test Directorate Field Command.
Rousseau, V., A. Leynaert, N. Daoud, and C. Lancelot. 2002. Diatom succession, silicification and silicic acid availability in Belgian coastal waters (southern North Sea). *Marine Ecology Progress Series* 236:61–73.
Ryther, J. H., D. W. Menzel, and N. Corwin. 1967. Influence of the Amazon River outflow on the tropical Atlantic. I: Hydrography and nutrient chemistry. *Journal of Marine Research* 25:69–83.

Sayles, F. L. 1981. The composition and diagenesis of interstitial solutions: II. Fluxes and diagenesis at the water–sediment interface in the high latitude North and South Atlantic. *Geochimica et Cosmochimica Acta* 45:1061–1086.

Sholkovitz, E. D. 1976. Flocculation of dissolved organic and inorganic matter during the mixing of river water and seawater. *Geochimica et Cosmochimica Acta* 40:831–845.

Siever, R. 1968. Sedimentological consequence of a steady-state ocean-atmosphere. *Sedimentology* 11:5–29.

Siever, R. 1971. Silicon: Abundance in rock forming minerals. Pp. 14-D1–D3 in *Handbook of geochemistry*, Vol. II-3, edited by K. H. Wedepohl. Berlin: Springer-Verlag.

Sillén, L. G. 1961. The physical chemistry of seawater. Pp. 549–581 in *Oceanography*, edited by M. Sears. Washington, DC: American Association for the Advancement of Science 67.

Soetaert, K., and P. Van Rijswijk. 1993. Spatial and temporal patterns of the zooplankton in the Westerschelde estuary. *Marine Ecology Progress Series* 97:47–59.

Uliana, E., C. B. Lange, B. Donner, and G. Wefer. 2001. Siliceous plankton productivity fluctuations in the Congo Basin over the past 460,000 years: Marine vs. riverine influence, ODP site 1077. *Proceedings of the Ocean Drilling Project, Scientific Results* 175:1–32.

van Bennekom, A. J., G. W. Berger, W. Helder, and R. T. P. de Vries. 1978. Nutrient distribution in the Zaire estuary and river plume. *Netherlands Journal of Sea Research* 12:296–323.

Van Cappellen, P., and L. Qiu. 1997. Biogenic silica dissolution in sediments of the Southern Ocean. I. Solubility. *Deep-Sea Research II* 44:1109–1128.

van der Zee, C., and L. Chou. 2005. Seasonal cycling of phosphorus in the Southern Bight of the North Sea. *Biogeosciences* 2:27–42.

Wilson, T. R. S. 1975. Salinity and major elements in sea water. Pp. 301–363 in *Chemical oceanography*, 2nd ed., Vol. 2, edited by J. P. Riley and G. Skirrow. London: Academic Press.

Wollast, R. 1978. Modelling of biological and chemical processes in the Scheldt estuary. Pp. 63–77 in *Hydrodynamics of estuaries and fjords*, edited by J. C. J. Nihoul. Oceanography Series. Amsterdam: Elsevier.

Wollast, R. 1988. The Scheldt estuary. Pp. 183–193 in *Pollution of the North Sea: An assessment*, edited by W. Salomon, B. Bayne, E. K. Duursma, and U. Forstner. Berlin: Springer Verlag.

Wollast, R., and F. De Broeu. 1971. Study of the behavior of dissolved silica in the estuary of the Scheldt. *Geochimica et Cosmochimica Acta* 35:613–620.

9

Physiological Ecology of Diatoms Along the River–Sea Continuum

Pascal Claquin, Aude Leynaert, Agata Sferratore, Josette Garnier, and Olivier Ragueneau

This chapter is a synthesis of diatom ecology and physiology along the river–sea continuum, where diatoms are ubiquitous and specifically adapted to various environmental conditions. Instead of an exhaustive literature review, we focus on the major factors that influence the growth characteristics of diatoms and their elemental composition, with special emphasis on the Si. We discuss the aspects of major changes observed in diatom assemblages, the Si sources, the physiological regulations (i.e., metabolisms, uptakes) and their effects on growth and stoichiometry, the bloom developments, and the successions of populations.

Diatoms, from Freshwater to the Marine Ecosystem

Diatoms are unicellular algae belonging to the Heteronkontophyta clade. There are 285 known genera (Round et al. 1990) and 10,000–12,000 recognized species according to Norton et al. (1996), but many of them remain to be described. Diatoms occur as single cells or chains; individual cells are in the size range of 5–500 μm, and multicellular chains can reach several millimeters in length. Diatoms occurring in the sea and freshwater can be planktonic, benthic, periphytic, or epizoic (Graham and Wilcox 2000), and a few species also occur in terrestrial moisture, aerial habitats, or soils.

Literature about river diatoms is scarce. Most studies are on phytoplankton physiology and dynamics in lakes and oceans. Among freshwater phytoplankton, some main differences occur because of the prevailing conditions: Rivers are permanently mixed, and nutrients are replenished from headwaters (soil leaching and anthropogenic sources). Also, water residence time is shorter (Garnier et al. 1995). Stream communities of diatoms usually are benthic and epiphytic species; planktonic genera are rare (Ruth 1977). The diatom flora differs greatly along a stream. In headwaters, species are generally cool adapted, whereas downstream, where the current is stronger, rheophil diatoms such as *Achnanthes* and *Cocconeis* can grow on substrates, and when the flow is reduced, many species are found in the sediments (Ruth 1977). Only predators that have a short lifespan (e.g., Rotifera) have enough time to develop as they move along the river, before they reach the estuary (Pourriot et al. 1982; Testard et al. 1993). In some ways the dynamics of plankton in a river may be comparable to that in shallow lakes with wind-induced mixing (Reynolds 1988). Important biomass of epipelagic communities consists of many species of *Gyrosigma, Navicula, Nitzschia, Pleurosigma,* and *Amphora,* which are adapted to the brackish water in the estuaries.

Sources of Dissolved Silica

Generally rock and soil weathering are considered the major source of dissolved silica (DSi) in rivers (Chapter 2, this volume). Wastewater treatment plants have been shown to be another source, although not a major one (Chapter 10, this volume). Few studies have been carried out on the biogeochemical cycle of silicon (Bartoli 1983), particularly on the role of biogenic silica (BSi) as a possible source of DSi. Plants absorb Si from soils and precipitate it in their tissues as phytoliths (opal particles), in a proportion that varies between a few parts per million to 15 wt% (Alexandre et al. 1997; see also Chapter 3, this volume). It is well known that gramineous species (such as oats) contain ten to twenty times more Si than leguminous species (Russell 1961) and that Si helps keep the plants erect, counteract manganese toxicity, and ward off fungal and insect attacks (Jones and Handreck 1967). Silicon is also present in pastoral plants, which are grazed by ruminants. Most of the Si ingested comes back to the soil unchanged, so the BSi stock present in the soil, coming from animal feces or directly from plant decomposition, is available for dissolution.

Rain forests are among the most efficient vegetation types in recycling Si (Lucas et al. 1993; Oliva et al. 1999). In particular, research in a small watershed in Cameroon studying the effect of organic matter on chemical weathering shows that DSi concentration in the waters of the swamp zone is well above quartz saturation; this excess of DSi (to the expected equilibrium with quartz) is attributed to the dissolution of phytoliths (Oliva et al. 1999) and confirms the important role of BSi (in addition to the lithogenic Si) as a source of DSi.

Although DSi is abundant in freshwater, it can become a limiting factor during spring diatom blooms in rivers enriched with N and P (Garnier et al. 1995). Seawater is everywhere undersaturated in DSi (Tréguer et al. 1995), and limitation of diatom silicification rates has been observed at least once in almost every ocean environment examined, whether in fertile coastal waters or oligotrophic open ocean gyres (Brzezinski and Nelson 1996; Brzezinski et al. 1997; Martin-Jézéquel et al. 2000; Leynaert et al. 2001). Consequently, DSi availability has important effects on the distribution and abundance of diatoms (Egge and Asknes 1992).

Silicon Metabolism and Diatom Stoichiometry

A particular feature of diatoms is their silicified cell wall, called a frustule, which consists of hydrated amorphous silica. As a consequence, silicon metabolism plays a fundamental role in diatoms; vegetative division cannot occur without the formation of the valves of the daughter cells, and cell growth cannot occur without girdle band formation (Volcani 1981). Growth condition can strongly affect the Si metabolism and consequently the Si content per cell. Silicon metabolism in diatoms is strictly linked to the cell cycle (Martin-Jézéquel et al. 2000).

The cell cycle is classically divided into four phases: G1, S, G2, and M. The DNA is replicated during the S phase, M corresponds to the period of mitosis and cell division, and G1 and G2 refer to gaps in the cycle during which most cell growth takes place (Mitchison 1971). Silicon uptake and deposition appear to be associated with the formation of new siliceous valves just before cell division and thus to be confined mainly to the G2 and M period . . . between cytokinesis and daughter cell separation (Hildebrand 2000). It has been suggested that an increase in G2 phase length should entail an increase in Si uptake (for example, Martin-Jézéquel et al. 2000), a hypothesis confirmed by Claquin et al. (2002) on a continuous culture of the marine diatom *Thalassiosira pseudonana.* They observed that the increase of the G2 phase length, which was caused by a decrease in the growth rate under light or nitrogen or phosphorus limitation, entailed an augmentation of cell silicification (Figure 9.1).

Thus it appears that the cellular Si content and frustule thickness (i.e., BSi per cell surface) are regulated by the total amount of Si uptake, directly driven by the length of the cell cycle (i.e., the growth rate). Consequently, the Si content variation appears to be linked not to the type of limitation but rather to the intensity. Under Si limitation, these results cannot be obtained, and the cellular amount of BSi decreases with the growth rate (Martin-Jézéquel et al. 2000).

Other factors such as salinity, pH, or metals (aluminium, germanium) can also influence the Si metabolism and deposition (Gensemer 1990; Vrieling et al. 1999a, 1999b). Recently, Milligan et al. (2004) proved the influence of pCO_2 on Si metabolism and content. They showed an increase of Si content with low pCO_2, which results in the lowering of the Si losses (i.e., efflux and dissolution). Moreover, the basic pH of seawater is corrosive to BSi. The frustule of diatoms is protected by an organic coating that prevents a direct exposure to seawater (Martin-Jézéquel et al. 2000). If the organic covering is removed by chemical, enzymatic, or bacterial attacks, the rate of BSi dissolution increases because of the greater surface area exposed to the water (Bidle and Azam 1999, Van Cappellen et al. 2002).

DSi uptake does not require photosynthetic energy, but it does require energy from respiration (Sullivan 1980; Raven 1983); consequently, DSi uptake can occur in the dark or in the presence of light (Chisholm 1981; Martin-Jézéquel et al. 2000). The strict link between Si metabolism and the cell cycle associated with a low respiratory energy need independent of photosynthesis (Raven 1983; Martin-Jézéquel et al. 2000)

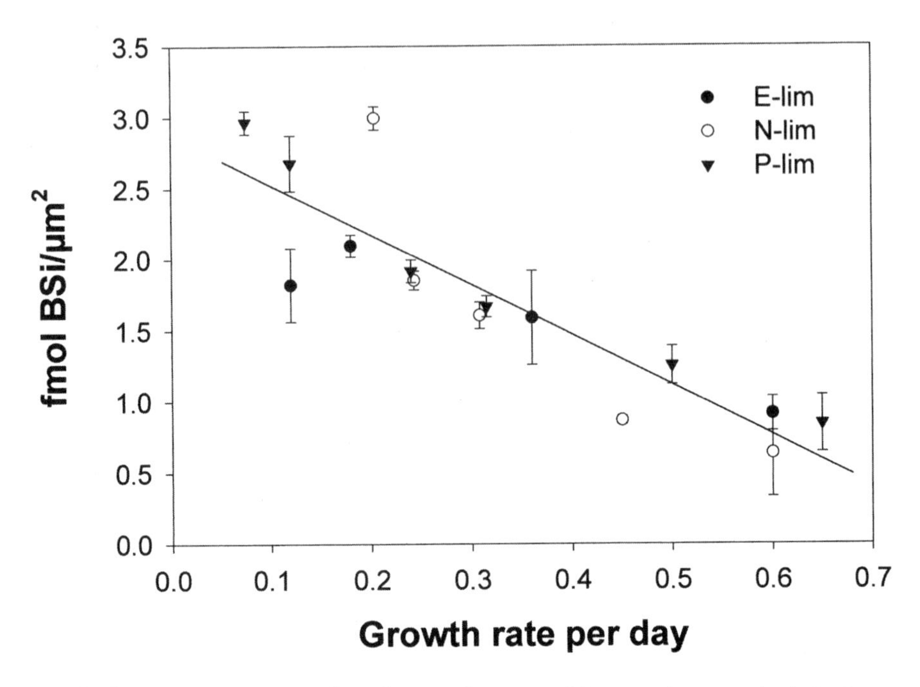

Figure 9.1. BSi content per cell surface as a function of the growth rate under light (E-lim), nitrogen (N-lim), and phosphorus (P-lim) limitations. Linear regression for the 3 limitations, $y = 2.87 - 3.52x$, $r^2 = 0.78$, $p < .001$ (from Claquin et al. 2002).

explains in part the uncoupling between C and Si metabolisms (Claquin et al. 2002). This controls the frequently observed variations of Si:C or Si:N ratios in response to changes in growth conditions caused by light intensity, temperature, or nutrient limitation (Paasche 1980b; Brzezinski 1985; Claquin et al. 2002).

A linear and positive correlation between Si content and diatom biovolume has been determined for both freshwater and marine diatoms (Conley et al. 1989). In marine species, one order of magnitude less BSi per unit of biovolume (0.00050 ± 0.00047 pmol/µm^3) is present than in freshwater ones (0.00558 ± 0.00400 pmol/µm^3). Similar findings emerge from studies on the species *Cyclotella meneghiniana* (Tuchman et al. 1984) and *Thalassiosira pseudonana* (Olsen and Paasche 1986), where cells growing at high salinity have lower Si contents than the ones growing at low salinity. A Si:C molar ratio of 0.79 ± 0.43 has been found for twelve freshwater species (Sicko-Goad et al. 1984), whereas a value of 0.13 ± 0.04 is reported for twenty-seven marine species (Brzezinski 1985).

Several possible reasons have been suggested for the difference in Si content between marine and freshwater diatoms (Conley et al. 1989). They include the following:

- *Sinking strategy.* A lower Si content in the frustule allows marine diatoms not to sink rapidly out of the photic zone (resuspension from great depths is unlikely). On the other hand, sinking is a strategy adopted by freshwater diatoms under nutrient limitations (Sommer and Stabel 1983); resuspension in this environment is generally possible.
- *DSi availability.* DSi concentration is higher in rivers than in the ocean. This may have led to selection of the species present in the two environments: Species with highly silicified frustules live in rivers, where DSi is more available, whereas marine diatoms are adapted to the ocean, where DSi is less abundant.
- *Salinity.* Although there is a relationship between salinity and Si content in diatoms, it is not yet clear whether this results from a direct interaction with salt or from other factors such as osmotic pressure (Olsen and Paasche 1986). Salinity can also affect valve morphology (Paasche et al. 1975). These authors showed on the brackish water plankton diatom *Skeletonoma subsalsum* that the valve faces were flat and the connecting processes much shorter at 3 psu or less, whereas at 5 psu or more the valves had a dome shape, and the connecting processes were longer. They obtained the same results by replacing salt with sucrose; therefore, they concluded that the morphological changes were caused by osmotic pressure variations and not directly by salinity. Vrieling et al. (1999b) found the same type of regulation on *Thalassiosira weissflogii* and *Navicula salinarum.* They observed that the decrease in salinity entailed an increase in BSi per cell and showed that the concentration of salts is an important factor that affects the Si polymerization.

Dissolved Silica Uptake and Growth Rate Parameters

Numerous studies have shown that the specific rate of DSi uptake (V) and the specific Si-dependent cell division rate (μ) of natural communities and diatoms in culture fit reasonably well to Michaelis–Menten saturation functions. Typically, the half saturation for growth (K_μ) is less than that for uptake (K_s). This can be explained by the fact that to determine the growth rate, the increase in cell numbers is used most often. But diatoms are able to decrease the Si content of their frustule when extracellular $Si(OH)_4$ concentrations limit Si uptake (Brzezinski et al. 1990). Thus, a doubling of the number of cells does not necessarily imply a doubling of the biomass.

Physiological parameters for growth (K_μ and μ_{max}) are determined from batch or continuous culture of a given species. They have been measured for many marine diatom species (review in Martin-Jézéquel et al. 2000). For freshwater species, data are scarce, making any comparison difficult. However, there is a large range of variation, from 0.02 to 8.60 μmol/L and from 0.6/d to 9.3/d, respectively.

Uptake parameters can be determined either in culture for a specific species (such as the parameters for growth) or in situ for short-term experiments (a few hours) by using a radioisotope of silicon (^{32}Si). In this latter case, they do not refer to a specific species (except in the case of a monospecific bloom) and can be biased by the fact that not all the cells actively take up Si during an experiment. In this context, Brzezinski (1992) argues that the Si uptake systems of diatoms could have lower affinity and greater maximum capacity than some current kinetic parameters indicate. However, these parameters still give interesting information on the functioning of the natural community that is present at the time of sampling.

K_s and V_{max} for Si uptake have been measured for many species and in many environments. Martin-Jézéquel et al. (2000) summarize the parameters obtained in culture for specific diatom species, and Table 9.1 reports data from in situ natural diatom assemblages.

As mentioned earlier, along the river–sea continuum large changes in environmental conditions, such as hydrodynamics, salinity, temperature, and nutrient availability, control the structure and elemental composition of phytoplankton communities. One of these factors that may influence DSi uptake directly is the DSi concentration in situ, which varies by more than an order of magnitude between rivers and the open ocean. We could expect the affinity to be higher when ambient DSi concentration is low. However, even if the highest K_s values usually are found in areas where DSi concentrations are high, the plot of K_s as a function of ambient DSi concentrations does not show any obvious relationship (Figure 9.2).

A relationship between K_s and ambient DSi concentrations probably would become visible if we take into account only data from systems in a quasistationary state (i.e., when physical, chemical, and biological parameters show very low variability at the time of sampling). In the equatorial upwelling, for example, because of a quasistationary upwelling driven by trade winds and associated with the surface divergence, the system works like a chemostat (Frost and Franzen 1992). The DSi uptake parameters measured in this region indicate that in situ diatom populations are able to regulate their uptake rate at about half the maximum, with a half saturation constant close to the ambient DSi concentration (Dugdale and Wilkerson 1998; Leynaert et al. 2001). However, the coastal environment is very dynamic, with important variation in DSi concentration on a very short time scale. The DSi concentration measured at a given time is the result of the difference between supply (e.g., dissolution, benthic, or river fluxes) and uptake. For instance, diatoms could use DSi released from BSi dissolution before we are able to measure an increase in water column concentration.

Thus, not only the concentration but also the patchiness of the nutrient may influence phytoplankton succession and community structure. The physiological differences in nutrient uptake ability between the different assemblages allow optimal use of the limiting nutrient in a particularly patchy regime. Turpin and Harrison (1979) demonstrate that optimization in a patchy environment takes place through an increase of the maximal uptake rate (high V_{max}), whereas optimization in a homogeneous environment appears to take place through an increase in substrate affinity (i.e., low K_s). In this context, it is obvious that the relationship between K_s or V_{max} and the ambient DSi concentration cannot be straightforward.

Extremely high values for K_s sometimes are reported (Brzezinski et al. 1998; Nelson

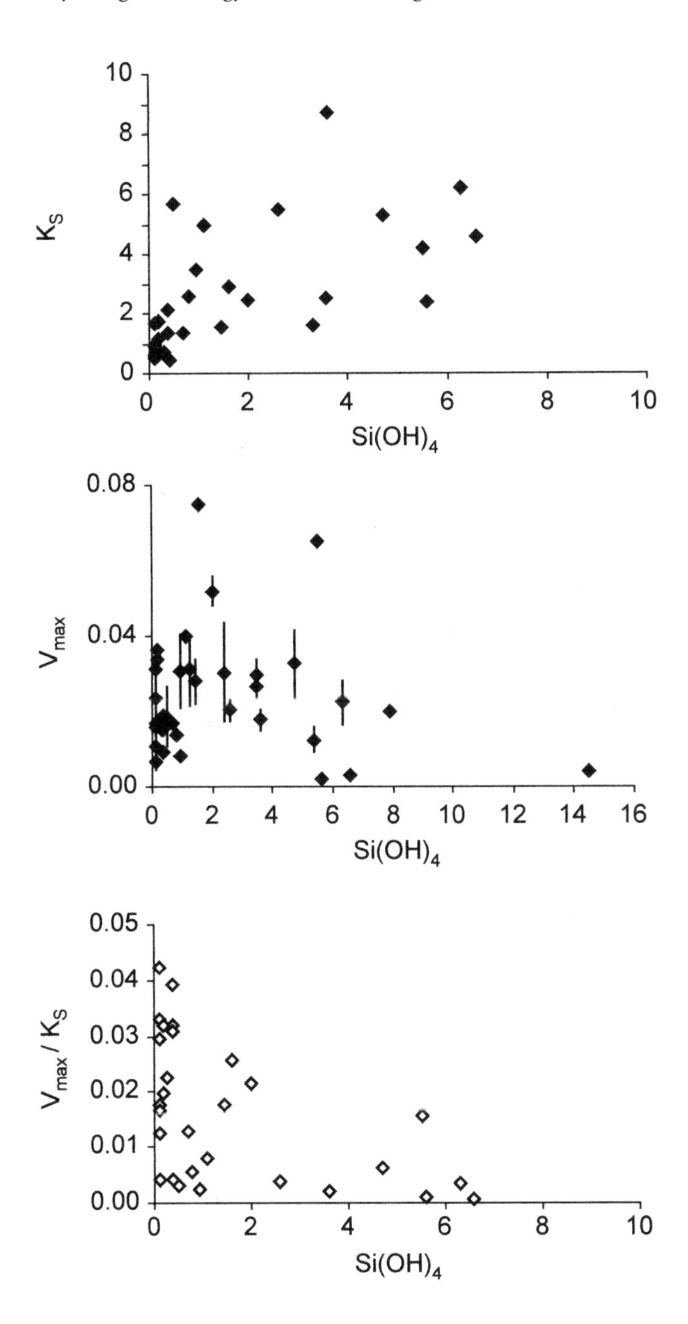

Figure 9.2. K_s and V_{max} values reported from the literature (see Table 9.1) for natural diatom assemblages, measured in situ in different environments as a function of ambient DSi concentrations.

Table 9.1. Review of K_s and V_{max} values from the literature.

Site	$Si(OH)_4$ (μmol/L)	K_s (μmol/L)	*SE*	V_{max} (per h)	*SE*	Reference
Equatorial Pacific						Leynaert et al. (2001)
0°, 180°W	2.00	2.42	0.53	0.052	0.004	
3°S, 180°W	1.45	1.57	1.32	0.028	0.006	
Sargasso Sea						Nelson and Brzezinski (1990)
Gulf Stream	0.10	0.53	0.55	0.007	0.0021	
warm core ring						
1981–1982	0.10	0.90	0.56	0.016	0.0035	
	0.10	0.56	0.58	0.024	0.0084	
	0.10	0.57	0.22	0.017	0.0022	
	0.10	0.66	0.5	0.011	0.0031	
				0.024	0.0041	
				0.035	0.0037	
Sargasso Sea						
Bermuda Atlantic	1.60	0.55	0.023	0.003		Brzezinski and Nelson (1996)
Time Series 1991						
Sargasso Sea						
Bermuda Atlantic	0.80	2.60	1.5	0.014	0.001	
Time Series 1993						

Table 9.1. Review of K_s and V_{max} values from the literature. (*continued*)

Site	$Si(OH)_4$ (µmol/L)	K_s (µmol/L)	*SE*	V_{max} (per h)	*SE*	Reference
Mississippi River plume		5.30				Nelson and Dortch (1996)
	0.41	0.48	0.06	0.019	0.0005	
	0.29	0.69	0.12	0.016		
	0.40	0.49	0.06	0.016		
	0.39	0.50	0.07	0.016		
	0.19	1.71	0.29	0.034		
	0.13	0.95	0.1	0.032		
	0.19	1.14	0.12	0.037		
	5.35			0.013	0.0032	
	3.48			0.030	0.0042	
	2.42			0.030	0.013	
	4.72	5.29	3.79	0.033	0.009	
	0.93			0.031	0.0099	
	1.24			0.031	0.0096	
Ross Sea	6.60	4.58	1.42	0.003	0.0001	Nelson and Tréguer (1992)
	5.60	2.37	0.61	0.002	0.0001	
Peru upwelling	1.61	2.93		0.075		Goering et al. (1973)
			3.70			Nelson et al. (1981)
Baja California coastal waters	3.30	1.59				Azam and Chisholm (1976)
		3.58	2.53			

Table 9.1. Review of K_s and V_{max} values from the literature. (*continued*)

Site	$Si(OH)_4$ (µmol/L)	K_s (µmol/L)	*SE*	V_{max} (per h)	*SE*	Reference
Northwest Mediterranean Sea	0.95	3.46		0.008		Leblanc et al. (2003)
1999–2000	1.11	4.97		0.040		
Southern Norway	6.30	6.20	4.1	0.022	0.0059	Kristiansen et al. (2000)
Oslo fjord	3.60	8.70	2.5	0.018	0.0027	
	2.60	5.50	1.7	0.020	0.0029	
	0.50	5.70	0.5	0.019	0.008	
	0.10	1.70	0.5	0.007	0.0005	
	0.40	2.10	0.4	0.009	0.0005	
California Monterey Bay		14.50			0.004	White and Dugdale 1997
	7.93			0.020		
	5.53	4.20		0.065		
Bay of Brest	0.70	1.34		0.017		Bonnet 2001
	0.39	1.38		0.160		
		3.20	0.6	0.040	0.003	Del Amo 1993
	1.40	0.1	0.041	0.001		
	2.10	0.3	0.044	0.002		

Table 9.1. Review of K_s and V_{max} values from the literature. (*continued*)

Site	$Si(OH)_4$ (µmol/L)	K_s (µmol/L)	*SE*	V_{max} (per h)	*SE*	Reference
Central North Pacific	2.53	1.97	0.008			Brzezinski et al. 1998
	2.09	1.06	0.010			
	0.55	0.26	0.008			
Southern Ocean	1.56		0.013			Nelson et al. 2001
Average	2.48	0.026				
SD	1.93	0.025				

et al. 2001), usually when the relationship between the specific uptake rate (V, per h) and $Si(OH)_4$ fits poorly to the Michaelis–Menten hyperbola or when the kinetics are assessed at very high $Si(OH)_4$ concentrations. This suggests that some other mechanism may exist. A new hypothesis proposes that uptake is controlled by the ratio of bound to unbound Si inside the cell (Thamatrakoln and Hildebrand 2005). When intracellular Si binding components are in excess, uptake occurs, and when levels of unbound Si are higher than bound Si, uptake is inhibited or efflux is induced. As DSi concentrations outside the cell rise, increasing surge uptake would occur, increasing the levels of unbound intracellular Si and inducing efflux to a greater extent. If this hypothesis is confirmed, the definition of physiological parameters for DSi uptake should be clarified.

Another variable, salinity, is especially important along the river–sea continuum. It does not affect the growth rate but can influence the Si metabolism. Indeed, salinity seems to play a major role in DSi uptake and Si content. Paasche (1980a) showed that marine diatoms have a half saturation constant for DSi that is one order of magnitude lower than that of freshwater diatoms. Olsen and Paasche (1986) observed a variation of DSi kinetic parameters in *Thalassiosira pseudonana* (Bacillariophyceae) in response to changes in the chemical composition of the growth medium. Half saturation constant for growth increased from 0.04 μmol Si/L to 8.6 μmol Si/L in *Thalassiosira pseudonana* grown at high salinity (marine medium) and at low salinity (freshwater medium), respectively. The overall means for these parameters are 2.48 ± 1.93 μmol/L and 0.63 ± 0.60/d, respectively.

It is interesting to note that K_s values used in biogeochemical models usually are in the higher range for river and lake models, whereas they are in the lower range for coastal biogeochemical models (Table 9.2). All values used for μ_{max} are clearly above the calculated average for V_{max}.

When Does the Bloom Occur? The Diatom Succession

The development of diatom blooms in coastal waters and more generally in seawater is linked to light availability (i.e., opposite of turbidity) (Cloern 1987), water temperature, hydrodynamics (Cloern 1996), salinity (Kirst 1995), and nutrient availability (Egge and Asknes 1992). Other parameters such as trace elements (i.e., Fe), grazing rate, and pCO_2 can also play a role (Kirst 1995; Gobler et al. 2002). The interactions between these factors are responsible for the growth and succession of diatoms and other phytoplankton taxa. In rivers, phytoplankton blooms are controlled by roughly the same parameters (Lalli and Parsons 1993) as in seawater, but there is some specificity. Several authors have pointed out a negative correlation between discharge and phytoplankton bloom in spring. This phenomenon has been reported for the Sacramento River (Greenberg 1964), the Thames River (Lack 1971), the Loire River (Champ 1980), the Seine River (Garnier et al. 1995), and others, where a bloom has been observed just after the spring discharge. Depending on the hydrological conditions (wet or dry), the occurrence of the diatom bloom can shift from early to late spring. The bloom usually takes place not in

Table 9.2. Diatom physiological parameters used in ecological models.

Reference	μ_{max} (per d)	K_s (μmol Si/L)	System
Garnier et al. (1995)	1.2	4.2	River
Chen et al. (2002)	1.2–1.6	5.0	Lake
Chai et al. (2002)	3.0	3.0	Ocean
Dugdale and Wilkerson (1998)	3.0		Ocean
Le Pape and Menesguen (1997)	0.6 at 0°C	1.0	Coast
Lancelot et al. (2002)	3.7	1.0	Coast

the upper reaches, where residence time is short, but in the downstream areas of the river, when the phytoplankton growth rate exceeds the dilution rate (Billen et al. 1994).

In the case of the Seine River, diatom blooms develop from fifth-order streams downward, with increasing biomass toward the estuary. This is essentially true for both spring and summer blooms, when diatoms constitute up to 76 and 51 percent of total phytoplankton biomass, respectively. However, whereas spring blooms are controlled from the bottom up (hydrology, nutrients, light, and temperature), summer blooms experience an intermittent decline in the seventh-order stretch because of biological top-down control (zooplankton grazing, virus lysis) before redevelopment toward the estuary (Garnier et al. 1995).

Conclusion

DSi availability along the river–sea continuum is not homogeneous. In contrast to marine diatoms, which are often limited by Si, river diatoms might occasionally be limited under high anthropogenic P and N inputs. Physiological differences between marine and freshwater diatoms appear to reflect the abundance of DSi in the two environments. However, few studies have analyzed the relationship between Si content, Si metabolism, and the role of environmental factors such as salinity, pH, and osmotic pressure. In general, phytoplankton blooms along the continuum are controlled mainly by similar factors. In rivers, discharge is an important additional factor controlling the development of blooms. Diatom successions have been widely studied in marine environments, but there is a lack of data concerning in situ river diatom assemblages. More physiological experiments are needed to better explain the link between environmental factors and Si metabolism and to elucidate the cellular regulations involved in these processes.

Literature Cited

Alexandre, A., J.-D. Meunier, F. Colin, and J.-M. Koud. 1997. Plant impact on the biogeochemical cycle of silicon and related weathering processes. *Geochimica et Cosmochimica Acta* 61:677–682.

Azam, F., and S. W. Chisholm. 1976. Silicic acid uptake and incorporation by natural marine phytoplankton populations. *Limnology and Oceanography* 21:427–433.

Bartoli, F. 1983. The biogeochemical cycle of silicon in two temperate forest ecosystems. *Ecological Bulletin* 35:469–476.

Bidle, K. D., and F. Azam. 1999. Accelerated dissolution of diatom silica by marine bacterial assemblages. *Nature* 397:508–512.

Billen, G., J. Garnier, and P. Hanset. 1994. Modelling phytoplankton development in whole drainage networks: The RIVERSTRAHLER model applied to the Seine River system. *Hydrobiologia* 289:119–137.

Bonnet, S. 2000. *Etude de l'écosystème Rade de Brest: Rôle du silicium dans la dynamique phytoplanctonique.* Rapport de stage de maîtrise. Université de Bretagne Occidentale, Brest, France.

Brzezinski, M. A. 1985. The Si:C:N ratio of the marine diatoms: Interspecific variability and the effect of some environmental variables. *Journal of Phycology* 21:347–357.

Brzezinski, M. A. 1992. Cell-cycle effects on the kinetics of silicic acid uptake and resource competition among diatoms. *Journal of Plankton Research* 14:1511–1539.

Brzezinski, M. A., and D. M. Nelson. 1996. Chronic substrate limitation of silicic acid uptake rates in the western Sargasso Sea. *Deep-Sea Research II* 43:437–453.

Brzezinski, M. A., R. J. Olson, and S. W. Chisholm. 1990. Silicon availability and cell cycle progression in marine diatoms. *Marine Ecology Progress Series* 67:83–96.

Brzezinski, M. A., D. A. Phillips, F. P. Chavez, G. E. Friederich, and R. C. Dugdale. 1997. Silica production in the Monterey, California, upwelling system. *Limnology and Oceanography* 42:1694–1705.

Brzezinski, M. A., T. A. Villareal, and F. Lipschultz. 1998. Silica production and the contribution of diatoms to new and primary production in the central North Pacific. *Marine Ecology Progress Series* 167:89–104.

Chai, F., R. C. Dugdale, T.-H. Peng, F. P. Wilkerson, and R. T. Barber. 2002. One-dimensional ecosystem model of the equatorial Pacific upwelling system. Part I: Model development and silicon and nitrogen cycle. *Deep-Sea Research II* 49:2713–2745.

Champ, P. 1980. Biomasse et production primaire du phytoplancton d'un fleuve: La Loire au niveau de la centrale nucleaire de Saint-Laurent-des-Eaux. *Acta Ecologica* 1:111–130.

Chen, C., J. Rubao, D. J. Schwab, D. Beletsky, G. L. Fahnenstiel, M. Jiang, T. H. Johengen, H. Vanderploeg, B. Eadie, J. Wells Budd, M. H. Bundy, W. Gardner, J. Cotner, and P. J. Lavrentyev. 2002. A model study of the coupled biological and physical dynamics in Lake Michigan. *Ecological Modelling* 152:145–168.

Chisholm, S. W. 1981. Temporal patterns of cell division in unicellular algae. In *Physiological bases of phytoplankton ecology*, edited by T. Platt. *Canadian Bulletin of Fisheries and Aquatic Sciences* 210:150–181.

Claquin, P., V. Martin-Jézéquel, J. C. Kromkamp, M. J. W. Veldhuis, and G. W. Kraay. 2002. Uncoupling of silicon compared to carbon and nitrogen metabolism, and role of the cell cycle, in continuous cultures of *Thalassiosira pseudonana* (Bacillariophyceae) under light, nitrogen and phosphorus control. *Journal of Phycology* 38:922–930.

Cloern, J. E. 1987. Turbidity as a control on phytoplankton biomass and productivity in estuaries. *Continental Shelf Research* 7:1367–1381.

Cloern, J. E. 1996. Phytoplankton bloom dynamics in coastal ecosystems: A review with some general lessons from sustained investigation of San Francisco Bay, California. *Review of Geophysics* 34:127–168.

Conley, D. J., S. S. Kilham, and E. Theriot. 1989. Differences in silica content between marine and freshwater diatoms. *Limnology and Oceanography* 34:205–213.

Del Amo, Y. 1993. *Dynamique et structure des communautés phytoplanctoniques en écosystème côtier perturbé; cinétique de l'incorporation de silicium par les diatomées.* Université de Bretagne Occidentale, Brest, France.

Dugdale, R. C., and F. P. Wilkerson. 1998. Silicate regulation of new production in the equatorial Pacific. *Nature* 391:270–273.

Egge, J. K., and D. L. Asknes. 1992. Silicate as a regulating nutrient in phytoplankton competition. *Marine Ecology Progress Series* 38:281–290.

Frost, B. W., and N. C. Franzen. 1992. Grazing and iron limitation in the control of phytoplankton stock and nutrient concentration: A chemostat analogue of the Pacific equatorial upwelling zone. *Marine Ecology Progress Series* 83:291–303.

Garnier, J., G. Billen, and M. Coste. 1995. Seasonal successions of diatoms and Chlorophyceae in the drainage network of the Seine River: Observations and modelling. *Limnology and Oceanography* 40:750–765.

Gensemer, R. W. 1990. Role of aluminium and growth rate on changes in cell size and silica content of silica-limited populations of *Asterionnella ralfsii* var. *americana* (Bacillariophyceae). *Journal of Phycology* 26:250–258.

Gobler, C. J., J. R. Donat, J. A. Consolvo, and S. A. Sanudo-Wilhelmy. 2002. Physicochemical speciation of iron during coastal algal blooms. *Marine Chemistry* 77:71–89.

Goering, J. J., D. M. Nelson, and J. A. Carter. 1973. Silicic acid uptake by natural populations of marine phytoplankton. *Deep Sea Research* 20:777–789.

Graham, L. E., and L. W. Wilcox. 2000. *Algae.* Upper Saddle River, NJ: Prentice Hall.

Greenberg, A. E. 1964. Plankton of the Sacramento River. *Ecology* 45:40–49.

Hildebrand, M. 2000. Silicic acid transport and its control during cell wall silicification in diatoms. Pp. 171–188 in *Biomineralization of nano- and micro-structures*, edited by E. Bauerlein. Weinheim, Germany: Wiley-VCH.

Jones, L. H. P., and K. A. Handreck. 1967. Silica in soils, plants and animals. *Advances in Agronomy* 19:107–149.

Kirst, G. O. 1995. Influence of salinity on algal ecosystems. Pp. 123–142 in *Algae, environment and human affairs*, edited by C. Starr. Bristol, UK: Biopress.

Kristiansen, S., T. Farbrot, and L. J. Naustvoll. 2000. Production of biogenic silica by spring diatoms. *Limnology and Oceanography* 45:472–478.

Lack, T. J. 1971. Quantitative studies on the phytoplankton of the Rivers Thames and Kennet at Reading. *Freshwater Biology* 1:213–224.

Lalli, C., and T. R. Parsons. 1993. *Biological oceanography: An introduction.* Oxford: Pergamon.

Lancelot, C., J. Staneva, D. Van Eeckhout, J.-M. Beckers, and E. Stanev. 2002. Modelling the Danube-influenced north-western continental shelf of the Black Sea. II: Ecosystem response to changes in nutrient delivery by the Danube River after its damming in 1972. *Estuarine, Coastal and Shelf Science* 54:473–499.

Leblanc, K., B. Quéguiner, N. Garcia, P. Rimmelin, and P. Raimbault. 2003. Silicon cycle in the northwestern Mediterranean Sea: A seasonal study of a coastal oligotrophic site. *Oceanologica Acta* 26:339–355.

Leynaert, A., P. Tréguer, C. Lancelot, and M. Rodier. 2001. Silicon limitation of biogenic silica production in the equatorial Pacific. *Deep-Sea Research I* 48:639–660.

Le Pape, O., and A. Menesguen. 1997. Hydrodynamic prevention of eutrophication in the Bay of Brest (France), a modelling approach. *Journal of Marine Systems* 12:171–186.

Lucas, Y., F. J. Luizao, J. Rouiller, and D. Nahon. 1993. The relationship between the biological activity of the rain forest and the mineral composition of the soils. *Science* 260:521–523.

Martin-Jézéquel, V., M. Hildebrand, and M. A. Brzezinski. 2000. Silicon metabolism in diatoms: Implications for growth. *Journal of Phycology* 36:821–840.

Milligan, A. J., D. E. Varela, M. A. Brzezinski, and F. M. M. Morel. 2004. Dynamics of silicon metabolism and silicon isotopic discrimination in a marine diatom as a function of pCO_2. *Limnology and Oceanography* 49:322–329.

Mitchison, J. M. 1971. *The biology of the cell.* Cambridge: Cambridge University Press.

Nelson, D. M., and M. A. Brzezinski. 1990. Kinetics of silicic acid uptake by natural diatom assemblages in two Gulf Stream warm-core rings. *Marine Ecology Progress Series* 62:283–292.

Nelson, D. M., M. A. Brzezinski, D. E. Sigmon, and V. M. Franck. 2001. A seasonal progression of Si limitation in the Pacific sector of the Southern Ocean. *Deep-Sea Research II* 48:3973–3995.

Nelson, D. M., and Q. Dortch. 1996. Silicic acid depletion and silicon limitation in the plume of the Mississippi River: Evidence from kinetic studies in spring and summer. *Marine Ecology Progress Series* 136:163–178.

Nelson, D. M., J. J. Goering, and D. W. Boisseau. 1981. Consumption and regeneration of silicic acid in three coastal upwelling systems. Pp. 242–256 in *Coastal upwelling,* edited by F. A. Richards. Washington, DC: American Geophysical Union.

Nelson, D. M., and P. Tréguer. 1992. Role of silicon as limiting nutrient to Antarctic diatoms: Evidences from kinetics studies in the Ross Sea ice-edge zone. *Marine Ecology Progress Series* 80:255–264.

Norton, T. A., M. Melkonian, and R. A. Andersen. 1996. Algal biodiversity. *Phycologia* 35:308–325.

Oliva, P., J. Viers, B. Dupre, J. P. Fortune, F. Martin, J. J. Braun, D. Nahon, and H. Robain. 1999. The effect of organic matter on chemical weathering: Study of a small tropical watershed: Nsimi-Zoetele site, Cameroon. *Geochimica et Cosmochimica Acta* 63:4013–4035.

Olsen, S., and E. Paasche. 1986. Variable kinetics of silicon-limited growth in *Thalassiosira pseudonana* (Bacillariophyceae) in response to changed chemical composition of the growth medium. *British Phycological Journal* 21:183–190.

Paasche, E. 1980a. Silicon. Pp. 259–284 in *The physiological ecology of phytoplankton,* edited by I. Morris. Oxford: Blackwell.

Paasche, E. 1980b. Silicon content of five marine plankton diatom species measured with a rapid filter method. *Limnology and Oceanography* 25:474–480.

Paasche, E., S. Johansson, and D. L. Evensen. 1975. An effect of osmotic pressure on the valve morphology of the diatom *Skeletonema subsalsum* (A. Cleve) Bethge. *Phycologia* 14:205–211.

Pourriot, R., D. Benest, P. Champ, and C. Rougier. 1982. Influence de quelques facteurs du milieu sur la composition et la dynamique saisonnière du zooplancton de la Loire. *Acta Ecologica* 3:353–371.

Raven, J. A. 1983. The transport and function of silicon in plants. *Biological Review* 58:178–207.

Reynolds, C. S. 1988. Potamoplankton: Paradigms, paradoxes and prognoses. Pp. 285–311 in *Algae and the aquatic environment,* edited by F. E. Round. Bristol: Biopress.

Round, F. E., R. M. Crawford, and D. G. Mann. 1990. *The diatoms: biology and morphology of the genera.* Cambridge: Cambridge University Press.

Russell, E. W. 1961. *Soil conditions and plant growth*, 9th ed. New York: Wiley.

Ruth, P. 1977. Ecology of freshwater: Diatoms and diatom communities. Pp. 284–332 in *The biology of diatoms*, edited by D. Werner. London: Blackwell Scientific.

Sicko-Goad, L., C. L. Schelske, and E. F. Stroermer. 1984. Estimation of intracellular carbon and silica content of diatoms from natural assemblages using morphometric techniques. *Limnology and Oceanography* 29:1170–1178.

Sommer, U., and H. H. Stabel. 1983. Silicon consumption and population density change of dominant planktonic diatoms in Lake Constance. *Journal of Ecology* 71:119–130.

Sullivan, C. W. 1980. Diatom mineralization of silicic acid. V. Energetic and macromolecular requirements for $Si(OH)_4$ mineralization events during the cell cycle of *Navicula pelliculosa. Journal of Phycology* 16:321–328.

Testard, P., R. Pourriot, A. Miquelis, and C. Rougier. 1993. Fonctionnement de l'écosystème fluvial: Organisation et role du zooplancton. Pp. 1–34 in *PIREN–Seine synthesis report*, edited by G. Billen and J. Allardi. Paris: Université Pierre et Marie Curie.

Thamatrakoln, K., and M. Hildebrand. 2005. Approaches for functional characterization of diatom silicic acid transporters. *Journal of Nanoscience and Nanotechnology* 5:158–166.

Tréguer, P., D. M. Nelson, A. J. Van Bennekom, D. J. DeMaster, A. Leynaert, and B. Quéguiner. 1995. The silica balance in the world ocean: A reestimate. *Science* 268:375–379.

Tuchman, M. L., E. F. Stroemer, and E. Theriot. 1984. Sodium chloride effects on the morphology and physiology of *Cyclotella meneghiniana. Archiv für Protistenkunde* 128:319–326.

Turpin, D. H., and P. J. Harrison. 1979. Limiting nutrient patchiness and its role in phytoplankton ecology. *Journal of Experimental Marine Biology and Ecology* 39:151–166.

Van Cappellen, P., S. Dixit, and J. van Beusekom. 2002. Biogenic silica dissolution in the oceans: Reconciling experimental and field-based dissolution rates. *Global Biogeochemical Cycles* 16:1075.

Volcani, B. E. 1981. Cell wall formation in diatoms: Morphogenesis and biochemistry. Pp. 157–200 in *Silicon and siliceous structures in biological systems*, edited by T. L. Simpson and B. E. Volcani. New York: Springer.

Vrieling, E. G., W. W. C. Gieskes, and T. P. M. Beelen. 1999a. Silicon deposition in diatoms: Control by the pH inside the silicon deposition vesicle. *Journal of Phycology* 35:548–559.

Vrieling, E. G., L. Poort, T. P. M. Beelen, and W. W. C. Gieskes. 1999b. Growth and silica content of the diatoms *Thalassiosira weissflogii* and *Navicula salinarum* at different salinities and enrichments with aluminium. *European Journal of Phycology* 34:307–316.

White, K. K., and R. C. Dugdale. 1997. Silicate and nitrate uptake in the Monterey Bay upwelling system. *Continental Shelf Research* 17:455–472.

10

Modeling Silicon Transfer Processes in River Catchments

Josette Garnier, Agata Sferratore, Michel Meybeck, Gilles Billen, and Hans Dürr

Less attention has been paid to the fluxes of Si transferred from terrestrial systems to surface waters than is the case for P and N fluxes (Howarth et al. 1996; Nixon et al. 1996). Although river dissolved silica (DSi) has been analyzed by geochemists for decades (Clarke 1924), it was not included in water quality surveys until very recently. As a result, whereas P and N concentrations in surface water can be found in the literature or in the databases of water agencies for the last 30 years, DSi values are found only for the last 15 years and even today remain scarce.

DSi inputs to surface water originate mostly from rock weathering and generally are considered not to be influenced by human activity or land use. However, eutrophication can lead to increased retention in the upstream sectors of the drainage network and its stagnant annexes (from small ponds to deep lakes), through increasing diatom biomass production (and the associated uptake of DSi and the formation of biogenic silica [BSi]) under nutrient (N, P) enrichment (Schelske 1985; Schelske et al. 1985; Conley et al. 1995; Campy and Meybeck 1995). More recently, the pools of BSi originating from both aquatic and terrestrial (phytolith) ecosystems have been shown to play a significant role on a global scale (Conley 2002). Redfield ratio (Si:N, Si:P) changes are known to be major indicators of ecological changes in coastal ecosystems (Officer and Ryther 1980; Conley et al. 1993; Billen and Garnier 1997; Humborg et al. 1997; Rabalais and Turner 2001).

A number of recent modeling approaches are devoted to the representation of nutrient transfer through the aquatic continuum (see Vannote et al. 1980), taking into account input from the terrestrial systems of the watershed and the biological uptake and transformation processes in the river system. The RIVERSTRAHLER model (Billen et al. 1994, 1998, 2001, 2005; Garnier et al. 1995, 1998, 2002a, 2002b, 2006; Billen and Garnier 1999; Garnier and Billen 2002) is one example of such an approach.

The RIVERSTRAHLER model offers a general framework for the study of the biogeochemical functioning of river systems and is used here to compare silica behavior in river networks that differ in their hydrological regime and climate and in the human activities in their watersheds. The rivers Seine and Danube in Europe and the Red River in Vietnam are the case studies chosen here. Special attention is paid to the factors controlling diffuse sources of silica to surface water. Following the previous work by Meybeck (1984, 1986, 1987, 2003), we investigated factors such as lithology, temperature, and latitude within a gradient of conditions represented by our three case studies. Once validated, the model has been used to explore a range of scenarios differing in the constraints imposed by land use and urban population.

Contrarily to multiregression models leading to annual or pluri-annual means (Kroeze and Seitzinger 1998; Green et al. 2004), the RIVERSTRAHLER model calculates seasonal fluxes delivered to the coastal zone and Redfield ratios. The model takes into account in-stream processes along the aquatic continuum, such as nutrient cycling and retention, and major hydrological features (i.e., reservoirs, canalization). The modular structure and resolution of the model allow us to explicitly consider the nutrient inputs (point or diffuse source) in a realistic way.

RIVERSTRAHLER: A Modeling Approach at the Basin Scale

Description

RIVERSTRAHLER is a generic model of the ecological functioning and nutrient cycling of large drainage networks as a function of the properties of their watersheds, taking into account the geomorphological characteristics of the watershed, land use, and urban effluent distribution. It is based on the idea that kinetics of the basic ecological processes in the functioning of the aquatic ecosystem are the same from headwaters to downstream sectors, and the hydrometeorological and morphological constraints and the point and nonpoint sources of material from the watershed modulate their expression. The RIVERSTRAHLER model thus results from the coupling of a unique model of processes (RIVE, Figure 10.1) and a hydrological model (HYDROSTRAHLER), describing in an idealized way the water fluxes in the drainage network, represented by a regular scheme of confluence of tributaries of increasing stream order with mean characteristics (see Strahler's concept of stream order; Strahler 1957).

The constraints necessary to build the hydrological model therefore are of geomorphological type which, with the rainfall–discharge relationship (Bultot and Dupriez 1976), are the basis for calculating the discharge in each stream order (Billen et al. 1994; Figure 10.2).

Discharge comprises two components; the surface and base flows are simulated at a ten-day step. This model simulates twenty-two variables for the water quality at the outlet of any subbasin considered, including the major biological compartment (phytoplankton, bacteria, and zooplankton) and the main nutrients (N, P, and Si). Ecological

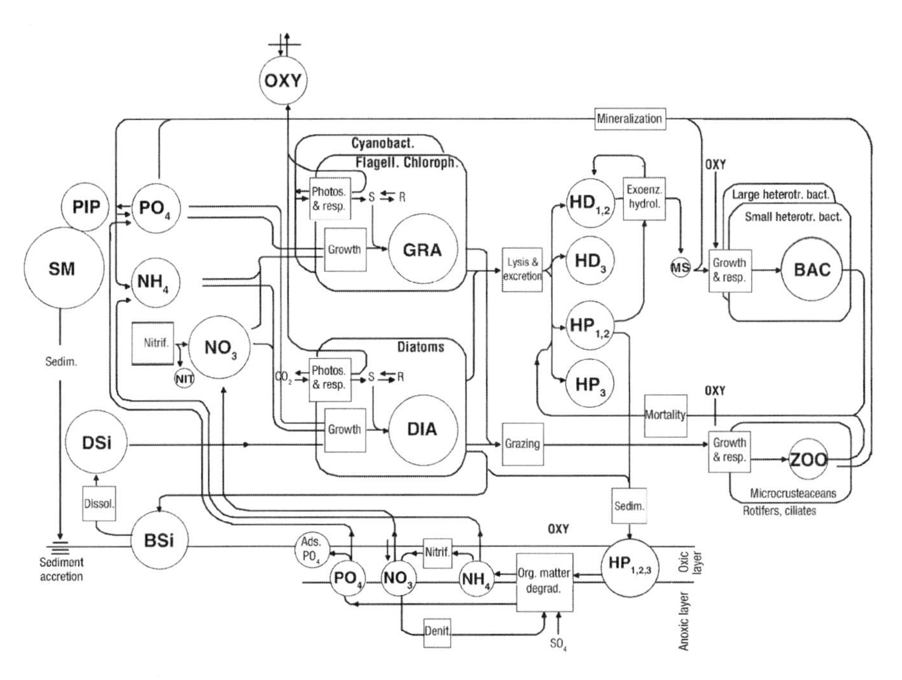

Figure 10.1. Representation of the RIVE model showing the complex interactions between the main biological compartments in the water column, at the water interface sediment, and the stocks of nutrients (PIP = particulate inorganic phosphorus, PO_4 = adsorbed orthophosphates, NH_4 = ammonium, NO_3 = nitrate), suspended matter (SM), oxygen (OXY), and organic matter ($HD_{1,2}$, HD_3 = dissolved organic carbon under 3 classes of biodegradability; $HP_{1,2}$, HP_3 = particulate organic carbon under 3 classes of biodegradability; MS = molecular substrates). Biological compartments are phytoplankton (GRA = green algae, DIA = diatoms, Cyanobact. = cyanobacteria), heterotrophic bacterioplankton (heterotr. bact. or BAC) as large and small bacteria, nitrifying bacteria (NIT), and zooplankton (ZOO) as microcrustaceans, rotifers, and/or ciliates. Processes taken into account are sedimentation of suspended matter (sedim.), nitrification of ammonium (nitrif.), BSi dissolution (dissol.), photosynthesis (photos.), of small metabolites (S) and large molecule reserves (R) respiration (resp.), protein synthesis (growth), lysis and excretion of phytoplankton, exoenzymatic hydrolysis (exoenz. hydrol.), mineralization, growth, and respiration (resp.), mortality of bacteria, and grazing. All compartments are submitted to sedimentation (sedim.) as particulate organic matter ($HP_{1,2,3}$) at the sediment interface is degraded (org. matter degrad.), NH_4 is nitrified in the oxic layer (nitrif.), and NO_3 is denitrified (denit.) in the anoxic layer.

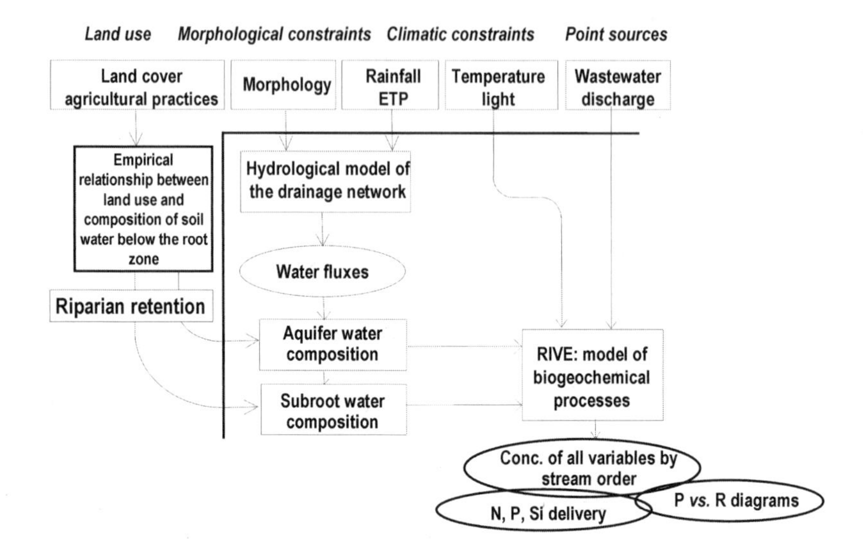

Figure 10.2. Data needed to build the RIVERSTRAHLER model. *Top to bottom*: Constraint data controlling hydrology and the RIVE model and output of the model. Riparian retention is taken into account according to Billen and Garnier (1999).

processes are calculated at less than 1 hour resolution and averaged over 24 hours. Taking into account silica as a state variable, similarly to N or P, implies quantifying silica inputs from point and diffuse sources. Although silica is contained in domestic effluents, it is principally of diffuse origin, from surface and base flows. Siliceous algae, the diatoms, represent one of the major groups in rivers (Garnier et al. 1995; Everbecq et al. 2001) that consume DSi in a ratio to P and N, defined by their physiology (Redfield 1958; Redfield et al. 1963; Conley and Kilham 1989). Dissolution of particulate BSi, both in the water column and in the sediments after deposition, is also taken into account, on the basis of experimental studies (Garnier et al. 2002b).

Field of Application

Although the RIVERSTRAHLER model was first developed on the Seine (Billen et al. 1994; Garnier et al. 1995), it was conceived as a generic model and was soon applied to several other rivers that offered a gradient of basin area, hydrological regime, and human pressure. The Seine, the Danube, and the Red River, selected here as case studies, differ mainly in their hydrological regimes and climates (Table 10.1). Population density is a good indicator of human activity within the watershed, as well as of land use (percentage in arable land, forest, and meadow).

Table 10.1. General characteristics of the hydrographic network chosen here for application of the RIVERSTRAHLER model. The rivers differ in the surface of their watersheds, the annual average discharge and specific discharge, climate type, population density, and land use.

Characteristics	Seine	Red River	Danube
Watershed surface (km^2)	65,000	169,000	817,000
Annual average and specific discharge (m^3/s [$L/s/km^2$])	500 [7.7]	4,000 [23.7]	6,400 [7.8]
Climate type	Temperate (oceanic)	Subtropical and monsoon	Temperate (continental)
Population density (inhabitants/km^2)	195	180	90
Land use (% forest, meadow, arable land)	35, 10, 55	34, 24, 21*	45, 15, 45

*Including 8% rice culture.

Whereas the constraints are river-specific and must be analyzed for the case of each investigated river, the kinetics of the processes and the corresponding parameters have been shown to be valid for any of the systems (Billen and Garnier 1997, 1999; Garnier et al. 1999b, 2001a, 2002a; Trifu 2002; Billen et al 2005). Although the RIVER-STRAHLER model is designed to represent the whole of the drainage network as a simplified scheme of confluences, where all tributaries of the same stream order are considered identical, the geographic resolution can be much improved by splitting the drainage network into several homogeneous subbasins, characterized by different sets of morphological, land use, and human discharge constraints. Thus, the Seine and Red rivers are divided into four and three subbasins, respectively, connected to one main branch, whereas the Danube is divided into nine main subbasins connected to one very long major branch.

The Seine River, down to the entry of its estuarine zone at Poses, has a drainage area of 65,000 km^2 (77,000 km^2 when the estuarine part is included). It is characterized by a pluvio-oceanic regime, with mean annual flow around 500 m^3/s (at Poses, the limit of the estuary for the last 10 years) and minimum summer flows of 120 m^3/s. The construction of three large reservoirs (725 Mm^3) in the upstream part of the basin sustains summer flow above 100 m^3/s. Upstream from Paris, the main subbasins considered are the upstream Seine, issued from the Langres Plateau, and the Marne. Downstream from Paris, the Seine receives the Oise and 150 km downstream the Eure, immediately downstream from the Poses station, where a sluice lock separates the lower Seine from the freshwater estuary. The results of the RIVERSTRAHLER model are given within the estuary at Caudebec, where the limit of the saline intrusion is found. Upstream from Paris, the river and its tributaries drain intensive agricultural areas, causing heavy nitrate load. Most of the population of the watershed is concentrated in the central Parisian area, and the associated wastewater discharges, treated at the Achères wastewater treatment plant (6.5 million inhabitant equivalents), are responsible for a severe increase in organic matter, phosphates, and ammonium loadings (Garnier et al. 2001b). Forests and meadows occupy a small part (35 percent and 10 percent, respectively) of the upstream Seine subbasins, and arable land occupies 55 percent of the basin.

The Red River originates in Yunnan Province, in the mountainous region of South China. The river comprises three main subbasins: the Lo, Da, and Thao rivers, joining at Viet Tri. From this confluence, the main branch (called Hong River) reaches the China Sea at Ba Lat through a large delta (Dang Anh Tuan 2000). The basin area is 169,000 km^2, of which 82,400 km^2 (about 48.8 percent) is in China, 85,100 km^2 (about 50.3 percent) in Vietnam, and 1,500 km^2 (0.9 percent) in Laos. The total length of the Red River is approximately 1,140 km (910 km from headwater to Viet Tri). The results of the RIVERSTRAHLER model are given immediately upstream from the delta, at Hanoi, and do not include the impact of the city, which discharges its effluents into a branch of the delta. The discharge is maximal in July, during the monsoon rainy season, and averages 4,000 m^3/s. Occasionally, discharge values greater than 30,000 m^3/s can be observed in the flood season (37,800 m^3/s in 1971, the maximum for the last 100 years). Forests and meadows occupy the majority of the upstream Red River subbasins (34 percent and 24 percent, respectively), 8 percent for the rice culture and 13 percent for industrial crops; the remaining part is occupied mainly by rocky mountain terrain.

With a catchment basin of 817,000 km^2 and a main course 2,860 km long, the Danube River flows through many different geological facies and types of land cover. From its source in the Black Forest (in Germany) to its mouth on the Black Sea, the Danube River receives on its right bank the alpine tributaries (Inn, Drava, and Sava) and the Velika Morava, flowing from the Balkans. On the left bank, the major tributaries are the Morava and the Tisza, which drain a large area of the Hungarian plain, and the Olt, Siret, and Prut rivers, which originate in the Carpathians. To achieve a compromise between fine spatial resolution and model flexibility, we divided the Danube network into the nine subbasins corresponding to the main tributaries and the main branch of the Danube, from its junction with the Inn to its mouth. The average discharge of the Danube River into the delta amounts to 6,400 m^3/s, with values ranging from 2,000 to 12,500 m^3/s. The results of the RIVERSTRAHLER model are given at Reni, immediately upstream from the three main branches of the delta. Because of high precipitation in the upstream alpine sector, discharge is already high in Vienna. The snow cover on the watershed lasts from more than 200 days in the highest mountain regions to only about 10 days on the Black Sea coast. Arable land occupies large areas of the Hungarian plain in the middle of the basin and represents nearly half of the total watershed area. Large forests exist in the southwestern part of the basin and in the Transylvanian Alps and Carpathian regions. Grassland represents about 15 percent of the whole watershed;

whereas forest occupies a larger part of the upstream subbasins (45 percent), cultivated soils dominate in the downstream subbasins (45 percent).

Sources of Silica from the Terrestrial Systems of the Catchment

Lithology-Based Diffuse Dissolved Silica

Quartz, amorphous silica, and silicate minerals are the only sources of DSi in pristine rivers. Physical denudation and chemical weathering of silicate minerals are associated with the presence of DSi in freshwater, which varies according to the rock type, temperature, and runoff in a basin (Meybeck 1986; Berner and Berner 1996; Drever 1997; Gaillardet et al. 1999). General surface water DSi values for more than 1,000 rivers in pristine condition are 0.5 and 10 mg DSi/L for 1st and 99th percentiles, respectively (Global Environment Monitoring System–Global Registration of Land–Ocean River Inputs [GEMS-GLORI] database: Meybeck and Ragu 1997), and 3 and 23 mg/L in groundwater (Davis et al. 2002), although cases with higher values of silica concentration are reported, particularly under hydrothermal influence.

The review by Meybeck and Ragu (1997) showed that the sixty largest rivers in the world drain about 80 percent of the water from continents; therefore, the large watershed approach may be enough to estimate major element fluxes to the ocean (Meybeck

Table 10.2. Average DSi concentrations calculated per ocean basin from the Global Environment Monitoring System–Global Registration of Land–Ocean River Inputs database (250 rivers) and the associated documented area.

	DSi (mg/L)	Documented Area (10^6 km^2)
Arctic Ocean	1.6	12.6
Baltic Sea	0.9	0.6
Black Sea	1.8	2.0
Hudson Bay	0.8	2.0
Indian Ocean	5.2	7.0
Mediterranean Sea*	3.3	3.1
Σ North Atlantic Ocean	3.3	16.3
Σ South Atlantic Ocean	5.5	9.2
Σ Pacific Ocean	4.1	10.6
Global discharge-weighted average	4.25	63.4

*With natural Nile discharge.

1988). On the basis of the GEMS-GLORI database, average DSi concentrations per ocean basin can be determined (Table 10.2).

The ocean and regional sea drainage basins display a large variation of DSi, by more than a factor of six, which reflects the overall controls on DSi at such a scale. Cold, lake-covered glaciated shield regions as the Arctic, Baltic, and Hudson show minimum DSi, whereas warm, humid, and volcanic regions have the maximum values.

To determine the diffuse source characteristics for each of the chosen basins or sub-basins, a small watershed approach that combines the effect of lithology and homogeneous climatic conditions (Meybeck 1987; White et al. 1998; Oliva et al. 1999, 2003; Miretzky et al. 2001; Asano et al. 2003) is more appropriate. The global scale lithology map recently developed by Dürr (2003) has allowed us to determine the percentage area of the main rock types in each of the cases studied and, following the work initiated by Meybeck (1987), a silica concentration of the headstream waters (without in-stream river cycling) was associated with each rock type (Table 10.3).

In addition, DSi surface flux is positively correlated with temperature (Meybeck 1979; White and Blum 1995), which directly enhances Si mineral dissolution (Stallard 1995; Berner and Berner 1996; Van Cappellen and Qiu 1997) or has an indirect influence through latitudes or altitudes. The importance of silica fluxes at tropical latitudes was reported by Meybeck (1979) and White and Blum (1995) (Figure 10.3).

On the basis of the studies by Schelske (1985) and Schelske et al. (1985) on the increase of permanent sedimentation in the Great Lakes, DSi deliveries at the outlet of large watersheds must take into account such in-stream retention. The effect of a lake

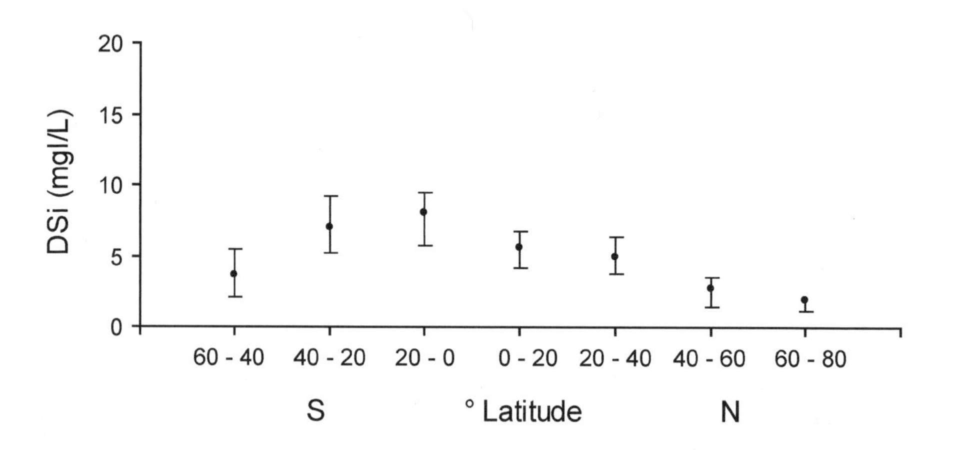

Figure 10.3. Relationship between DSi concentrations in world rivers (Global Environment Monitoring System–Global Registration of Land–Ocean River Inputs [GEMS-GLORI] database) and latitude (average and *SD* for each 20° latitude class).

Table 10.3. Lithological characteristics of the Seine, Danube, and Red River watersheds as determined from the lithological world map (Dürr 2003): percentage of surface area in each watershed occupied by lithological types and DSi concentration found for a monolithological basin with an average annual temperature of 10°C (based on Meybeck 1986).

Lithology	Seine (%)	Danube (%)	Red River (%)	DSi Concentration at 10°C (mg DSi/L)
Polar ice and glaciers				
Plutonic basic–ultrabasic		0.6		5.6
Plutonic acid	2.9	2.6	1.7	3.9
Volcanic basic		2.1		6.7
Volcanic acid				5.0
Shield and basement (Precambrian)			16.9	3.4
Metamorphic rocks				3.4
Complex lithologies		13.5	34	3.9
Siliciclastic sedimentary consolidated		11.3	15.1	4.5
Mixed sedimentary consolidated			15.2	2.8
Carbonate rocks*	66.6	12.6	17.1	2.2*
Quaternary evaporites				
Semiconsolidated to unconsolidated sedimentary	30.5	28		2.8
Alluvial deposit		13		2.8
Loess		16.3		3.4
Dunes and shifting sand		0		

*When carbonate rocks are chalk with microfossils, a DSi concentration of 5.6 is taken into account (case of the Seine Basin).

or reservoir is not important in any of these catchments in terms of permanent retention because of their low residence time, ranging from days to months.

As a result of lithology and temperature dependence (Figure 10.4), DSi concentrations of 4.7, 2.6, and 7.6 mg Si/L were found as an annual river average for the two components of discharge, surface and base flow, for the Seine, Danube, and Red River, respectively.

Where data are available, these values are refined on the basis of the lithology

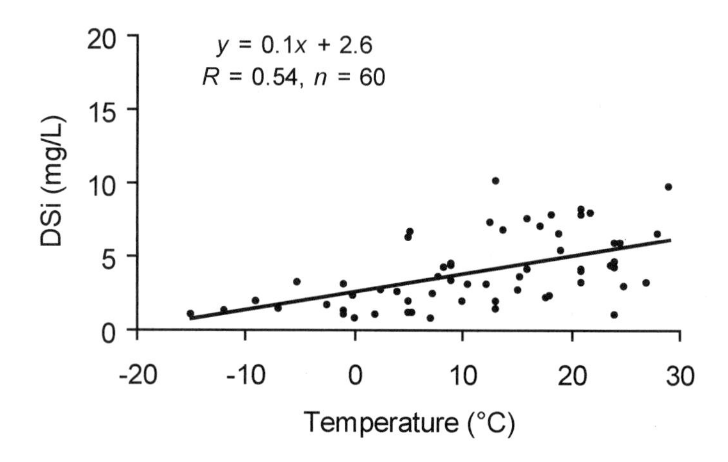

Figure 10.4. Relationship between DSi concentrations in world rivers (GEMS-GLORI database) and temperature.

specific to each subbasin considered within any watershed. In addition, DSi concentrations of aquifers, generally slightly higher than those of surface waters, can be taken into account in base flow when such data are available.

Finally, Oliva et al. (2003) show that besides the role of temperature in chemical weathering in granitic environments, high runoff (particularly when greater than 1,000 mm/yr) favors silica dissolution by increasing the effective contact time and exchange area of minerals with water. Because runoff is less than 1,000 mm/yr for the three rivers (240, 280, and 750 mm/yr for the Seine, the Danube, and the Red River, respectively), it is not explicitly considered here.

Anthropogenic Input of Dissolved and Biogenic Silica

Silica in present domestic raw and treated effluents has been measured in the Achères wastewater treatment plant, which serves 6.5 million inhabitants per day in the Parisian region. On the basis of a silica concentration of 4.5 mg Si/L in tap water and values of 7.1 and 6.4 mg Si/L in raw and treated water, respectively, a specific (per capita) additional Si load is estimated at 0.8 and 0.6 g Si/inhabitant/d (raw and treated), half in biogenic form (measured according to Conley and Schelske 2001). This new source of silica may result partly from the replacement of polyphosphates in household detergents by new additives containing silica. The value of 0.8–0.6 g Si/inhabitant/d fits well with the data obtained by Billen et al. (2001) from statistics on the domestic consumption of synthetic detergents over a 50-year period (Figure 10.5).

Thus, compared with the Redfield ratios (Si:N = 5.5 and Si:P = 41 in g/g, this new

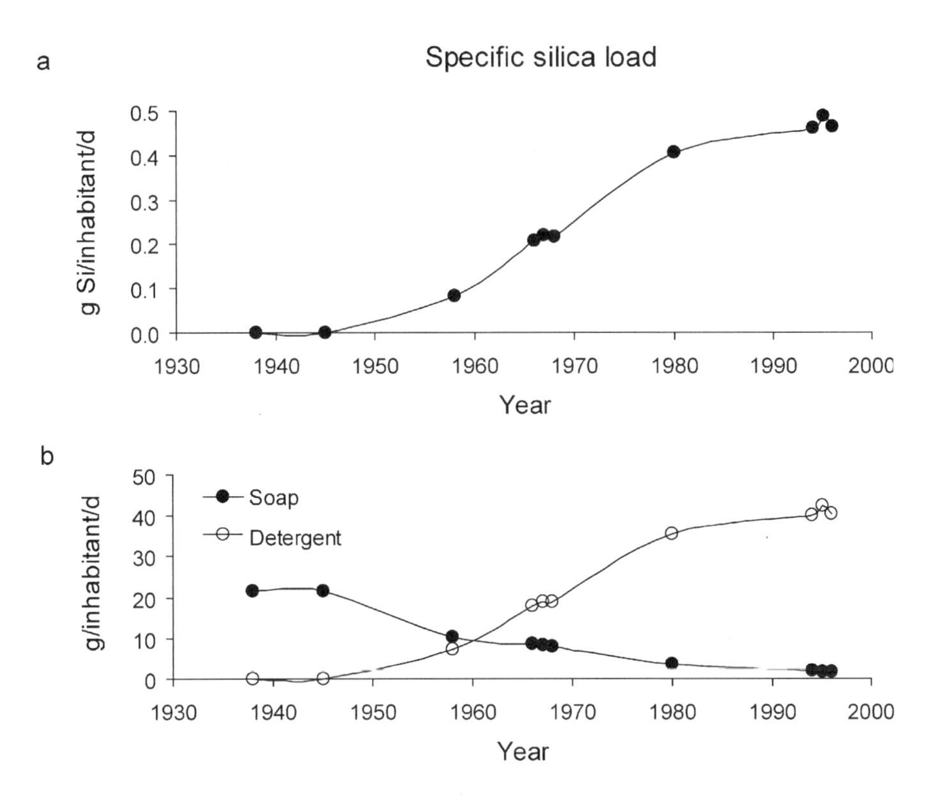

Figure 10.5. *(a)* Increase of the per capita Si load from 1930 to 2000 in developed countries. *(b)* Concomitant increase in detergent use and decrease in soap use.

specific (per capita) silica load is low compared with per capita loads in treated sewage of 1 and 10 g/inhabitant/d for P and N, respectively (Garnier et al. 2006). For raw wastewater, the Si:P and Si:N ratios would be even lower, with P and N specific load amounts of 2 and 12.5 g/inhabitant/d, respectively, for raw domestic wastewater. This very low silica load from wastewater led us to neglect silica point sources in the model.

Modeling Nutrient Dynamics (Si, N, P) in the Drainage Network

Hydrology and Nutrient Concentration

As seen in Table 10.1, the Seine River, with low discharge in summer and high discharge in winter, has a typical oceanic regime, whereas the Danube, influenced by snowmelt,

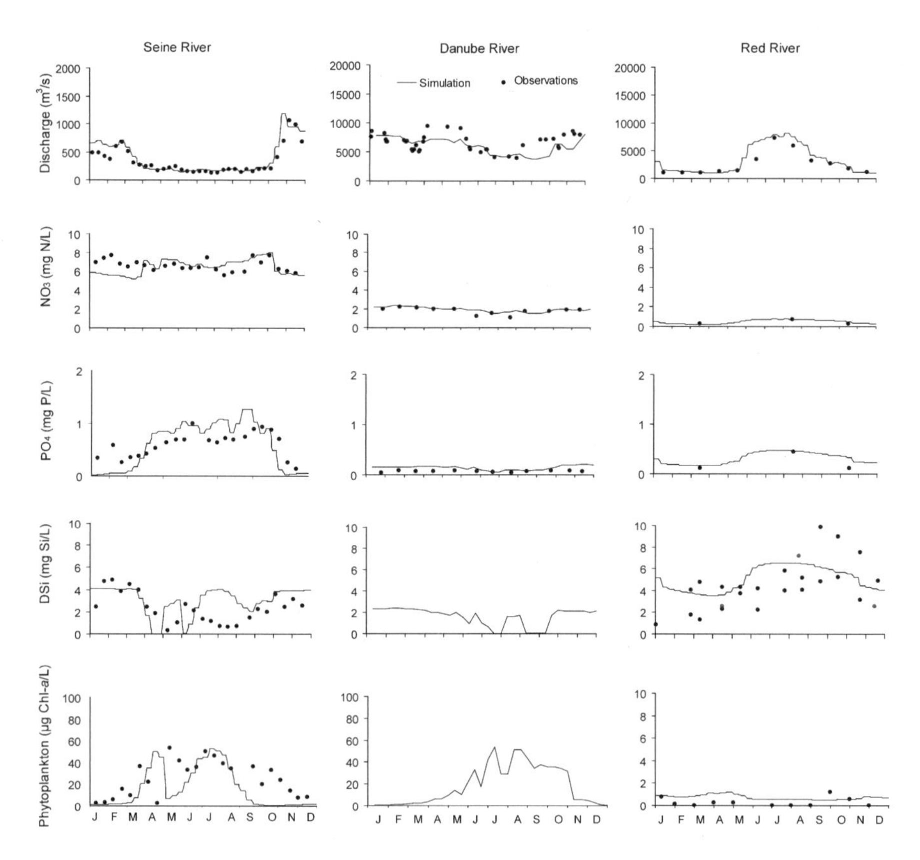

Figure 10.6. Observed and simulated seasonal variations of discharge, nitrates, phosphates (mg P/L), DSi, and phytoplankton biomass, expressed as chlorophyll-*a* concentrations, in the Seine River (in 1996), Danube River (in 1995), and Red River (average of 1995–1998 and additional observations in 2001).

has a maximum discharge in spring (Figure 10.6). The Red River (of monsoon type) experiences flooding in summer.

Because of differences in watershed size, comparisons are made between specific N, P, and Si fluxes (i.e., fluxes per square kilometer).

The model simulation of discharge and of the main variables of water quality shows a good agreement with observations. Although the seasonal variations in water quality variables are not always phased with the observations, the level of the concentrations is (Figure 10.6). Discrepancies may result from limitations in the quality of observations and constraints on data such as domestic and industrial point sources.

Among the three rivers, particularly striking are the differences in the concentrations of nitrates, much higher for the Seine by a factor of four to eight compared with the Danube and the Red rivers, respectively, because of intensive agriculture in the French watershed and the use of fertilizers characteristic of Western European countries (Figure 10.6). Phosphate concentrations are about ten times higher in the Seine, and such high values result mainly from point sources at the outlet of the whole basin, although diffuse sources are equally important in the upstream basins (Némery 2003; Garnier et al. 2005; Némery et al. 2005). It is interesting to note that the levels of silica for tropical waters are twice those of the temperate countries, in accordance with the general tendency observed on a world scale. Important variations observed in the Seine are a typical response to eutrophication (Figure 10.6). In a nonlimiting phosphate environment, diatom blooms exhaust silica concentration, mainly when dry hydrological conditions and a high residence time allow the algae to build high biomass (Garnier et al. 1995, 2002b). The poor simulation of DSi in summer calls for more studies both on the timing of algal dynamics and species succession and on silica dissolution rates (Garnier et al. 2002b).

The model has been run on the Seine River, the most eutrophic of the cases studied, for a scenario of drastic abatement of P by wastewater treatment plants (a further 85 percent reduction beyond the abatement of 65 percent that has already been observed in the last 10 years; Garnier et al. 2005). Under these conditions, phytoplankton growth is reduced, and silica is less depleted (Figure 10.7).

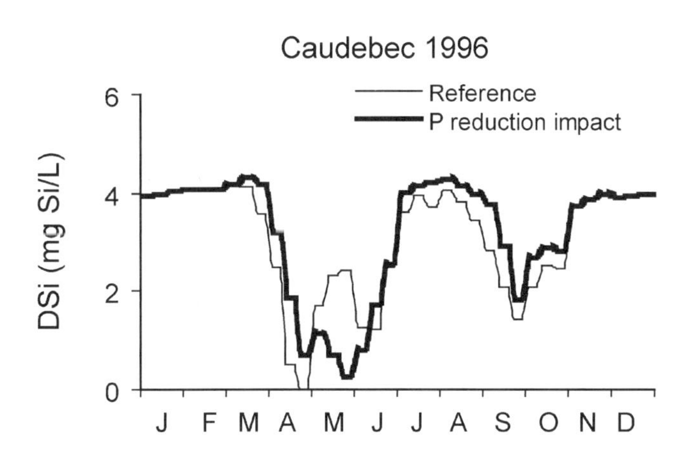

Figure 10.7. Response of the RIVERSTRAHLER model, in terms of DSi concentrations, to a reduction in phosphorus in the Seine River at the limit of saline intrusion in the estuary (Caudebec).

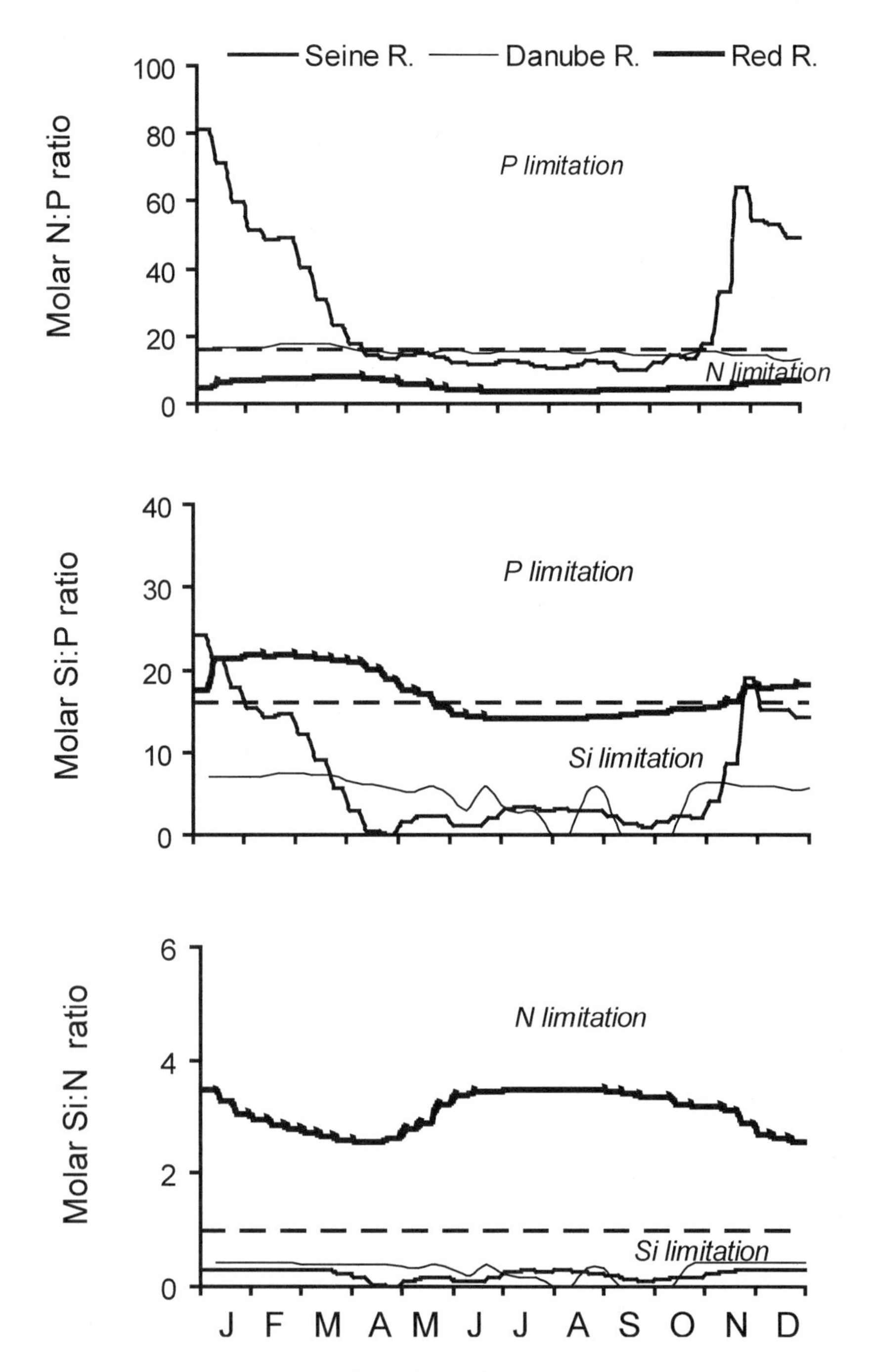

Figure 10.8. Seasonal variations of N:P, Si:P, and Si:N in the Seine River, Danube River, and Red River calculations from the simulation results (see Figure 10.6).

The seasonal variations in the molar ratio N:P:Si, calculated by the model for each watershed, reveal great seasonal changes for the Seine River, shifting from P limitation in spring to N limitation in summer, with Si limitation invariably observed in regard to either P or N (Figure 10.8).

For the Danube, silica limitation is also observed, with the N:P ratio being close to the Redfield value of algal requirement. With concentrations of silica about twice those of the other rivers, silica generally is not limiting in the Red River compared with P or to N. The Red River nutrient load appears to be low in N rather than in P, which is explained by the large amount of diffuse P, whereas N fertilizers are not used as much as in European countries, at least in the upstream basins.

Silica Retention in Reservoirs

Silica retention in the reservoirs of all three river systems considered here is taken into account in the model. We already demonstrated that silica retention in the diverted reservoirs of the Seine River (residence time of 0.5 year) is low on an annual scale, amounting to about 3 percent of the total retention in the drainage network and 0.3 percent of the silica flux at the outlet of the fluvial estuary (Garnier et al. 1999b, 2000, 2002b). Diverging from Humborg et al. (1997), we found silica retention in the large reservoirs of the Danube (Iron Gates) to be low because of the short residence time of the water masses (a few days). Our studies suggested a greater retention in the drainage network itself (Garnier et al. 2002a; Trifu 2002). Similarly, for the Red River, comparing the model results before and after damming showed that only low silica retention occurred in reservoirs (Garnier and Billen 2002).

Exploring Human Impact on N, P, Si Fluxes

The N, P, and Si specific fluxes calculated at the outlet of the three watersheds on the basis of the simulation presented in Figure 10.6 are compared in Figure 10.9.

In addition to the reference situations, two hypothetical scenarios have been tested for each river system: a "pristine" scenario considering the watershed entirely covered with natural vegetation (e.g., forest), without any point source of wastewater and without hydrological regulation, and a scenario called "no population," corresponding to the present land use and hydrological regulation but without wastewater release.

Specific fluxes of N are much higher for the Seine River than for the other two basins (Figure 10.9). The smaller difference between the reference and the no-population scenario than between the reference and pristine scenarios indicates that, for all three watersheds, total N is mostly of diffuse origin and related to agriculture.

Regarding P, the highest specific fluxes are observed in the tropical Red River (Figure 10.9). The results of the scenarios indicate the predominantly diffuse origin of P fluxes in this basin. Erosion of agricultural soils in tropical regions and even of forested

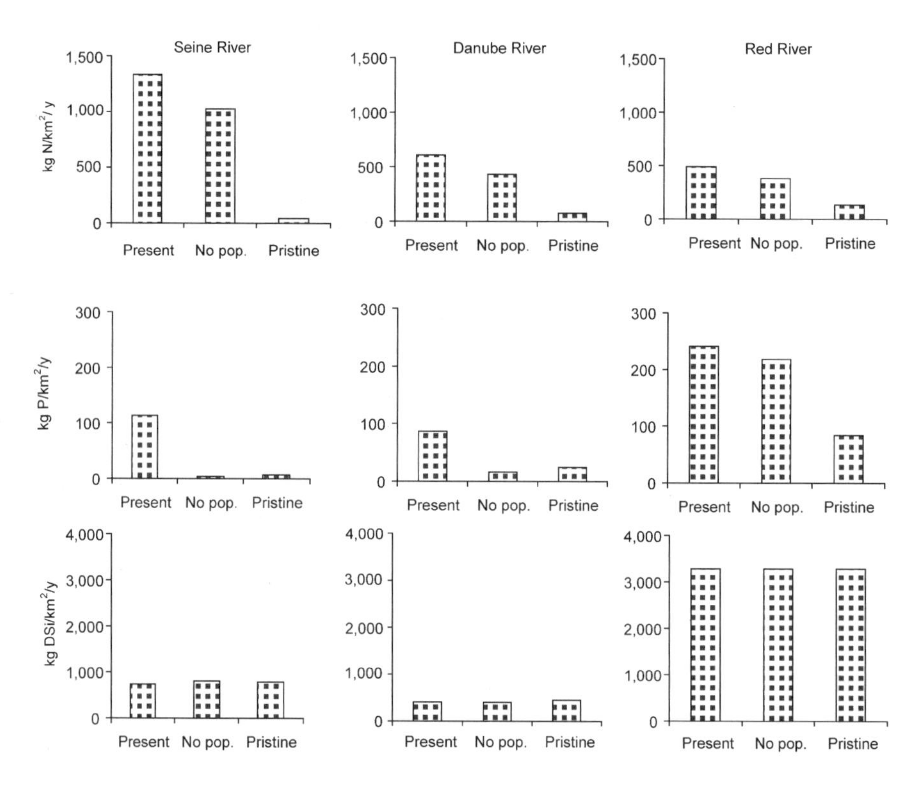

Figure 10.9. Response of the model to exploration of unrealistic scenarios: *(a)* specific fluxes of nitrates, *(b)* of phosphate, and *(c)* of DSi, under the conditions of the validation *(present)*, without domestic input (*no pop.*, i.e., all domestic effluents are treated), and without domestic input and with natural (forested) vegetation and no fertilization *(pristine)*.

areas is known to be a major source of suspended matter to which P is associated (Meybeck 1982); diffuse P sources are taken into account in the model (Garnier and Billen 2002). In the temperate systems, on the other hand, P fluxes are much lower and mainly of urban origin at the outlet of the watersheds because population is concentrated along the lower, larger rivers. For the Seine and the Danube the majority of domestic and industrial effluents is collected but only partially treated in wastewater treatment plants, usually without P removal.

Because Si is not directly related to human activity, there are no notable differences for the various conditions explored. Particularly striking are the much greater specific Si fluxes in the Red River (Figure 10.9). This is explained both by the lithology of the upstream Red River basin, composed of very heterogeneous Si-rich formations (e.g., volcanic rocks, sedimentary rocks), and by the wet tropical climate conditions (higher runoff) enhancing rock weathering.

Conclusion

Modeling the silica biogeochemical cycle at a multiregional scale, explicitly taking into account the mechanisms of the transfer and transformation processes within the catchment area and in the drainage network, is an original approach that needs further development in order to correctly simulate the absolute and relative value of the nutrient fluxes delivered to the coastal zones. Such a deterministic approach is needed if realistic scenarios are to be established for preventing or reducing coastal eutrophication, which closely depends on the ratios of riverine N, P, and Si inputs to the coastal environment (Cugier et al. 2005). Until now, most modeling approaches of riverine N and P delivery to the coastal ocean at a multiregional or global scale have been based on statistical regression approaches (Cole and Caraco 2001; Seitzinger et al. 2002a, 2002b; Green et al. 2004) instead of on a mechanistic representation of the biogeochemical processes involved.

The approach suggested in this chapter is still imperfect. In particular, it takes into

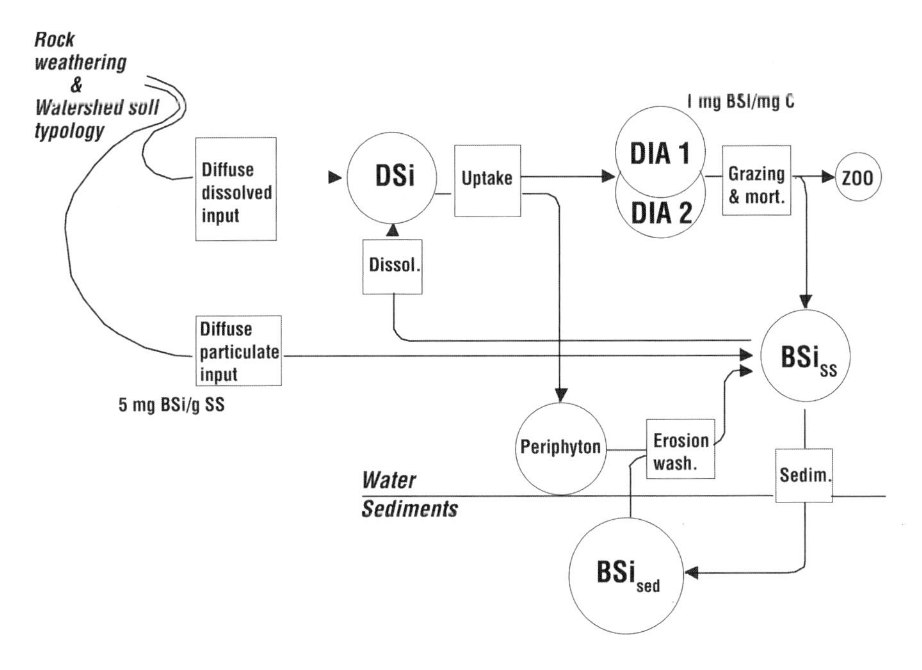

Figure 10.10. Future representation of Si in the RIVERSTRAHLER model, taking into account diffuse particulate BSi associated with suspended solids (BSiSS) and additional compartments of planktonic diatoms (DIA 1 and DIA 2) and benthic diatoms (periphyton). The processes of silica uptake and dissolution (dissol.) as well as diatom grazing, mortality (mort.), sedimentation (sedim.) and erosion/washing (wash.) are also represented.

account only the diffuse sources of DSi, and particulate sources such as BSi are neglected (Figure 10.10).

Although BSi input from the watershed represents only a small fraction (less than 2 percent) of total diffuse DSi and BSi sources in the Seine Basin (Garnier et al. in press), such a proportion could be much higher in low-relief regions where the terrestrial silica cycle and the vegetation silica uptake are more active (e.g., in tropical soils). The importance of phytoliths (opal particles) was first shown by Bartoli (1983). However, it was well known that gramineous species (e.g., oat) contain ten to twenty times more Si than leguminous species (Russell 1961) and that silica helps to keep the plants erect, counteract manganese toxicity, and prevent fungal and insect attacks (Jones and Handreck 1967). More recent works mention that plants absorb DSi from soils and precipitate it in their tissues as phytoliths, in a proportion up to 15 wt% (Alexandre et al. 1997).

From another point of view (river chemistry), the role of vegetation in DSi inputs to surface water could also be better represented in the model, which takes into account only the direct role of lithology; indeed, Humborg et al. (Chapter 5, this volume) showed a clear positive correlation between DSi and total organic carbon in small rivers draining differently vegetated catchments for the same lithology and climate.

A better representation of the aquatic biological uptake of silica (Figure 10.10) would be another improvement to the model. At present, one group of planktonic diatoms is considered. In analysis of the seasonal evolution of diatom successions in the Seine River (Garnier et al. 1995) and the pattern of seasonal successions in freshwater ecosystems (Sommer et al. 1986), at least two groups of diatoms should be taken into account, one for spring bloom and another one for late summer. This would probably improve the simulation of both seasonal phytoplankton and DSi variation levels. In addition, because the RIVERSTRAHLER model is more suited to simulating eutrophication in large rivers, development of benthic diatoms in small upstream rivers is not yet represented, although it could influence the silica biogeochemical cycle in headwaters (Flipo et al. 2004). Benthic algal development can be a sink for DSi in spring in headwaters but also can be a source of BSi by washing out at high discharges (Dessery et al. 1984; Barillier et al. 1993). To our knowledge the role of macrophytes in silica sequestration is not well documented, but it should not be excluded.

In the three river systems studied in this chapter, we found limited silica retention in reservoirs. On a global scale, however, damming is known to influence silica fluxes to the ocean because most of the reservoirs have much higher residence times than those studied here. However, residence times in river channels can reach 3 months, compared with 20–30 days in natural conditions (Vörösmarty et al. 1997, 2003). But as silica retention in both reservoirs and hydrographic networks increases with eutrophication, short-term silica retention, not significant in an annual budget, can damage the ecological functioning of the coastal zone of temperate systems, when silica depletion in late spring leads to a shift from diatoms to blooms of undesirable nondiatoms (e.g., *Phaeocystis*, Lancelot 1995; or *Dinophysis*, Cugier et al. 2005). Silica transformation and retention in estuaries and deltas are much less documented, but, as in reservoirs, the intensity of the transformations depends on residence time. For example, in the Seine estuary, BSi content of suspended solids (SS) is much higher in the turbidity maximum (residence time of a few months at low summer discharge), compared with the BSi content of headwaters (20 mg BSi/g SS and 5 mg BSi/g SS, respectively).

Although Si has long been studied by geochemists as a major ion, studies of its combination with the nutrient cycle and its importance in water quality date back only to the 1980s, so Si studies do not sufficiently cover various aspects of its cycle (e.g., transformations such as dissolution rates of BSi and lithogenic silica). Its circulation or retention along an aquatic continuum, from land to sea, is still poorly understood, especially under human impact. The "Anthropocene" Si cycle should be affected by changes in climate (temperature, runoff, and vegetation) and in N, P, and Si inputs (increase of fertilizers, improvement of N and P treatment, new use of Si in detergents) and by

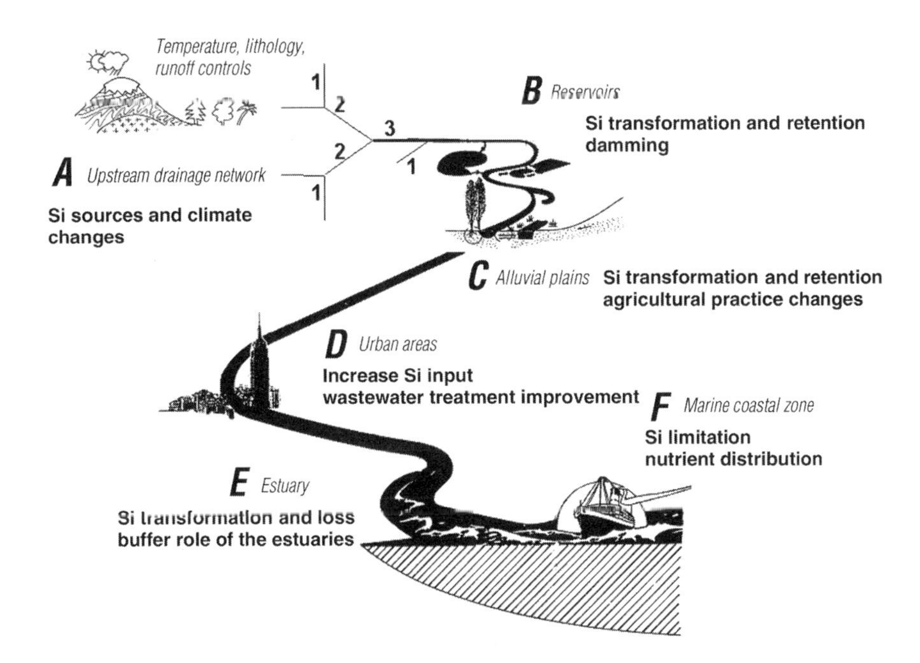

Figure 10.11. "Anthropocene" Si transfers from land to sea. Routing of riverine Si and Si cycling and retention along the aquatic continuum. *(a)* Strahler ordination of the headwater and diffuse Si sources, *(b)* reservoir Si transformation and retention, *(c)* large wetland Si retention, *(d)* large city input of DSi and BSi, *(e)* estuarine Si transformation and accumulation of BSi in the turbidity maximum, and *(f)* DSi limitation in the coastal zone, BSi recycling at the sediment interface, and redistribution within the water column.

impoundments (increased retention time of surface waters), necessarily leading to a redistribution of coastal nutrients (Figure 10.11).

Our study shows that regional specificities (i.e., climate, lithology, human impact) should be considered explicitly to elucidate the present river Si transfers at the global scale.

Literature Cited

Alexandre, A., J.-D. Meunier, F. Colin, and J.-M. Koud. 1997. Plant impact on the biogeochemical cycle of silicon and related weathering processes. *Geochimica et Cosmochimica Acta* 61:677–682.

Asano, Y., T. Uchida, and N. Ohte. 2003. Hydrologic and geochemical influences on the dissolved silica concentration in natural water in a steep headwater catchment. *Geochimica et Cosmochimica Acta* 67:1973–1989.

Barillier, A., J. Garnier, and M. Coste. 1993. Experimental reservoir water release: Impact on the water quality on a river 60 km downstream (upper Seine River, France). *Water Research* 27:635–643.

Bartoli, F. 1983. The biogeochemical cycle of silicon in two temperate forest ecosystems. *Ecological Bulletins* 35:469–476.

Berner, E. K., and R. Berner. 1996. *Global environment, water, air and geochemical cycles.* Englewood Cliffs, NJ: Prentice Hall.

Billen, G., and J. Garnier. 1997. The Phison River plume: Coastal eutrophication in response to changes in land use and water management in the watershed. *Aquatic Microbial Ecology* 13:3–17.

Billen, G., and J. Garnier. 1999. Nitrogen transfers through the Seine drainage network: A budget based on the application of the RIVERSTRAHLER model. *Hydrobiologia* 410:139–150.

Billen, G., J. Garnier, A. Ficht, and C. Cun. 2001. Modelling the response of water quality in the Seine estuary to human activity in its watershed over the last 50 years. *Estuaries* 24(6):977–993.

Billen, G., J. Garnier, and P. Hanset. 1994. Modelling phytoplankton development in whole drainage networks: The RIVERSTRAHLER model applied to the Seine River system. *Hydrobiologia* 289:119–137.

Billen, G., J. Garnier, and M. Meybeck. 1998. Les sels nutritifs: L'ouverture des cycles. Pp. 531–565 in *La Seine en son bassin: Fonctionnement écologique d'un système fluvial anthropisé*, edited by M. Meybeck, G. De Marsily, and F. Fustec. Paris: Elsevier.

Billen, G., J. Garnier, and V. Rousseau. 2005. Nutrient fluxes and water quality in the drainage network of the Scheldt Basin over the last 50 years. *Hydrobiologia* 540:47–67.

Bultot, F., and G. Dupriez. 1976. Conceptual hydrological model for an average-sized catchment area. *Journal of Hydrology* 39:251–292.

Campy, M., and M. Meybeck. 1995. Les sédiments lacustres. Pp. 184–226 in *Limnologie générale*, edited by R. Pourriot and M. Meybeck. Paris: Masson.

Clarke, F. W. 1924. Data of geochemistry. *U.S. Geological Survey Bulletin* 770.

Cole, J. J., and N. F. Caraco. 2001. Emissions of nitrous oxide (N_2O) from a tidal, freshwater river, the Hudson River, New York. *Environmental Science and Technology* 35:991–996.

Conley, D. 2002. Terrestrial ecosystems and the global biogeochemical silica cycle. *Global Biogeochemical Cycles* 16:1121–1129.

Conley, D. J., and S. S. Kilham. 1989. Differences in silica content between marine and freshwater diatoms. *Limnology and Oceanography* 34(1):205–213.

Conley, D. J., and C. L. Schelske. 2001. Biogenic silica. Pp. 281–293 in *Tracking environmental change using lake sediments*, edited by J. P. Smol, H. J. B. Birks, and W. M. Last. Dordrecht: Kluwer.

Conley, D., C. Schleske, and E. Stoemer. 1993. Modification of the biogeochemical cycle of silica with eutrophication. *Marine Ecology Progress Series* 101:179–192.

Conley, D. J., W. M. Smith, J. C. Corwell, and T. R. Fisher. 1995. Transformation of particle-bound phosphorus at the land–sea interfaces. *Estuarine, Coastal and Shelf Science* 40:161–176.

Cugier, P., G. Billen, J. F. Guillaud, J. Garnier, and A. Ménesguen. 2005. Modelling the eutrophication of the Seine bight (France) under historical, present and future riverine nutrient loading. *Journal of Hydrology* 304:381–396.

Dang Anh Tuan. 2000. *The Red River Delta: The cradle of the nation* (in Vietnamese). Hanoi: National University of Hanoi.

Davis, C. C., H. W. Chen, and M. Edwards. 2002. Modeling silica sorption to iron hydroxide. *Environmental Science and Technology* 36:582–587.

Dessery, S., C. Dulac, J. M. Laurenceau, and M. Meybeck. 1984. Evolution du carbone organique particulaire "algal" et détritique dans trois rivières du bassin parisien. *Archiv für Hydrobiologie* 100(2):235–260.

Drever, J. I. 1997. *The geochemistry of natural waters: Surface and groundwater environments*, 3rd ed. Upper Saddle River, NJ: Prentice Hall.

Dürr, H. H. 2003. *Towards a typology of global river systems: Some concepts and examples at medium resolution.* Thèse de doctorat, Université Paris VI, Pierre et Marie Curie.

Everbecq, E., V. Gosselain, L. Viroux, and J. P. Descy. 2001. Potamon: A dynamic model for predicting phytoplankton composition and biomass in lowland rivers. *Water Research* 35:901–912.

Flipo, N., S. Even, M. Poulin, M.-H. Tusseau-Vuillemin, T. Améziane, and A. Dauta. 2004. Biogeochemical modelling at the river scale: Plankton and periphyton dynamics (Grand Morin case study, France). *Ecological Modelling* 176:333–347.

Gaillardet, J., B. Dupre, P. Louvat, and C. J. Allegre. 1999. Global silicate weathering and CO_2 consumption rates deduced from the chemistry of large rivers. *Chemical Geology* 159:3–30.

Garnier, J., and G. Billen. 2002. The RIVERSTRAHLER modelling approach applied to a tropical case study (The Red-Hong River, Vietnam): Nutrient transfer and impact on the coastal zone. *SCOPE, Collection of Marine Research Works* 12:51–65.

Garnier, J., G. Billen, and M. Coste. 1995. Seasonal succession of diatoms and Chlorophyceae in the drainage network of the river Seine: Observations and modelling. *Limnology and Oceanography* 40:750–765.

Garnier, J., G. Billen, E. Hannon, S. Fonbonne, Y. Videnina, and M. Soulie. 2002a. Modeling transfer and retention of nutrients in the drainage network of the Danube River. *Estuarine, Coastal and Shelf Science* 54:285–308.

Garnier, J., G. Billen, P. Hanset, P. Testard, and M. Coste. 1998. Développement algal et eutrophisation. Pp. 593–626 in *La Seine en son bassin: Fonctionnement écologique d'un système fluvial anthropisé*, edited by M. Meybeck, G. De Marsily, and F. Fustec. Paris: Elsevier.

Garnier, J., G. Billen, and L. Palfner. 1999a. Understanding the oxygen budget of the Mosel drainage network with the concept of heterotrophic/autotrophic sequences: The RIVERSTRAHLER approach. *Hydrobiologia* 410:151–166.

Garnier, J., G. Billen, N. Sanchez, and B. Leporcq. 2000. Ecological functioning of the Marne Reservoir (Upper Seine Basin, France). *Regulated Rivers: Research & Management* 16:51–71.

Garnier, J., A. d'Ayguesvives, G. Billen, D. Conley, and A. Sferratore. 2002b. Silica dynamics in the hydrographic network of the Seine River. *Océanis* 28:487–508.

Garnier, J., O. Dufayt, G. Billen, and M. Roulier. 2001a. Eutrophisation et gestion des apports de phosphore dans le bassin de la Seine. Colloque "Scientifiques & Décideurs: Agir Ensemble pour une Gestion Durable des Systèmes Fluviaux," Lyon, France.

Garnier, J., L. Laroche, and S. Pinault. 2006. Characterization of the effluents of two large wastewater plants of the city of Paris (France): How to comply with the European directive. *Water Research* 2006.

Garnier, J., B. Leporcq, N. Sanchez, and X. Philippon. 1999b. Biogeochemical budgets in three large reservoirs of the Seine basin (Marne, Seine & Aube reservoirs). *Biogeochemistry* 47:119–146.

Garnier, J., J. Némery, G. Billen, and S. Théry. 2005. Nutrient dynamics and control of eutrophication in the Marne River system: Modelling the role of exchangeable phosphorus. *Journal of Hydrology* 304:397–412.

Garnier, J., P. Servais, G. Billen, M. Akopian, and N. Brion. 2001b. The oxygen budget in the Seine estuary: Balance between photosynthesis and degradation of organic matter. *Estuaries* 24(6):964–977.

Green, P., C. J. Vörösmarty, M. Meybeck, J. Galloway, and B. Peterson. 2004. Pre-industrial and contemporary fluxes of nitrogen through rivers: A global assessment based on typology. *Biogeochemistry* 68:71–105.

Howarth, R. J., G. Billen, D. Swaney, A. Townsend, N. Jaworski, K. Lajtha, J. A. Downing, R. Elmgren, N. Caraco, T. Jordan, F. Berendse, J. Freney, V. Kudeyarov, P. Murdoch, and Z. Zhao-Liang. 1996. Regional nitrogen budgets and riverine N and P fluxes for the drainages of the North Atlantic Ocean: Natural and human influences. *Biogeochemistry* 35:141–180.

Humborg, C., V. Ittekkot, A. Cosiascu, and B. von Bodungen. 1997. Effect of Danube River dam on Black Sea biogeochemistry and ecosystem structure. *Nature* 386:385–388.

Jones, L. H. P., and K. A. Handreck. 1967. Silica in soils, plants and animals. *Advances in Agronomy* 19:107–149.

Kroeze, C., and S. P. Seitzinger. 1998. Nitrogen inputs to rivers, estuaries and continental shelves and related nitrous oxide emissions in 1990 and 2050: A global model. *Nutrient Cycling in Agroecosystems* 52:195–212.

Lancelot, C. 1995. The mucilage phenomenon in the continental coastal waters of the North Sea. *Science of the Total Environment* 165:83–102.

Meybeck, M. 1979. Concentrations des eaux fluviales en éléments majeurs et apports en solution aux océans. *Revue de Géologie Dynamique et de Géographie Physique* 21:215–246.

Meybeck, M. 1982. Carbon, nitrogen, and phosphorus transport by world rivers. *American Journal of Science* 282:401–450.

Meybeck, M. 1984. *Les fleuves et le cycle géochimique des éléments.* Thèse de doctorat d'etat, Université Paris 6.

Meybeck, M. 1986. Composition chimique naturelle des ruisseaux non pollués en France. *Sciences Géologiques Bulletin* 39:3–77.

Meybeck, M. 1987. Global chemical weathering of surficial rocks estimated from river dissolved loads. *American Journal of Science* 287:401–428.

Meybeck, M. 1988. How to establish and use world budgets of riverine materials. Pp. 247–272 in *Physical and chemical weathering in geochemical cycles*, edited by A. Lerman and M. Meybeck. Dordrecht: Kluwer.

Meybeck, M. 2003. Global occurrence of major elements in rivers. Pp. 207–224 in *Treatise on geochemistry*, Vol. 5, edited by H. D. Holland and K. K. Turekian. Oxford: Pergamon.

Meybeck, M., and A. Ragu. 1997. Presenting the GEMS-GLORI, a compendium for world river discharge to the oceans. *International Association of Hydrology Scientific Publication* 243:3–14.

Miretzky, P., V. Conzonno, and A. Fernandez Cirelli. 2001. Geochemical processes controlling silica concentrations in groundwaters of the Salado River drainage basin, Argentina. *Journal of Geochemical Exploration* 73:155–166.

Némery, J. 2003. *Origine et devenir du phosphore dans le continuum aquatique de la Seine, des petits basins à l'estuaire. Rôle du phosphore échangeable sur l'eutrophisation.* Thèse Université Paris 6.

Némery, J., J. Garnier, and C. Morel. 2005. Phosphorus budget in the marine watershed (France): Urban vs. diffuse sources, dissolved vs. particulate forms. *Biogeochemistry* 72(1):35–66.

Nixon, S. W., J. W. Ammerman, L. P. Atkinson, V. M. Berounsky, G. Billen, W. C. Boicourt, W. R. Boynton, T. M. Church, D. M. Di Toro, R. Elmgren, J. H. Garber, A. E. Giblin, R. A. Jahnke, N. J. P. Owens, M. E. Q. Pilson, and S. P. Seitzinger. 1996. The fate of nitrogen and phosphorus at the land–sea margin of the North Atlantic Ocean. *Biogeochemistry* 35:141–180.

Officer, C. B., and J. H. Ryther. 1980. The possible importance of silicon in marine eutrophication. *Marine Ecology Progress Series* 3:83–91.

Oliva, P., J. Viers, and B. Dupre. 2003. Chemical weathering in granitic environments. *Chemical Geology* 202:225–256.

Oliva, P., J. Viers, B. Dupre, J. P. Fortune, F. Martin, J. J. Braun, D. Nahon, and H. Robain. 1999. The effect of organic matter on chemical weathering: Study of a small tropical watershed: Nsimi-Zoetele site, Cameroon. *Geochimica et Cosmochimica Acta* 63:4013–4035.

Rabalais, N. N., and R. E. Turner (eds.). 2001. Coastal hypoxia. *Coastal and Estuarine Studies Series* 58. Washington, DC: American Geophysical Union.

Redfield, A. C. 1958. The biological control of chemical factors in the environment. *American Scientist* 46:205–222.

Redfield, A. C., B. H. Ketchum, and F. A. Richards. 1963. The influence of organisms on the composition of sea-water. Pp. 12–37 in *The sea*, edited by M. N. Hill. New York: Wiley.

Russell, E. W. 1961. *Soil conditions and plant growth*, 9th ed. London: Longman.

Schelske, C. L. 1985. Biogeochemical silica mass balances in Lake Michigan and Lake Superior. *Biogeochemistry* 1:197–218.

Schelske, C. L., D. J. Conley, and W. F. Warwick. 1985. Historical relationships between phosphorus loading and biogenic silica accumulation in Bay of Quinte sediments. *Canadian Journal of Fisheries and Aquatic Science* 42:1401–1409.

Seitzinger, S. P., C. Kroeze, A. F. Bouwman, N. F. D. Caraco, and R. V. Styles. 2002a. Global patterns of dissolved inorganic and particulate nitrogen inputs to coastal systems: Recent conditions and future projections. *Estuaries* 25:640–655.

Seitzinger, S. P., R. V. Styles, E. W. Boyer, R. B. Alexander, G. Billen, R. W. Howarth, B. Mayer, and N. Van Breemen. 2002b. Nitrogen retention in rivers: Model development and application to watersheds in the northeastern U.S.A. *Biogeochemistry* 57:199–237.

Sommer, U., Z. M. Gliwicz, W. Lampert, and A. Duncan. 1986. The PEG-model of seasonal succession of planktonic events in fresh waters. *Archiv für Hydrobiologie* 106:433–471.

Stallard, R. F. 1995. Relating chemical and physical erosion. In *Chemical weathering rates of silicate minerals*, edited by A. F. White and S. L. Brantley. Mineralogical Society of America. *Reviews in Mineralogy* 31:543–562.

Strahler, A. H. 1957. Quantitative analysis of watershed geomorphology. *Transactions–American Geophysical Union* 38:913–920.

Trifu, M. C. 2002. *Transfer des nutriments dans le Bassin du Danube et les apports à la Mer Noire.* Thèse, Université Paris 6, Paris.

Van Cappellen, P., and L. Qiu. 1997. Biogenic silica dissolution in sediments of the Southern Ocean. I. Solubility. *Deep-Sea Research Part II* 44:1109–1128.

Vannote, R. L., G. W. Minshall, K. W. Cummins, J. R. Sedell, and C. E. Cushing. 1980. The river continuum concept. *Canadian Journal of Fisheries and Aquatic Science* 37:130–137.

Vörösmarty, C. J., M. Meybeck, B. Fekete, and K. Sharma. 1997. The potential impact of neo-Castorization on sediment transport by the global network of rivers. Pp. 261–273 in *Human impact on erosion and sedimentation.* Proceedings of the Rabat Symposium. International Association of Hydrological Sciences Publication 245.

Vörösmarty, C. J., M. Meybeck, B. Fekete, K. Sharma, P. Green, and J. Syvitski. 2003. Anthropogenic sediment retention: Major global-scale impact from the population of registered impoundments. *Global and Planetary Changes* 39:169–190.

White, A. F., and A. Blum. 1995. Effects of climate on chemical weathering in watersheds. *Geochimica et Cosmochimica Acta* 47:805–816.

White, A. F., A. E. Blum, M. S. Schulz, D. V. Vivit, D. A. Stonestrom, M. Larsen, S. F. Murphy, and D. Eberl. 1998. Chemical weathering in a tropical watershed, Luquillo Mountains, Puerto Rico: I. Long-term versus short-term weathering fluxes. *Geochimica et Cosmochimica Acta* 62:209–226.

11

Role of Diatoms in Silicon Cycling and Coastal Marine Food Webs

Olivier Ragueneau, Daniel J. Conley, Aude Leynaert, Sorcha Ni Longphuirt, and Caroline P. Slomp

The importance of studying the marine biogeochemical cycle of Si arises from both an ecological and a biogeochemical perspective and is related to the importance of diatoms in the global C cycle. Diatoms form the basis of some of the most productive food chains and play a fundamental role in the export of C to higher trophic levels (Cushing 1989) and to the deep sea (Goldman 1993, Buesseler 1998). They need dissolved silica (DSi) for growth. When DSi becomes limiting (Figure 11.1), it can cause shifts from diatoms to nonsiliceous algae (Officer and Ryther 1980), which increases the likelihood of harmful algal blooms (HABs) in coastal waters (review in Smayda 1990) and decreases export to the open sea (Dugdale et al. 1995).

This chapter concentrates on the Si cycle and the ecology of diatoms in coastal waters. The role of diatoms in the biogeochemical cycle of C in the open ocean has been reviewed recently by Smetacek (1999). The continental margins are of great importance for the global Si and C cycles for two major reasons. First, continental margins may be the site of the proposed missing sink of biogenic silica (DeMaster 2002). This idea has arisen with the downward revision of the importance of the Southern Ocean in the global accumulation of biogenic silica (BSi), and a new sink is needed to balance the global Si budget. The missing sink may take the form not only of BSi but also of lithogenic silica (LSi) after the conversion of diatoms into new silicate mineral phases through reverse weathering reactions (Michalopoulos and Aller 1995; Michalopoulos et al. 2000). Second, continental margins filter DSi inputs from the land to the ocean. This is essential because DSi has been recognized as a limiting nutrient in several biogeochemical provinces of the ocean (Dugdale and Wilkerson 1998; Pondaven et al. 1998; Wong and Matear 1999), with important consequences for the efficiency of the biological pump, be it in today's ocean or during the last glacial maximum (Smetacek 1999; Ragueneau et al. 2000; Archer et al. 2000).

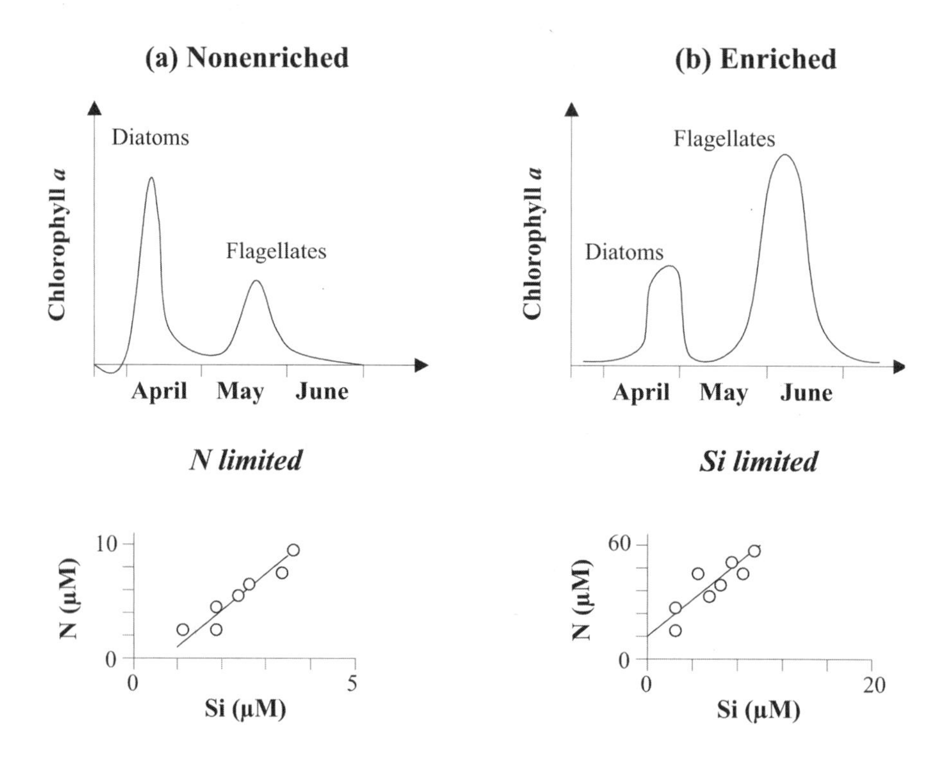

Figure 11.1. Classic sequence of phytoplankton dynamics in temperate waters of *(a)* unperturbed and *(b)* perturbed, nutrient-enriched coastal areas. Inspired by Billen et al. (1991). Note the switch from N to Si limitation under conditions of excessive N inputs *(b)*.

In this review we examine several paradigms and evaluate their status relative to field and laboratory and mesocosm evidence, including the following: Diatoms are the best food for grazers (Jonasdottir et al. 1998; Irigoien et al. 2002); DSi limitation controls the shift from diatoms to nonsiliceous species (Officer and Ryther 1980); and diatoms always dominate when DSi concentrations are more than 2 μM (Egge and Aksnes 1992). What have we learned in the past two decades that can confirm or invalidate these paradigms? First, we concentrate on the importance of diatoms in the functioning of coastal ecosystems. We start by examining the importance of diatoms in the diet of pelagic and benthic grazers, mostly copepods and bivalves. We continue by discussing the role of diatoms during HABs. We then review our knowledge of the mechanisms that control the Si cycle in coastal waters, focusing on the relative importance of the external (rivers, atmosphere, groundwater, and oceanic inputs) and internal (recycling in the water column and at the sediment–water interface) supply of DSi in sustaining the diatom demand.

Importance of Diatoms in Coastal Ecosystem Functioning

It is widely believed that diatoms form the basis of the shortest, most economically desirable food webs, leading to fish via copepods or to shellfish without intermediate trophic levels (Ryther 1969; Cushing 1989; Figure 11.2).

This paradigm is based on numerous field observations, with two important aspects that are difficult to contest: Diatoms form the basis of the food web that characterizes the most productive regions and sustains the most important fisheries on the planet, and most of the ecological problems related to eutrophication have been encountered with nondiatom species of phytoplankton. Have we learned anything in the last two decades that could modulate this paradigm?

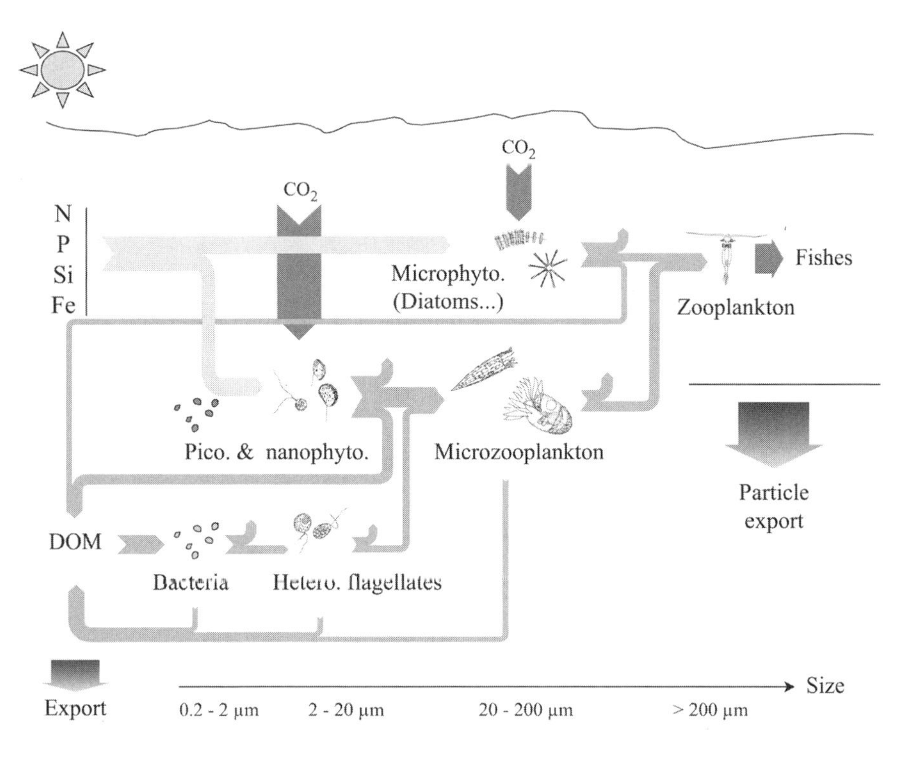

Figure 11.2. Schematic view of a pelagic food chain, inspired by the BIOGEN model (Lancelot et al. 2000), illustrating the direct, short link between diatoms and the higher trophic level, compared with the microbial network. DOM = dissolved organic matter, hetero. = heterotrophic, microphyto. = microphytoplankton, nanophyto. = nanophytoplankton, pico. = picoplankton.

Diatoms and Pelagic Food Webs

Diatoms clearly exhibit the appropriate size range for direct grazing by microzooplankton (Frost 1972). The herbivorous copepods feed directly on cells larger than 5 μm, and the food chain from bacteria to protozoa, ciliates, and copepods is affected by energetic losses at each step (Cushing 1989; Figure 11.2). The quantity of food necessary for copepods to grow has been explored, and ecological models can now be parameterized with appropriate ingestion curves for several zooplankton taxa (Hansen et al. 1997). However, the quality of the food is as important as the quantity and remains to be investigated (Touratier et al. 1999). In this respect, there is little evidence that diatoms represent a better diet than other species of similar size. In fact, dinoflagellates have a higher volume-specific organic content than diatoms under the same growth conditions (Hitchcock 1982) and are increasingly recognized as being important in the diet of *Calanus* (review in Kleppel 1993). In addition, the ability to switch from ambush feeding and motile prey to suspension feeding when food becomes abundant relates to the prey-switching theory described by Kiørboe et al. (1996). It shows that although diatoms are good, they are not the only good food: Diversity in the diet must help copepods survive in the absence of diatom blooms or between blooms. Other groups of zooplankton are also considered nonselective. Salps can ingest 100 percent of primary production per day without performing any selection (Dubischar and Bathmann 1997). Krill shows a nonselective feeding behavior, too.

Diatoms and Benthic Food Webs

In temperate and boreal coastal waters, the most prominent event in the annual flux of organic material to the benthos usually is the spring diatom bloom (Smetacek 1984; Wassmann 1991; Graf 1992; Olesen and Lundsgaard 1995). This bloom may reach the sea floor intact without being ingested by zooplankton (see Smetacek 1985 for review; Alldredge and Gotschalk 1989). Seasonally sedimented phytoplankton blooms are a major source of nutrients that are processed rapidly through the benthic system in open coastal areas (Graf et al. 1982; Gili and Coma 1998). Benthic suspension feeders are among the main contributors to the biomass of benthic communities of coastal and estuarine ecosystems worldwide (Grall and Chauvaud 2002); they benefit directly from pelagic primary production in the overlying water column (Graf et al. 1982; Christiansen and Kanneworf 1986) and are responsible for a large share of the energy flow from the pelagic to the benthic system (Figure 11.3), in addition to secondary production in benthic environments (Petersen and Black 1987; Gili and Coma 1998). What is the importance of diatoms in the diet of benthic suspension feeders?

To our knowledge, there is very little information on preference for diatoms over other phytoplankton for suspension feeders. Food uptake depends on water flow, on the size, shape, and load of the particles, and on their organic composition. The mechanisms involved have been studied in detail, and excellent reviews have been published by Jør-

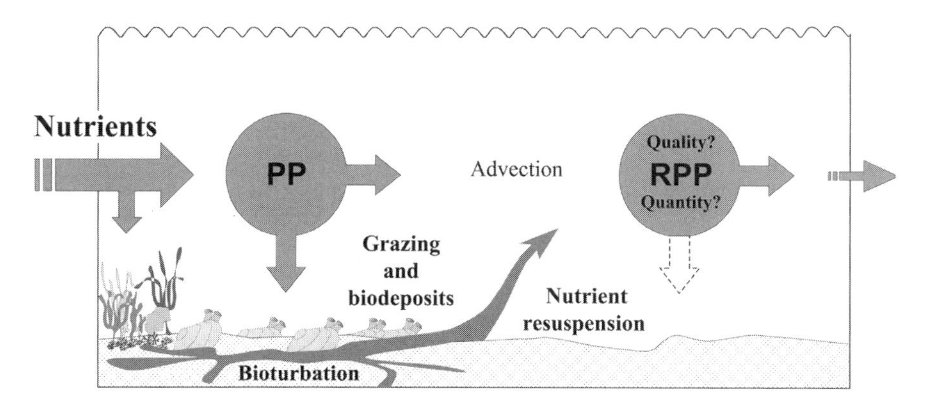

Figure 11.3. Schematic representation of nutrient fluxes and pelagic primary production dynamics in a suspension feeder–dominated ecosystem. PP = primary production, RPP = regenerated primary production. From Grall and Chauvaud (2002).

gensen (1990), Dame (1996), and Wildish and Kristmanson (1997). These mechanisms are complex because selection can occur at different stages of particle processing (Shumway et al. 1985): retention of particles on gills, preingestive selection on gills or labial palps, and differential absorption in the gut (i.e., postingestive selection). Aside from particle sorting by phytoplankton classes, there are controversies over the importance of size, shape, or site of selection (Cognie et al. 2003 and references therein). What seems well established is that suspension feeders are able to ingest a wide spectrum of particle sizes and to sort inorganic and organic particles, preferentially rejecting inorganic particles as pseudofeces (Newell et al. 1989).

As noted by Bougrier et al. (1997), studies on feeding behavior using flow cytometry to determine different populations of particles in mixed samples are scarce (Chrétiennot-Dinet et al. 1991; Cucci et al. 1985). Results do not demonstrate any preference for diatoms compared with other phytoplankters. Even more interestingly, especially compared with zooplankton diet, several studies have shown preferential consumption of a dinoflagellate relative to a diatom of similar size and preferential rejection of these diatoms in pseudofeces (Shumway et al. 1985; Bougrier et al. 1997). Bougrier et al. (1997) hypothesized that diatoms may appear as a mineral particle because of their frustule and be rejected when particles are being sorted at the labial palps level.

Care must be taken in generalizing such results. Indeed, grazing pressure on different species of algae depends on many parameters, linked not only to the characteristics of the prey, as discussed before, but also to the age of the animal studied, the time of day, and the composition of the mixture (Chrétiennot-Dinet et al. 1991). Even among diatoms of the same genus, some species have been shown to be preferentially retained and others rejected simply because of their relative biochemical composition (Robert et al. 1989).

Diatoms and Harmful Algal Blooms

Many coastal eutrophication problems are related to the development of nondiatom species, appearing in the same size range as diatoms but producing HABs. Although only about 20 of more than 1,000 species of dinoflagellates are toxic, their economic impact can be extensive (Steidinger and Baden 1984). Planktonic toxic dinoflagellates, their benthic resting stages, and their toxins can accumulate in filter-feeding animals such as clams and oysters and cause neurotoxic, paralytic, and diarrhetic shellfish poisoning; this can lead to illness or death of consumers such as birds, marine mammals, and humans. There are other ways in which blooms can be considered harmful. For example, some species, such as the prymnesiophyte *Phaeocystis pouchetii*, forms large foam banks on North Sea beaches (Lancelot et al. 1987); it also has deleterious effects on the marine food web, making a poor diet for copepods, causing low egg production and high mortality of nauplii (Tang et al. 2001).

There is abundant literature on HABs. Smayda (1990) revealed a global epidemic and related the increasing frequency and magnitude of these blooms to long-term decreasing Si:N and Si:P nutrient ratios. In a recent review, Anderson et al. (2002) detailed the various mechanisms leading to HABs, showing that nutrient balance is only one of the many factors controlling the appearance of such blooms. This important aspect is discussed further in Chapter 12. We concentrate on diatoms here and examine what we have learned recently that could modulate the previous view of diatoms as not being harmful.

Diatoms can also cause problems for grazers. The diatom *Pseudo-nitzschia australis* produces a neurotoxin, domoic acid, that killed planktivorous fish and sea lions along the central California coast (Scholln et al. 2000). Interestingly, the production of domoic acid has been linked to nutrient depletion, especially DSi (Pan et al. 1996a), P, and possibly Fe (Pan et al. 1996b). Some antiproliferative compounds have also been found in some diatoms (*Thalassiosira rotula*, *Skeletonema costatum*, and *Pseudo-nitzschia delicatissima*), leading to reduced egg hatching rates in copepods and inhibited cleavage of sea urchin embryos (Ban et al. 1997; Miralto et al. 1999). Such results are still debated in the literature (Irigoien et al. 2000). Another means by which diatoms may not be readily available for grazers is in the formation of aggregates. Under conditions that are not completely clear (Passow et al. 2001; Thornton 2002), diatoms may form large aggregates that sink rapidly to the bottom, allowing them to escape grazing by copepods, which may represent a life strategy to avoid being grazed (Smetacek 1985). However, this process may also cause problems for benthic suspension feeders. In the Bay of Brest, it has been shown that massive sedimentation of diatom blooms can cause growth anomalies in *Pecten maximus*, either through gill clogging or oxygen depletion caused by the degradation of the organic matter (Chauvaud et al. 2001; Lorrain et al. 2000).

Importance of Diatoms in Pelagic and Benthic Food Chains

From this brief survey of the literature it appears that for a grazer, a diatom is important because it represents a pool of energy with the appropriate size, but only very little preference for diatoms has been shown. If it is not the only good food, why are diatoms so important?

Although this might appear tautological, diatoms are essential simply because they are often there. They tend to dominate whenever conditions become optimal for phytoplankton growth (Guillard and Kilham 1978). These conditions are met in spring blooms (Hulburt 1990), coastal upwelling plumes (Rojas de Mendiola 1981), equatorial divergences (Dugdale and Wilkerson 1998), river plumes (Nelson and Dortch 1996; Ragueneau et al. 2002b), macrotidal coastal ecosystems (Ragueneau et al. 1994, 1996), ice edge blooms (Wilson et al. 1986; Tréguer et al. 1991), transient open ocean blooms triggered by wind-mixing events (Marra et al. 1990), decay of ocean eddies (Nelson et al. 1989; Brzezinski et al. 1998), and atmospheric dust inputs (Young et al. 1991). Most of these situations constitute hydrodynamic singularities that tend to favor large cells (Margalef 1978; Legendre and Le Fèvre 1989). Large phytoplankton cells (e.g., diatoms and dinoflagellates) generally have a low surface to volume ratio, which leads to a need for a nutrient-rich habitat, in contrast to smaller picophytoplankton (e.g., prochlorophytes and cyanobacteria), whose higher surface to volume ratio allows more efficient exploitation of low nutrient concentrations (Chisholm 1992). Therefore, diatoms dominate in a number of regimes that offer high-nutrient and turbulent conditions. That diatoms often outcompete other algae of similar size could be related to the existence of their frustule, acting as a protection against grazers (Hamm et al. 2003), or their higher division rates under similar environmental conditions (Smetacek 1999).

Thus, the combination of the success of diatoms at marine ergoclines (Legendre et al. 1986) and the very efficient export of C to higher trophic levels when the food chain is based on large cells probably explains why diatoms form the basis of the most productive ecosystems on the planet. In addition, there seem to be only few instances of exclusion of diatoms in a grazer diet.

Silica Cycle in Coastal Waters

Numerous factors control the contribution of diatoms to total primary production in coastal waters. These include hydrodynamics, nutrient concentrations, and grazing by pelagic and benthic animals. In this section we concentrate on nutrients only, particularly on DSi. Physical and direct biological controls and the importance of nutrient balance are discussed in Chapter 12. The primary question asked here is, "Where do diatoms get their DSi from?"

Coastal diatoms often are seen as being influenced only by river inputs. This is a gross oversimplification of the way diatoms satisfy their demand for DSi. The geographic position of coastal waters at the interface between land and ocean implies that diatoms can use Si coming from land, via rivers, groundwater, or the atmosphere, and from the open ocean, through advection of surface waters and upwelling. These will be called external inputs of DSi (Figure 11.4). Diatoms can also use DSi from recycling in the

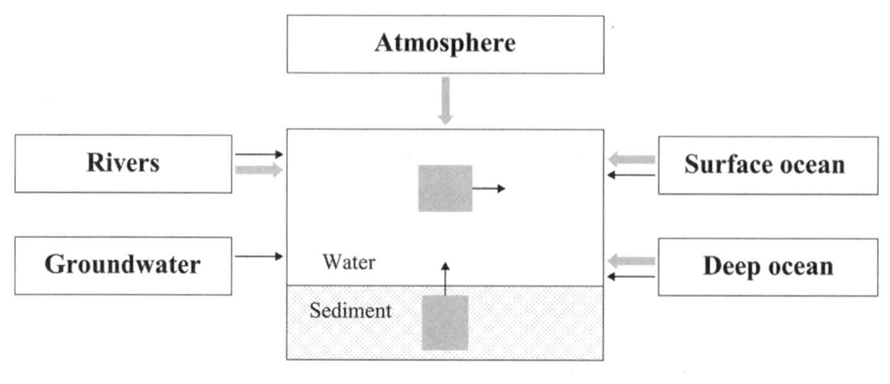

Figure 11.4. Schematic depiction of the potential sources of Si for coastal diatoms. Coastal diatoms use external sources of Si from rivers, groundwater, the atmosphere, and the ocean; these inputs take place either directly in the form of DSi *(black arrows)* or in particulate form, as BSi or LSi *(gray arrows)*, which then dissolve in coastal waters. Diatoms also use DSi coming from the recycling of BSi produced at that site, defined as internal recycling and including water column dissolution and recycling at the sediment–water interface.

water column, at the sediment–water interface and deeper in the sediments. These will be called internal recycling.

External Inputs

River Fluxes

Rivers are responsible for almost 80 percent of the Si entering the global ocean (Tréguer et al. 1995). River inputs have long been seen as transporting only DSi, a product of continental rock weathering (Meybeck 1982). This view has been challenged recently because it has been shown that rivers also carry Si in various forms of particulate matter that can play a significant role as a source of Si for coastal diatoms (Conley 1997).

LSi, consisting of quartz particles, aluminosilicates, and other minerals, has long been thought to react and dissolve slowly and linearly (McKyes et al. 1974). However, many studies have demonstrated that silicate mineral dissolution proceeds in two steps, with an initial dissolution rate, higher and nonlinear, at the beginning of the process (Huertas et al. 1999). LSi concentration in coastal waters can be substantial (Ragueneau and Tréguer 1994). However, it seems reasonable to assume that aged silicate minerals being delivered by rivers will have lost this reactive phase, so the impact of their dissolution on coastal diatoms can be ignored on the time scale of biological processes. The question of the influence of recently physically weathered silicate minerals as a potential source of DSi for coastal diatoms remains open.

BSi, being amorphous, dissolves five orders of magnitude faster than silicate minerals (Hurd 1983), so its dissolution can constitute an important, albeit poorly known source of Si for coastal diatoms. What is the importance of BSi in Si river inputs? Technical progress has been made in the last 25 years, allowing chemical methods to distinguish between LSi and BSi in suspended matter of rivers and coastal waters (DeMaster 1981; Ragueneau and Tréguer 1994; Ragueneau et al. 2005b and references therein). This allowed Conley (1997) to estimate that 16 percent of the gross riverine Si load is delivered to the world ocean as BSi (Figure 11.5). This estimate hides enormous spatial and temporal variability because the partitioning between DSi and BSi depends on many factors, especially those controlling bloom formation along the river (e.g., turbidity, water residence time).

During blooms, BSi may constitute one third of the gross delivery of Si, as in the Danube (Ragueneau et al. 2002b); it may even exceed DSi inputs, as in the Rhine (Admiraal et al. 1990; Figure 11.5). Most of this BSi was thought to be in the form of freshwater diatoms because it was easily digestible, a characteristic of diatoms. This is not surprising because diatoms have been observed in the suspended matter of many rivers (Anderson 1986; Admiraal et al. 1990; Ragueneau et al. 2002b). More recently, Conley (2002) suggested that phytoliths may well constitute an important fraction of the BSi found in sediments behind dams and may also constitute an important fraction of the BSi carried by rivers (see Chapter 3, this volume).

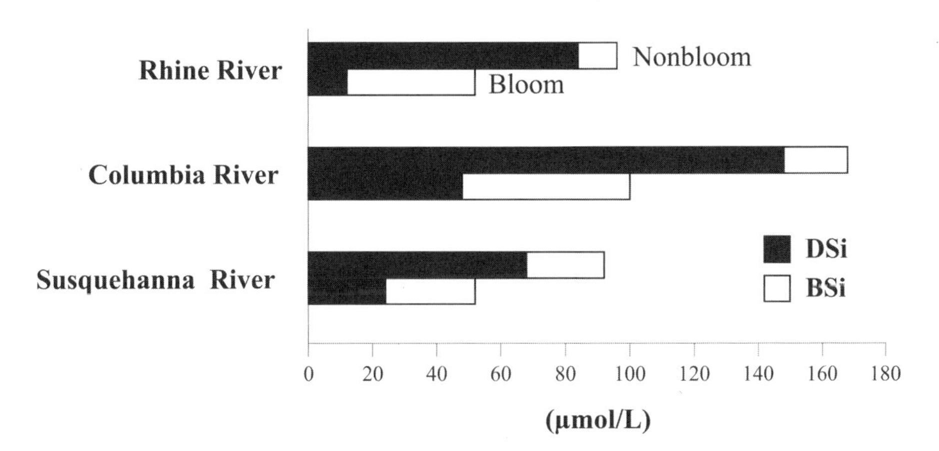

Figure 11.5. Concentrations of DSi and BSi for the Rhine (Admiraal et al. 1990), Columbia, and Susquehanna rivers during nonbloom periods and during periods when diatom blooms are present. The sum of DSi and BSi also decreases during blooms, suggesting sedimentation losses of BSi (Admiraal et al. 1990). Redrawn from Conley (1997).

Phytoliths and freshwater diatoms may constitute an important source of Si for coastal diatoms, provided they dissolve when diluted with coastal waters. What do we know about the fate of this material upon mixing with marine waters? Some fraction of Si (as the sum of DSi and BSi) is lost during transport in rivers (Admiraal et al. 1990), suggesting that sedimentation may be a sink for Si. In fact, freshwater diatoms and phytoliths can be buried in estuarine and marine sediments and constitute a source for reverse weathering reactions (Michalopoulos and Aller 1995). The fraction that is not sedimented can then either dissolve or be exported to the adjacent coastal, shelf, or ocean waters. Export of BSi to the open ocean has rarely been quantified (see DeMaster and Aller 2001 for an exception). Measurements of BSi dissolution in coastal waters have been reported (Brzezinski et al. 2003; Beucher et al. 2004), but none exist for freshwater diatoms and phytoliths. Such measurements are urgently needed to evaluate the importance of freshwater diatoms and phytoliths as a source of DSi for coastal diatoms.

Groundwater

Submarine groundwater discharge (SGD) is a potentially important but poorly quantified source of nutrients to the coastal ocean. The nutrient flux to coastal waters through SGD is generally determined as the product of the water flux and the nutrient concentrations in the groundwater. Both are often highly variable, both spatially and temporally, which makes them difficult to quantify (Burnett et al. 2001, 2003; Slomp and Van Cappellen 2004).

Most current estimates of the global flux of groundwater entering the ocean through SGD range from 0.01 percent to 10 percent of surface water runoff (Taniguchi et al. 2002). On a regional scale, however, SGD may account for a much larger proportion of the water flux. This is particularly the case in regions where coastal aquifers consist of permeable sand or limestone (Slomp and Van Cappellen 2004 and references therein).

Concentrations of N and P in SGD are controlled mostly by anthropogenic inputs, the water residence time, and the redox conditions in coastal aquifers and sediments; they can be orders of magnitude higher than in river water. This particularly holds true for N, which is generally removed from groundwater less efficiently than P (Slomp and Van Cappellen 2004). Sources of Si are largely natural and involve weathering of feldspars, micas, sedimentary BSi, and phytoliths (Conley 2002). The behavior of Si in groundwater is largely redox independent. Concentrations of Si in groundwater are generally on the same order of magnitude as those of river water (Table 11.1).

As a consequence of the high input of N relative to Si and limited removal relative to P, N:P and N:Si ratios in SGD are generally expected to exceed those in river water and be higher than the Redfield ratio. SGD of nutrients thus could contribute to a higher occurrence of HABs over diatom blooms and hence to a deterioration of water quality in the coastal ocean in the coming decades.

Table 11.1. DSi concentrations in coastal groundwater at 4 locations in the United States and in average river water.

Aquifer Type and Location	Si (μM)	Reference
Volcanic alluvium and calcareous rock (Hawaii)	175–621	Garrison et al. (2003)
Carbonate rock (Florida)	100	Corbett et al. (2002)
Granite, sandstone, and shale (California)	350–430	Oberdorfer et al. (1990)
Sand (Long Island, New York)	50–300	Montlucon and Sanudo-Wilhelmy (2001)
Average river water	150	Tréguer et al. (1995)

Atmospheric Inputs

Atmospheric deposition of macronutrients may modify the nutrient balance and play a major role in aquatic production by relieving nutrient limitation. There exists some literature on the role of atmospheric deposition of P, especially in lakes, where P is often limiting (Jassby et al. 1995), or during oligotrophic periods in coastal waters (Migon and Sandroni 1999). There are a number of studies regarding the importance of N deposition in coastal waters because of its possible role in eutrophication, hypoxia, and development of HABs (Paerl and Whitall 1999 for review). Literature concerning Si atmospheric deposition is far less abundant. It is discussed in detail by Tegen and Kohfeld (Chapter 7, this volume), mostly from an open ocean perspective. Such studies are of particular interest after the hypothesis of Harisson (2000), who suggested that glacial and interglacial changes in dust deposition may provide changes in DSi availability and induce switches from coccoliths to diatoms, with important implications for atmospheric CO_2 concentrations.

Closer to the coast, the importance of atmospheric deposition as a source of DSi for coastal diatoms must decrease because the diatom demand is larger and the other DSi inputs to sustain this demand overwhelm that of atmospheric deposition. Let us take an example from a site (Capo Carvallo, Corsica, north Mediterranean Sea) subject to Saharan dust deposition events (Losno 1989). There, the maximum flux of LSi to the surface waters has been estimated at 67.1 mmol Si/m^2/yr. The solubility of LSi is believed to range between 1 and 10 percent (Maring and Duce 1987). Even with a 20 percent solubility, the maximum daily flux of DSi from the atmosphere will be 0.04 mmol Si/m^2/d. This value, clearly representing a maximum for a site directly under the influence of the Sahara and an exaggerated solubility, is one fifth the minimum values of BSi production encountered in oligotrophic regions (Nelson et al. 1995).

Thus, from the limited information available on Si atmospheric deposition in coastal waters, it seems that such a source can be neglected in coastal budgets. More data will be welcome to challenge this view, especially data on the composition of dust particles and their dissolution kinetics. Dust particles may indeed contain recently eroded LSi particles, which may still contain a reactive surface layer; they may also be constituted of phytoliths (Folger et al. 1967; Romero et al. 2003), transported by the burning of tropical vegetation and wind erosion of soils (Abrantes 2003).

Oceanic Inputs

The high productivity of continental margins results from large inputs of nutrients, not only from the continents via rivers but also from the transfer of nutrient-rich deep water across the shelf break (Wollast 2002 and references therein). Obviously, its importance depends on the relative contribution of river and upwelling fluxes, both varying seasonally for a given system. Embayments with limited exchange with the ocean will be less affected than open shelves; eastern ocean boundaries will also be more affected than western ocean boundaries, where upwelling is hindered by stratification.

Ragueneau et al. (1994) estimated that no more than 5 percent of spring diatom production was sustained by oceanic inputs of DSi in a semienclosed ecosystem of northwestern Europe. However, the oceanic influence can be much larger, as reported from other ecosystems. Calvert (1966) estimated that the oceanic DSi input to the Gulf of California was two orders of magnitude higher than the riverine DSi input, and the most recent Si budget for the Amazon shelf (DeMaster and Aller 2001) suggested that advection and upwelling of offshore surface and subsurface waters represented about one third of the main external sources of DSi, and the other two thirds came from the Amazon (Figure 11.6). These examples are given simply to highlight the heterogeneity of coastal zones, especially with respect to the relative influence of external sources of nutrients.

Distinguishing land-derived from upwelling-derived DSi is essential in understanding whether higher productivity on continental margins during the late glacial maximum resulted from increased upwelling or increased supply of riverborne nutrients (Abrantes 2000; Peterson et al. 2000). Unfortunately, in the 200 nutrient budgets of the Land–Ocean Interactions in the Coastal Zone (LOICZ) project (data.ecology.su.se/MNODE/), Si is rarely included, and clearly this should be changed (Ragueneau 2004).

Internal Recycling

Productivity does not depend only on external inputs of nutrients. Recycling, both in the water column and at the sediment–water interface, may play a crucial role in sustaining production, and the relative rates of Si, N, and P remineralization may affect phytoplankton dynamics to a large extent, especially the shift from diatom to nondiatom species (Officer and Ryther 1980), as we shall see in Chapter 12. In this section, we briefly discuss the factors that control Si recycling, both in the water column and at the sediment–water interface.

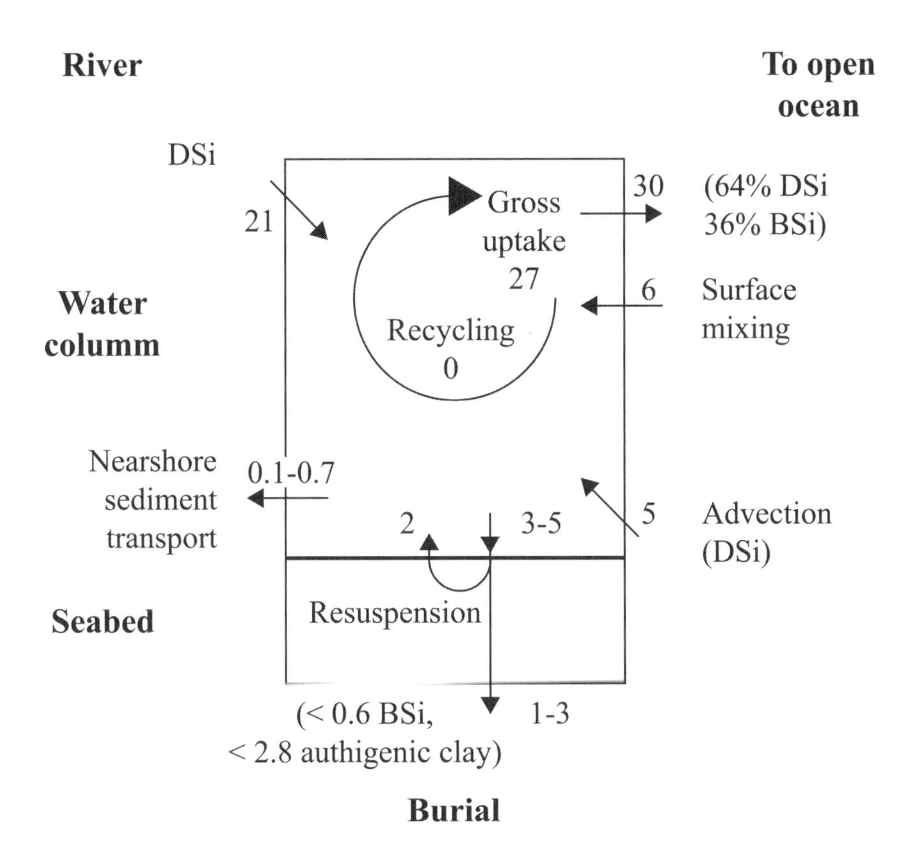

Figure 11.6. Biogeochemical Si fluxes on the Amazon shelf (modified after DeMaster and Aller 2001). Si fluxes are given in 10^8 mol Si/d. Illustration of the importance of oceanic inputs in total external DSi inputs, using one of the rare Si budgets in coastal waters published to date.

Dissolution of Biogenic Silica in Surface Waters

As often occurs with recycling, the direct measurement of BSi dissolution under in situ simulated conditions is very difficult; it involves the use of stable ^{29}Si and ^{30}Si isotopes (Nelson and Goering 1977a, 1977b; Corvaisier et al. 2005) and the measurement of their relative abundance compared to the most abundant ^{28}Si isotope, using mass spectrometry. To date, about 430 profiles of BSi production have been obtained (190 in coastal waters), particularly since the development of the radioactive ^{32}Si tracer (Tréguer et al. 1991); only 74 profiles have been obtained for dissolution, with only half of them concerning coastal waters (Beucher 2003). The dissolution to production (D:P) ratio in surface waters is very variable in coastal areas, with values as low as 0.02 in Monterey Bay (Brzezinski et al. 2003) and as high as 5.8 in the upwelling off northwest Africa

(Nelson and Goering 1977a). As observed for open ocean waters by Nelson et al. (1995), such variability cannot be explained simply in terms of latitude, temperature, or rate of production.

In fact, numerous factors affect the dissolution of biogenic silica in surface waters (reviews in Nelson et al. 1995; Ragueneau et al. 2000). These include physicochemical controls such as temperature (Lawson et al. 1978; Kamatani 1982), incorporation of trace elements such as Al (Van Bennekom et al. 1989), departure from equilibrium (Hurd 1972), or specific surface area (Hurd and Birdwhistell 1983). The influence of these factors has been reviewed recently by Van Cappellen et al. (2002). But biology also is essential for both dissolution and preservation mechanisms.

It is well known that dissolution is prevented until the organic matter coating opal surfaces has been removed (Kamatani 1982). The first evidence of a biological influence on BSi dissolution was provided by Patrick and Holding (1985) in lake studies. They demonstrated a clear increase in the solubilization of diatom frustules by natural bacterial populations, which produce hydrolytic enzymes that help to degrade the organic matter coating surrounding the frustule. Later, Bidle and Azam (1999) performed similar experiments and demonstrated bacteria-mediated increases in dissolution rates in marine waters.

More recently, it has been shown that dissolution rates of BSi can be reduced by a factor of two to three when diatoms are embedded in zooplankton fecal pellets (Schultes et al. in revision) and in aggregates (Moriceau et al. in press). These studies point to the importance of the way diatom blooms terminate. When diatoms are embedded in zooplankton fecal pellets or form aggregates, they sink very rapidly, with rates that can exceed that of single cells by two orders of magnitude (Fowler and Smaal 1972; Alldredge and Gotschalk 1989). Such fast sinking associated with reduced dissolution rates and very shallow coastal waters implies that diatoms probably spend very little time in the undersaturated surface waters of coastal areas. Thus, it is believed that most BSi dissolution in coastal areas takes place at the sediment–water interface and within the sediment column.

Dissolution of Biogenic Silica at the Sediment–Water Interface

Coastal waters, from estuaries to the shelf break, are shallow. This unique property induces strong interactions between pelagic and benthic compartments. The ecology and biogeochemistry of the sediment–water interface and upper sediment column depend strongly on pelagic production; in turn, pelagic production can be strongly affected by processes occurring on the seabed. The latter occurs directly, via filtering activities, and indirectly, via the recycling of nutrients caused by shallow depths and tidal mixing. Several excellent reviews demonstrate the importance of this interaction (Suess 1980; Heip et al. 1995; Jørgensen 1996; Marcus and Boero 1998; Soetaert et al. 2000; Cloern 2001; Grall and Chauvaud 2002; Middleburg and Soetaert 2003). In the next section we highlight the importance of the sediment–water interface as an essential source of DSi for pelagic diatoms. We then describe the processes that affect Si cycling at the sediment–water interface, with a major emphasis on three aspects as they modulate benthic fluxes on various time scales: reverse weathering reactions, BSi production by benthic diatoms, and biodeposition by invasive species.

Importance of DSi Benthic Fluxes

A number of methods are available to quantify Si recycling at the sediment–water interface, including direct measurements of DSi benthic fluxes by means of benthic chambers (Callender and Hammond 1982; D'Elia et al. 1983; Berelson et al. 1987) or core incubation (Ragueneau et al. 1994, 2002a). Fluxes can also be estimated indirectly, using the gradient of porewater DSi concentrations near the sediment–water interface. This approach has often been used in the deep sea (see McManus et al. 1995; Rickert 2000 for details of the calculations) and compared with direct measurements (e.g., Ragueneau et al. 2001). Such a comparison is very interesting because consistent estimates validate the technique used (Berelson et al. 1987), and differences between measured and calculated fluxes have been interpreted as increased dissolution of BSi at the sediment–water interface (Conley et al. 1988). In coastal waters, DSi fluxes are affected by processes such as bottom water currents and irrigation (Berner 1980; Boudreau 1996; Aller and Aller 1998). Therefore, the direct approaches are generally preferred, and the numbers given here have all been derived using these direct measurements.

Examples of DSi benthic fluxes encountered in various coastal environments are shown in Table 11.2. These fluxes typically range from negative values, when the flux is directed toward the sediment, to a few tens of mmol Si/m^2/d. This is typically the order of magnitude of BSi production in coastal waters (synthesis in Beucher 2003), except for coastal upwelling, where production rates can be much higher (e.g., Brzezinski and Philipps 1996).

The importance of DSi benthic fluxes as a source of coastal diatoms depends on the site, the degree of coupling between the pelagic and benthic ecosystems, and the season (DeMaster and Aller 2001; D'Elia et al. 1983). Clearly, benthic recycling plays a very important role for coastal diatoms, and this source cannot be overlooked in studies of the response of coastal ecosystems to any kind of perturbation (Chapter 12, this volume). We will now examine the mechanisms affecting DSi fluxes from the seabed.

Mechanisms Controlling DSi Benthic Fluxes

Several recent reviews have addressed the processes that control the thermodynamics and kinetics of BSi dissolution in sediments (see Ragueneau et al. 2000; Van Cappellen et al. 2002; Sarmiento and Gruber 2006). Here we discuss physical, chemical, and biological influences on the Si cycling in sediments that are peculiar to the coastal zone: the importance of sediment permeability, because permeable sediments represent some 70

Table 11.2. DSi benthic fluxes in various coastal ecosystems.

Site	DSi Flux (mmol Si/m^2/d)	Period	Reference
Lake Michigan	2.2–10.1	April–Aug (1983–1985)	Conley et al. (1988)
Bay of Brest	0.8–2.6	April–June (1992)	Ragueneau et al. (1994)
Bay of Brest	0.11–6.25	May–November (2000)	Ragueneau et al. (2002a)
San Nicolas Basin	0.9–1.3	August 1983–April 1985	Berelson et al. (1987)
San Pedro Basin	0.48–0.90	August 1983–April 1985	Berelson et al. (1987)
Potomac estuary	1–25	August 1979	Callender and Hammond (1982)
Chesapeake Bay	3.6–43.2	July 1980 and May 1981	D'Elia et al. (1983)
Skagerrak	0.55–3.97	March–September (1991–1994)	Hall et al. (1996)
Northwestern Black Sea	0.2–6.7	May 1997 and August 1995	Friedrich et al. (2002)
Long Island Sound	0.8–1.1	Winter	Aller and Benninger (1981)
Bering Sea, outer shelf	0.3–0.8	1979–1982	Banahan and Goering (1986)
Amazon shelf	0.13–1.25	August 1989–November 1991	DeMaster and Pope (1996)
Oosterschelde	7.2–112.8		Prins and Smaal (1994)
Pontevedra Ria	0.5–5.0	February–October 1998	Dale and Prego (2002)

percent of shelf sediments (Emery 1968); the importance of reverse weathering reactions, because of the importance of Al and detrital inputs to coastal sediments; the importance of benthic diatoms, because light often reaches the sediment–water interface; and the importance of biodeposition, because of the crucial role played by benthic suspension feeders in many ecosystems.

Sediment Permeability

Sandy sediments are abundant on continental shelves (de Haas et al. 2002). Sands are permeable, and because of bedform topography, high current velocities, and wave activities, exchange of water between sediments and the water column is greatly enhanced (Precht and Huettel 2003 and references therein). This favors the transfer of particles from the benthic boundary layer to sediments and increases influxes of oxidants (Middelburg and Soetaert 2003). This in turn induces high mineralization rates in such permeable sediments, as seen from nutrient profiles and fluxes in various shelf regions (Marinelli et al. 1998; Jahnke et al. 2000; Nedwell et al. 1993). Shum and Sundby (1996) suggest that increased organic matter turnover rates in permeable sediments may be linked to enhanced BSi dissolution in such sediments. Ehrenhauss and Huettel

(2004) confirmed recently that BSi and organic matter are rapidly degraded in permeable coastal sands. Several mechanisms may be involved. The rapid advective solute exchange reduces the accumulation of regenerated nutrients in porewater (Ehrenhauss et al. 2004), which may favor BSi dissolution from a thermodynamic point of view. The macrofaunal activity may help in flushing DSi from sediments (Ehrenhauss et al. 2004), although uptake of DSi by benthic diatoms may modulate DSi efflux from the sediment (Marinelli et al. 1998). Also, desorptive fluxes may be stimulated by the large and rapid variations in porewater concentrations in such permeable sediments, leading to enhanced DSi fluxes (Gehlen and van Raaphorst 2002).

Reverse Weathering Reactions

Geochemical reactions occurring in the sediment can also modulate the intensity of DSi benthic fluxes because some of the DSi released from diatom frustules does not diffuse out of the sediment but rather enters reverse weathering reactions. Recently, the transformation of diatom BSi to authigenic clay minerals in continental margin sediments was demonstrated (cf. Michalopoulos and Aller 1995, 2004; Michalopoulos et al. 2000) based on porewater stoichiometry, scanning electron microscope studies (Figure 11.7), and acid–base selective leaches in order to resolve whether clay minerals originate from BSi or terrestrial sources.

The BSi content of continental margin sediments is very low (about 1 wt%) and is attributed mostly to diatoms, with minor contributions from silicoflagellate skeletal material (mostly dissolved before burial), radiolarians (rare), and siliceous sponge spicules (limited to low accumulation rates). The contribution of phytoliths as a source of DSi for reverse weathering reactions is unknown. Scanning electron microscope studies confirm the transformation of diatom BSi to authigenic clays but not the extent. Clay mineral formation on the Amazon shelf (0.17 Tmol Si/yr or less) corresponds to 20 percent of the riverine Si discharge, but similar estimates must be obtained for other continental margins and on a global scale.

Biodeposition

Biodeposition in beds of suspension-feeding bivalves results from active filter feeding by a bivalve, which leads to nondigested material being excreted to the sediment surface as feces and pseudofeces (Norkko et al. 2001). Therefore, bivalves strongly affect physical, chemical, and biological properties near the sediment–water interface (review in Graf and Rosenberg 1997). In particular, these processes typically result in local deposition rates that exceed that of passive physical sedimentation (Dame 1993; Dobson and Mackie 1998) and create an enrichment of sediments in C and N (Kautsky and Evans 1987). Beyond their effect on sediment properties, bivalve mollusks strongly influence the cycling of several biogenic elements such as carbon (Doering et al. 1987), nitrogen (Dame et al. 1991), phosphorus (Asmus et al. 1995), sulfur (Hansen et al. 1996), and inorganic carbon (Chauvaud et al. 2003). Because of the intensity of benthic–pelagic

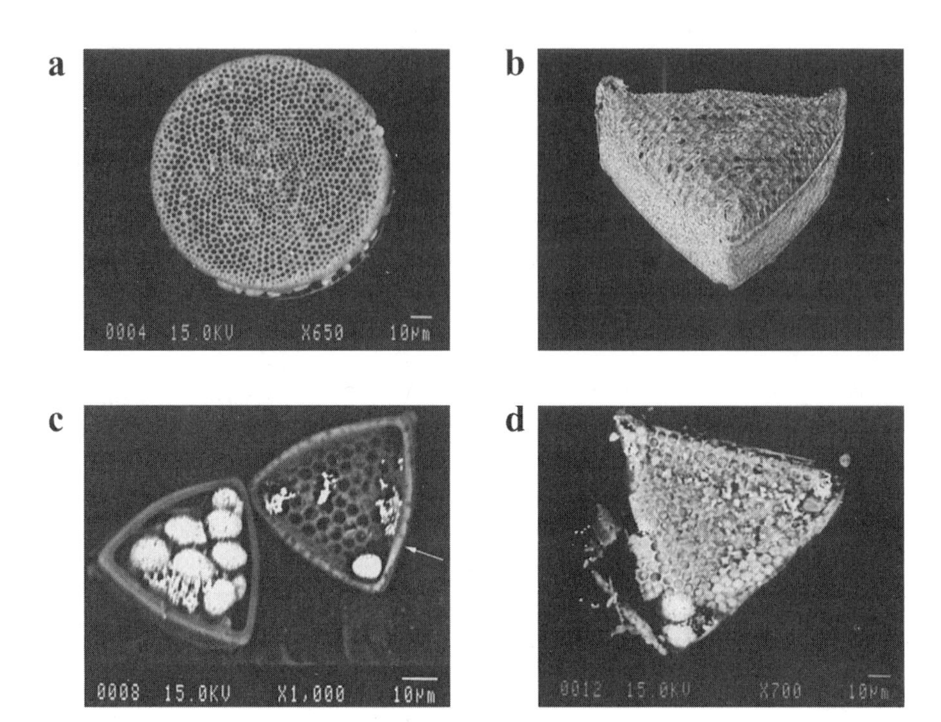

Figure 11.7. Representative scanning electron microscope images illustrating the range of preservation modes of distinct diatom cells (from Michalopoulos and Aller 2004). *(a)* Discoid diatom cell from station OST-2 (393–403 cm). Parts of the frustule aureoles are covered by aluminosilicate coatings. *(b)* Triagonal diatom, *Triceratium favus*, with nearly continuous authigenic aluminosilicate and metal-rich coating. *(c)* Valves of triagonal diatom opened to reveal internal framboidal pyrite fill. Parts of the frustule exhibit enrichments in K, Al, and Fe *(arrow)*, whereas the remainder is unaltered. *(d)* Diatom in advanced stages of alteration and initial fragmentation. The microarchitecture is recognizable but extensively altered. Aureoles are filled with aluminosilicate precipitate, and the cell lumen is occupied by framboidal pyrite (from OST-2, 425–435 cm).

coupling in coastal waters, they play an essential role in the functioning of coastal ecosystems in general (Alpine and Cloern 1992; Dame 1996; Wildish and Kristmanson 1997).

Diatoms are taken up by benthic suspension feeders in the process of feeding, with regeneration of DSi from the biodeposits in the sediments (Asmus et al. 1990). Very few investigations of bivalves and bivalve beds as sources and sinks of silicon exist in the literature (Asmus 1986; Doering et al. 1987; Dame et al. 1991; Prins and Smaal 1994). Recently, Chauvaud et al. (2000) proposed that benthic suspension feeders also affect the Si cycle, with potentially important ecological and biogeochemical implications.

Ragueneau et al. (2002b) explored the ecological implications of the proliferation of a benthic suspension feeder, *Crepidula fornicata*, in the Bay of Brest ecosystem. *C. fornicata* exerts primary control on DSi benthic fluxes (Figure 11.8a), with important implications for summer phytoplankton dynamics in this ecosystem (see also Ragueneau et

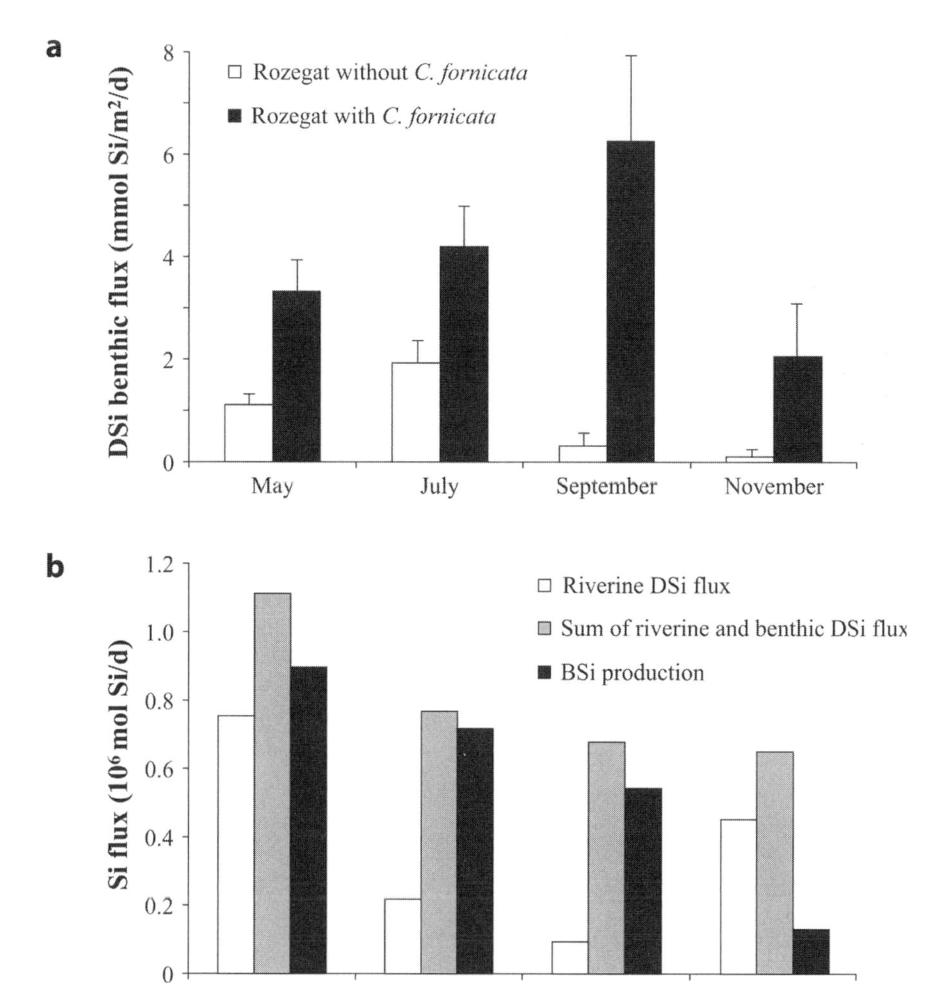

Figure 11.8. *(a)* Synthesis of DSi benthic fluxes measured at 2 contrasting sites during the productive period in the Bay of Brest. *Black bars*: site Rozegat with high densities of *Crepidula fornicata* (1,243 individuals/m^2, Thouzeau et al. 2000); *white bars*: site Rozegat without *C. fornicata* (<30 individuals/m^2, Thouzeau et al. 2000). *(b)* Seasonal budgets of DSi fluxes in the Bay of Brest (Ragueneau et al. 2002a). This figure highlights the importance of benthic fluxes in sustaining the diatom demand for DSi in summer in such ecosystems.

al. 2004). Biogeochemical implications are explored by Ragueneau et al. (2005a) and described in Chapter 12 as one example of anthropogenic perturbation (proliferation of invasive species) affecting the Si cycle in coastal waters.

Role of Microphytobenthos

Distinct benthic microalgal assemblages, consisting primarily of diatoms (McIntyre et al. 1996; Wulff et al. 1997; Facca et al. 2002), develop in most coastal areas where light reaches the sea floor (McIntyre et al. 1996) and in intertidal flats. They have been shown to act as a major control on nutrient fluxes at the sediment–water interface in myriad ways (Sundbäck and Granéli 1988; Sundbäck 1991; Epping 1996; Granéli and Sundbäck 1986; Rizzo 1990).

The combination of the concentrated microphytobenthos (MPBs) present at the sediment surface and the extra polymeric substances (Staats et al. 1999; Smith and Underwood 2000) they produce results in the formation of a biofilm, and the direct uptake of DSi by the MPBs either from the interstitial waters of the sediment or from the overlying pelagic zone can influence the release of nutrients from the underlying sediment (Srithongouthai et al. 2003; Sigmon and Cahoon 1997).

Diel variations of nutrient fluxes at the sediment–water interface in subtidal areas where MPBs are present illustrate that during periods of illumination significantly lower fluxes of DSi have emanated from the sediment than in darkened sediments (Sundbäck 1991; Sigmon et Cahoon 1997; Epping 1996). Explanations have been proposed in terms of DSi uptake being higher during the day (Syrett 1981), although DSi uptake has been recently shown to be coupled more to the cell cycle than to the day–night cycle (Martin-Jézéquel et al. 2000; Claquin et al. 2002) or, in terms of MPB active migration, shown to depend on the day–night cycle on intertidal mudflats (Blanchard et al. 1998; Serodio et al. 1997).

Benthic diatoms are highly silicified (Conley et al. 1994; Sigmon and Cahoon 1997), presumably because they live in a DSi-rich environment and most often in a low-light regime. The rate of primary production by benthic diatoms can exceed that of the pelagic production (Guarini et al. 2002). Because of this importance of benthic primary production and because benthic diatoms take up nutrients at a ratio that probably leaves less DSi behind, relative to dissolved inorganic nitrogen and phosphorus, benthic primary production has the potential to modify drastically the nutrient balance of the overlying waters. This impact must be evaluated at the scale of ecosystems, an exercise that has not been done so far for the Si cycle.

Conclusion

The importance of diatoms in coastal food webs, both pelagic and benthic, was discussed. Diatoms often are seen as the best food for grazers. From the evidence we have gathered, diatoms are very important in carbon transfer to higher trophic levels because of their relative size and because they are successful under conditions of high turbulence and high nutrients, not because they constitute a better diet for grazers. Given the importance of diatoms, it is essential to better understand the relative importance of the sources from which they get their nutrients.

There is a clear lack of DSi budgets in coastal waters with which to evaluate the relative importance of external and internal sources of DSi for diatom production, which varies seasonally and spatially. The limitation comes from the fact that too few laboratories measure the rate of DSi uptake directly by means of Si stable and radioactive isotopes, and even fewer measure water column recycling directly. In budgets built using the LOICZ approach (Gordon et al. 1996), Si has received much less attention than N and P, and this should be changed. Nevertheless, we have seen that rivers clearly play an important role, bringing Si in the form of dissolved but also particulate matter, both lithogenic and biogenic (freshwater diatoms, phytoliths). Among other external sources of DSi, atmospheric inputs seem to play a minor role, but data are scarce. A similar conclusion can be drawn for groundwater inputs, especially at global scale. The importance of groundwater at specific locations must be better assessed because it may modify the nutrient balance on a local scale. It is suggested that oceanic inputs may well constitute the most important source of DSi to sustain diatom productivity, especially in continental shelves subject to coastal upwelling.

When external sources decrease, internal recycling becomes very important in sustaining diatom production. More data are needed on recycling in the water column to better quantify the importance of this flux. More data exist concerning recycling at the sediment–water interface. These data show that it can sustain a very significant part of diatom production. Mechanisms that control the fate of BSi at the sediment–water interface have been described, with a particular emphasis on those specific to coastal sediments, such as their permeability, the importance of biodeposition, reverse weathering reactions, and benthic production. These mechanisms are complex, and much work is still needed before Si fluxes can be properly incorporated in models describing benthic–pelagic coupling (Soetaert et al. 2000), which plays an essential role in coastal ecosystem functioning.

Acknowledgments

This work is part of the Natural and Anthropogenic Modifications of the Si Cycle Along the Land–Ocean Continuum: Worldwide Ecological, Biogeochemical and Socio-economical Consequences (Si-WEBS) Research Training Network, funded by the European Union (HPRN-CT-2002-00218). We are very grateful to V. Ittekkot, C. Humborg, and T. Jennerjahn for inviting us to write this review chapter and organizing the fall 2003 meeting in Bremen.

Literature Cited

Abrantes, F. 2000. 200,000 yr diatom records from Atlantic upwelling sites reveal maximum productivity during LGM and a shift in phytoplankton community structure at 185,000 yr. *Earth and Planetary Science Letters* 179(1):7–16.

Abrantes, F. 2003. A 340,000 year continental climate record from tropical Africa: News from opal phytoliths from the equatorial Atlantic. *Earth and Planetary Science Letters* 209(1–2):165–179.

Admiraal, W., P. Breugem, D. M. Jacobs, and E. D. de Ruyter van Steveninck. 1990. Fixation of dissolved silicate in the lower river Rhine during diatom blooms. *Biogeochemistry* 9:175–185.

Alldredge, A. L., and C. C. Gotschalk. 1989. Direct observations of the mass flocculation of diatom blooms: Characteristics, settling velocities and formation of diatom aggregates. *Deep-Sea Research I* 36(2):159–171.

Aller, R. C., and J. Y. Aller. 1998. The effect of biogenic irrigation intensity and solute exchange on diagenetic reaction rates in marine sediments. *Journal of Marine Research* 56:905–936.

Aller, R. C., and L. K. Benninger. 1981. Spatial and temporal patterns of dissolved ammonium, manganese and silica fluxes from bottom sediments of Long Island Sound, U.S.A. *Journal of Marine Research* 39:295–314.

Alpine, A. E., and J. E. Cloern. 1992. Trophic interactions and direct physical effects control phytoplankton biomass and production in an estuary. *Limnology and Oceanography* 37:946–955.

Anderson, D. M., P. M. Glibert, and J. M. Burkholder. 2002. Harmful algal blooms and eutrophication: Nutrient sources, composition and consequences. *Estuaries* 25(4b):704–726.

Anderson, G. F. 1986. Silica, diatoms and a freshwater productivity maximum in Atlantic coastal plain estuaries, Chesapeake Bay. *Estuarine, Coastal and Shelf Science* 22:183–197.

Archer, D., A. Winguth, D. Lea, and N. Mahowald. 2000. What caused the glacial/interglacial atmospheric pCO_2 cycles? *Reviews of Geophysics* 38(2):159–189.

Asmus, H., R. M. Asmus, and K. Reise. 1990. Exchange processes in an intertidal mussel bed: A sylt-flume study in the Wadden Sea. *Berichte Biologische Anstalt Helgoland* 6:1–79.

Asmus, H., R. M. Asmus, and G. F. Zubillaga. 1995. Do mussel beds intensify the phosphorus exchange between sediment and tidal waters? *Ophelia* 41:37–55.

Asmus, R. 1986. Nutrient flux in short-term enclosures of intertidal sand communities. *Ophelia* 26:1–18.

Ban, S., C. Burns, J. Castel, Y. Chaudron, E. Christou, R. Escribano, S. F. Umani, S. Gasparini, F. G. Ruiz, M. Hoffmeyer, A. Ianora, H. K. Kang, M. Laabir, A. Lacoste, A. Miralto, X. Ning, S. Poulet, V. Rodriguez, J. Runge, J. Shi, M. Starr, S. Uye, and Y. Wang. 1997. The paradox of diatom–copepod interactions. *Marine Ecology Progress Series* 157:287–293.

Banahan, S., and J. J. Goering. 1986. The production of biogenic silica and its accumulation on the southeastern Bering Sea shelf. *Continental Shelf Research* 5(1–2):199–213.

Berelson, W. M., D. E. Hammond, and K. S. Johnson. 1987. Benthic fluxes and the cycling of biogenic silica and carbon in two southern California borderlands. *Geochimica et Cosmochimica Acta* 51:1345–1363.

Berner, R. A. 1980. *Early diagenesis, a theoretical approach.* Princeton Series in Geochemistry, edited by H. D. Holland. Princeton, NJ: Princeton University Press.

Beucher, C. 2003. *Production et dissolution de la silice biogène dans les écosystèmes marins: Mesure par spectrométrie de masse.* Thèse de doctorat, Université de Bretagne Occidentale, Brest.

Beucher, C., P. Tréguer, R. Corvaisier, A.-M. Hapette, and M. Elskens. 2004. Production and dissolution of biosilica, and changing microphytoplankton dominance in the Bay of Brest (France). *Marine Ecology Progress Series* 267:57–69.

Bidle, K. D., and F. Azam. 1999. Accelerated dissolution of diatom silica by marine bacterial assemblages. *Nature* 397:508–512.

Billen, G., C. Lancelot, and M. Meybeck. 1991. N, P and Si retention along the aquatic continuum from land to ocean. Pp. 19–44 in *Ocean margin processes in global change*, edited by R. F. C. Mantoura, J.-M. Martin, and R. Wollast. Chichester, UK: Wiley.

Blanchard, G. F., J.-M. Guarini, C. Bacher, and V. Huer. 1998. Control of the short-term dynamics of intertidal microphytobenthos by the exondation–submersion cycle. Académie des Sciences. Sciences de la vie. *Ecology* 321:501–508.

Boudreau, B. P. 1996. *Diagenetic models and their implementation.* New York: Springer.

Bougrier, S., A. J. S. Hawkins, and M. Héral. 1997. Preingestive selection of different microalgal mixtures in *Crassostrea gigas* and *Mytilus edulis*, analysed by flow cytometry? *Aquaculture* 150:123–134.

Brzezinski, M. A., J. Jones, K. Bidle, and F. Azam. 2003. The balance between biogenic silica production and silica dissolution in the sea: Insights from Monterey Bay, California, applied to the global data set. *Limnology and Oceanography* 48(5):1846–1854.

Brzezinski, M. A., and D. R. Phillips. 1996. Silica production rates in the Monterey Bay, California, upwelling system. Pp. 32–36 in *OPALEO: On the use of opal as a paleo-productivity proxy*, edited by O. Ragueneau, A. Leynaert, and P. Tréguer. Minutes of the first workshop. Brest, France.

Brzezinski, M. A., T. A. Villareal, and F. Lipschultz. 1998. Silica production and the contribution of diatoms to new and primary production in the central North Pacific. *Marine Ecology Progress Series* 167:89–104.

Buesseler, K. O. 1998. The decoupling of production and particulate export in the surface ocean. *Global and Biogeochemical Cycles* 12:297–310.

Burnett, W. C., H. Bokuniewicz, M. Huettel, W. S. Moore, and M. Taniguchi. 2003. Groundwater and porewater inputs to the coastal zone. *Biogeochemistry* 66:3–33.

Burnett, W. C., M. Taniguchi, and J. Oberdorfer. 2001. Measurement and significance of the direct discharge of groundwater into the coastal zone. *Journal of Sea Research* 46(2):109–116.

Callender, E., and D. E. Hammond. 1982. Nutrient exchange across the sediment–water interface in the Potomac river estuary. *Estuarine, Coastal and Shelf Science* 15:395–413.

Calvert, S. E. 1966. Accumulation of diatomaceous silica in the sediments of the Gulf of California. *Geological Society of America Bulletin* 77:569–596.

Chauvaud, L., A. Donval, G. Thouzeau, Y.-M. Paulet, and E. Nézan. 2001. Variations in food intake of *Pecten maximus* (L.) from the Bay of Brest (France): Influence of environmental factors and phytoplankton species composition. *Comptes Rendus de l'Académie des Sciences, Paris, Sciences de la Vie/Life Sciences* 324:1–13.

Chauvaud, L., F. Jean, O. Ragueneau, and G. Thouzeau. 2000. Long-term variation of the Bay of Brest ecosystem: Pelagic–benthic coupling revisited. *Marine Ecology Progress Series* 200:35–48.

Chauvaud, L., J. K. Thompson, J. E. Cloern, and G. Thouzeau. 2003. Clams as CO_2 generators: The *Potamocorbula amurensis* example in San Francisco Bay. *Limnology and Oceanography* 48(6):2086–2092.

Chisholm, S. W. 1992. Phytoplankton size. Pp. 213–238 in *Primary productivity and biogeochemical cycles in the sea*, edited by P. G. Falkowski and A. D. Woodhead. New York: Plenum.

Chrétiennot-Dinet, M. J., D. Vaulot, R. Galois, A. M. Spano, and R. Robert. 1991. Analysis of larval oyster grazing by flow cytometry. *Journal of Shellfish Research* 10:453–457.

Christiansen, H., and E. Kanneworf. 1986. Sedimentation of phytoplankton during a spring bloom in the Oresund. *Ophelia* 26:109–122.

Claquin, P., V. Martin-Jézèquel, J. C. Kromkamp, M. J. W. Veldhuis, and G. W. Kraay. 2002. Uncoupling of silica compared with carbon and nitrogen metabolisms and the role of the cell cycle in continuous cultures of *Thalassiosira pseudonana* Bacillariophyceae under light, nitrogen and phosphorus control. *Journal of Phycology* 38:922–930.

Cloern, J. E. 2001. Our evolving conceptual model of the coastal eutrophication problem. *Marine Ecology Progress Series* 210:223–253.

Cognie, B., L. Barillé, G. Massé, and P. G. Benninger. 2003. Selection and processing of large suspended algae in the oyster *Crassostrea gigas. Marine Ecology Progress Series* 250:145–152.

Conley, D. J. 1997. Riverine contribution of biogenic silica to the oceanic silica budget. *Limnology and Oceanography* 42:774–777.

Conley, D. J. 2002. Terrestrial ecosystems and the global biogeochemical silica cycle. *Global Biogeochemical Cycles* 16(4):1121.

Conley, D. J., M. A. Quigley, and C. L. Schelske. 1988. Silica and phosphorus flux from sediments: Importance of internal recycling in Lake Michigan. *Canadian Journal of Fisheries and Aquatic Sciences* 45:1030–1035.

Conley, D. J., P. V. Zimba, and E. Theriot. 1994. Silica content of freshwater and marine benthic diatoms. Pp. 95–101 in *Proceedings of the 11th Diatom Symposium*, San Francisco, CA.

Corbett, D. R., K. Dillon, W. Burnett, and G. Schaefer. 2002. The spatial variability of nitrogen and phosphorous concentration in a sand aquifer influenced by onsite sewage treatment and disposal systems: A case study on St. George Island, Florida. *Environmental Pollution* 117:337–345.

Corvaisier, R., P. Tréguer, C. Beucher, and M. Elskens. 2005. Determination of the rate of production and dissolution of biosilica in marine waters by thermal ionisation mass spectrometry. *Analytica Chimica Acta* 534:149–155.

Cucci, T. L., S. E. Shumway, R. C. Newell, and C. M. Yentsch. 1985. A preliminary study of the effects of *Gonyaulax tamarensis* on feeding in bivalve molluscs. Pp. 395–400 in *Toxic dinoflagellates*, edited by D. M. Anderson, A. W. White, and D. G. Baden. New York: Elsevier.

Cushing, D. H. 1989. A difference in structure between ecosystems in strongly stratified waters and in those that are only weakly stratified. *Journal of Plankton Research* 11:1–13.

Dale, A. W., and R. Prego. 2002. Physico-biogeochemical controls on benthic–pelagic coupling of nutrient fluxes and recycling in a coastal upwelling system. *Marine Ecology Progress Series* 235:15–28.

Dame, R. F. 1993. Bivalve filter feeders in estuarine and coastal processes. *NATO ASI Series G: Ecological Sciences* 33. Heidelberg: Springer-Verlag.

Dame, R. F. 1996. *Ecology of marine bivalves: An ecosystem approach.* Boca Raton, FL: CRC Press.

Dame, R. F., N. Dankers, T. Prins, H. Jongsma, and A. Smaal. 1991. The influence of mussel beds on nutrients in the western Wadden Sea and eastern Scheldt estuaries. *Estuaries* 14:130–138.

De Haas, H., T. C. E. van Weering, and H. de Stiger. 2002. Organic carbon in shelf seas: Sinks or sources, processes and products. *Continental Shelf Research* 22(5):691–717.

D'Elia, C. F., D. M. Nelson, and W. R. Boynton. 1983. Chesapeake Bay nutrient and plankton dynamics III. The annual cycle of dissolved silicon. *Geochimica et Cosmochimica Acta* 47:1945–1955.

DeMaster, D. J. 1981. The supply and accumulation of silica in the marine environment. *Geochimica et Cosmochimica Acta* 45:1715–1732.

DeMaster, D. J. 2002. The accumulation and cycling of biogenic silica in the Southern Ocean: Revisiting the marine silica budget. *Deep-Sea Research II* 49(16):3155–3167.

DeMaster, D. J., and R. C. Aller. 2001. Biogeochemical processes on the Amazon shelf: Changes in dissolved and particulate fluxes during river/ocean mixing. Pp. 328–357 in *The biogeochemistry of the Amazon Basin*, edited by M. E. McClain, R. L. Victoria, and J. E. Richey. New York: Oxford University Press.

DeMaster, D. J., and R. H. Pope. 1996. Nutrient dynamics in Amazon shelf waters: Results from AMASSEDS. *Continental Shelf Research* 16:263–289.

Dobson, E. P., and G. L. Mackie. 1998. Increased deposition of organic matter, polychlorinated biphenyls, and cadmium by zebra mussels (*Dreissena polymorpha*) in western Lake Erie. *Canadian Journal of Fisheries and Aquatic Sciences* 55:1131–1139.

Doering, P. H., J. R. Kelly, C. A. Oviatt, and T. Sowers. 1987. Effect of the hard clam *Mercenaria mercenaria* on benthic fluxes of inorganic nutrients and gases. *Marine Biology* 94:377–383.

Dubischar, C. D., and U. V. Bathmann. 1997. Grazing impact of copepods and salps on phytoplankton in the Atlantic sector of the Southern Ocean. *Deep-Sea Research II* 44(1–2):415–433.

Dugdale, R. C., and F. P. Wilkerson. 1998. Understanding the eastern equatorial Pacific as a continuous new production system regulating on silicate. *Nature* 391:270–273.

Dugdale, R. C., F. P. Wilkerson, and H. J. Minas. 1995. The role of a silicate pump in driving new production. *Deep-Sea Research* 42:697–719.

Egge, J. K., and D. L. Aksnes. 1992. Silicate as a regulating nutrient in phytoplankton competition. *Marine Ecology Progress Series* 83:281–289.

Ehrenhauss, S., and M. Huettel. 2004. Advective transport and decomposition of chain-forming planktonic diatoms in permeable sediments. *Journal of Sea Research* 52:179–197.

Ehrenhauss, S., U. Witte, F. Janssen, and M. Huettel. 2004. Decomposition of diatoms and nutrient dynamics in permeable North Sea sediments. *Continental Shelf Research* 24:721–737.

Emery, K. O. 1968. Relict sediments on continental shelves of the world. *American Association of Petroleum Geologists Bulletin* 52:445–464.

Epping, E. H. G. 1996. *Benthic phototrophic communities and the sediment–water exchange of oxygen, Mn-11, Fe-11 and silica acid.* Ph.D. thesis, University of Groningen.

Facca, C., A. Sfriso, and G. Socal. 2002. Temporal and spatial distribution of diatoms in the surface sediments of the Venice Lagoon. *Botanica Marina* 452:170–183.

Folger, D. W., L. H. Burckle, and B. C. Heezen. 1967. Opal phytoliths in a North Atlantic dust fall. *Science* 155:1243–1244.

Fowler, S. W., and L. F. Smaal. 1972. Sinking rates of euphausiid fecal pellets. *Limnology and Oceanography* 17:293–296.

Friedrich, J., C. Dinkel, G. Friedl, N. Pimenov, J. Wijsman, M. T. Gomoiu, A. Cociasu, L. Popa, and B. Wehrli. 2002. Benthic nutrient cycling and diagenetic pathways in the north-western Black Sea. *Estuarine, Coastal and Shelf Science* 54:369–383.

Frost, B. W. 1972. Effects of size and concentration of food particles on the feeding behaviour of the marine planktonic copepod *Calanus pacificus. Limnology and Oceanography* 17:805–815.

Garrison, G. H., C. R. Glen, and C. M. McMurtry. 2003. Measurement of submarine groundwater discharge in Kahana Bay, O'ahu, Hawai'i. *Limnology and Oceanography* 48(2):920–928.

Gehlen, M., and W. van Raaphorst. 2002. The role of adsorption–desorption surface reactions in controlling interstitial $Si(OH)_4$ concentrations and enhancing $Si(OH)_4$ turn-over in shallow shelf seas. *Continental Shelf Research* 22:1529–1547.

Gili, J. M., and R. Coma. 1998. Benthic suspension feeders: Their paramount role in littoral marine food webs. *Trends in Ecology and Evolution* 13(8):316–321.

Goldman, J. C. 1993. Potential role of large oceanic diatoms in new primary production. *Deep-Sea Research* 40:159–186.

Gordon, D. C. Jr., P. R. Boudreau, K. H. Mann, J.-E. Ong, W. L. Silvert, S. V. Smith, G. Wattayakorn, F. Wulff, and T. Yanagi. 1996. LOICZ biogeochemical modelling guidelines. *LOICZ Reports and Studies* 5. Texel, The Netherlands: LOICZ.

Graf, G. 1992. Benthic–pelagic coupling: A benthic view. *Oceanography and Marine Biology Annual Review* 30:149–190.

Graf, G., W. Bengtsson, U. Diesner, and H. Theede. 1982. Benthic response to sedimentation of a spring phytoplankton bloom: Process and budget. *Marine Biology* 67:201–208.

Graf, G., and R. Rosenberg. 1997. Bioresuspension and biodeposition: A review. *Journal of Marine Systems* 11:269–278.

Grall, J., and L. Chauvaud. 2002. Marine eutrophication and benthos: The need for new approaches and concepts. *Global Change Biology* 8:813–830.

Granéli, W., and K. Sundbäck. 1986. Can microphytobenthic photosynthesis influence below-halocline oxygen conditions in the Kattegat? *Ophelia* 26:195–206.

Guarini, J.-M., J.-E. Cloern, J. E. Edmunds, and P. Gros. 2002. Microphytobenthic potential productivity estimated in three tidal embayments of the San Francisco Bay: A comparative study. *Estuaries* 25(3):409–417.

Guillard, R. R. L., and P. Kilham. 1978. The ecology of marine planktonic diatoms. Pp. 372–469 in *The biology of diatoms*, edited by D. Werner. Berkeley: University of California Press.

Hall, P. O. J., S. Hulth, G. Hulthe, A. Landen, and A. Tengberg. 1996. Benthic nutrient fluxes on a basin-wide scale in the Skagerrak (north-eastern North Sea). *Journal of Sea Research* 35:123–137.

Hamm, C. E., R. Merkel, O. Springer, P. Jurkojc, C. Maier, K. Prechtel, and V. Smetacek. 2003. Architecture and material properties of diatom shells provide efficient mechanical protection. *Nature* 421:841–843.

Hansen, K., G. M. King, and E. Kristensen. 1996. Impact of soft-shell clam *Mya arenaria* on sulfate reduction in an intertidal sediment. *Aquatic Microbial Ecology* 10:181–194.

Hansen, P. J., P. K. Bjornsen, and B. W. Hansen. 1997. Zooplankton grazing and growth: Scaling within the 2–2,000 µm body size range. *Limnology and Oceanography* 42:687–704.

Harisson, K. G. 2000. Role of increased marine silica input on paleo-pCO_2 levels. *Paleoceanography* 15(3):292–298.

Heip, C. H. R., N. K. Goosen, P. M. J. Herman, J. Kromkamp, J. J. Middleburg, and K. Soetaert. 1995. Production and consumption of biological particles in temperate tidal estuaries. *Oceanography and Marine Biology: An Annual Review* 33:1–149.

Hitchcock, G. L. 1982. A comparative study of the size-dependent organic composition of marine diatoms and dinoflagellates. *Journal of Plankton Research* 4:363–377.

Huertas, F. J., L. Chou, and R. Wollast. 1999. Mechanism of kaolinite dissolution at room temperature and pressure, Part II: Kinetic study. *Geochimica et Cosmochimica Acta* 63(19/20):3261–3275.

Hulburt, E. M. 1990. Description of phytoplankton and nutrients in spring in the western North Atlantic Ocean. *Journal of Plankton Research* 12:1–28.

Hurd, D. C. 1972. Factors affecting solution rate of biogenic opal in seawater. *Earth and Planetary Science Letters* 15:411–417.

Hurd, D. C. 1983. Physical and chemical properties of the siliceous skeletons. Pp. 187–244 in *Silicon geochemistry and biogeochemistry*, edited by S. R. Aston. London: Academic Press.

Hurd, D. C., and S. Birdwhistell. 1983. On producing a general model for biogenic silica dissolution. *American Journal of Science* 283:1–28.

Irigoien, X., R. P. Harris, H. M. Verheye, P. Joly, J. Runge, M. Starr, D. Pond, R. Campbell, R. Shreeve, P. Ward, A. N. Smith, H. G. Dam, W. Peterson, V. Tirelli, M. Koski, T. Smith, D. Harbour, and R. Davidson. 2002. Copepod hatching success in marine ecosystems with high diatom concentrations. *Nature* 419:387–389.

Irigoien, X., R. N. Head, R. P. Harris, D. Cummings, and D. Harbour. 2000. Feeding selectivity and egg production of *Calanus helgolandics* in the English Channel. *Limnology and Oceanography* 45(1):44–54.

Jahnke, R. A., J. R. Nelson, R. L. Marinelli, and J. E. Eckman. 2000. Benthic flux of biogenic elements on the southeastern US continental shelf: Influence of porewater advective transport and benthic microalgae. *Continental Shelf Research* 20:109–127.

Jassby, A. D., C. R. Goldman, and J. E. Reuter. 1995. Long-term change in Lake Tahoe (California–Nevada, U.S.A.) and its relation to atmospheric deposition of algal nutrients. *Archive für Hydrobiologie* 135:1–21.

Jonasdottir, S. H., T. Kiorboe, K. W. Tang, M. St. John, A. W. Visser, E. Saiz, and H. G. Dam. 1998. Role of diatoms in copepod production: Good, harmless or toxic? *Marine Ecology Progress Series* 172:305–308.

Jørgensen, C. B. 1990. *Bivalve filter feeding: Hydrodynamics, bioenergetics, physiology and ecology*. Fredensborg: Olsen and Olsen Press.

Jørgensen, C. B. 1996. Bivalve filter feeding revisited. *Marine Ecology Progress Series* 142:287–302.

Kamatani, A. 1982. Dissolution rates of silica from diatoms decomposing at various temperatures. *Marine Biology* 68:91–96.

Kautsky, N., and S. Evans. 1987. Role of biodeposition by *Mytilus edulis* in the circulation of matter and nutrients in a Baltic coastal ecosystem. *Marine Ecology Progress Series* 38:201–212.

Kjorboe, T., E. Saiz, and M. Vitasalo. 1996. Prey switching behaviour in the planktonic copepod *Acartia tonsa*. *Marine Ecology Progress Series* 143:65–75.

Kleppel, G. S. 1993. On the diets of calanoid copepods. *Marine Ecology Progress Series* 99:183–195.

Lancelot, C., G. Billen, A. Sournia, T. Weisse, F. Colijn, M. J. W. Veldhuis, A. Davies, and P. Wassman. 1987. *Phaeocystis* blooms and nutrient enrichment in the continental coastal zones of the North Sea. *Ambio* 16:38–46.

Lancelot, C., E. Hannon, S. Becquevort, C. Veth, and H. J. W. de Baar. 2000. Modeling phytoplankton blooms and carbon export production in the Southern Ocean: Dominant controls by light and iron of the Atlantic sector in austral spring 1992. *Deep-Sea Research I* 47:1621–1662.

Lawson, D. S., D. C. Hurd, and H. S. Pankratz. 1978. Silica dissolution rates of decomposing assemblages at various temperatures. *American Journal of Science* 278:1373–1393.

Legendre, L., S. Demers, and D. Lefaivre. 1986. Biological production at marine ergoclines. Pp. 1–29 in *Marine interface ecohydrodynamics*, edited by J. C. J. Nihoul. Amsterdam: Elsevier.

Legendre, L., and J. Le Fèvre. 1989. Hydrodynamical singularities as controls of recycled versus export production in oceans. Pp. 49–63 in *Productivity of the ocean: Present and past*, edited by W. H. Berger, V. S. Smetacek, and G. Wefer. Chichester, UK: Wiley.

Lorrain, A., Y.-M. Paulet, L. Chauvaud, N. Savoye, E. Nézan, and L. Guérin. 2000. Growth anomalies in *Pecten maximus* from coastal waters (Bay of Brest, France): Relationship with diatom blooms. *Journal of Marine Biological Association U.K.* 80:667–673.

Losno, R. 1989. *Chimie d'éléments minéraux en trace dans les pluies Méditerranéennes.* Thèse de doctorat, Université de Paris 7, Paris.

Marcus, N. H., and F. Boero. 1998. Mini-review: The importance of benthic–pelagic coupling and the forgotten role of life cycles in coastal aquatic systems. *Limnology and Oceanography* 43(5):763–768.

Margalef, R. 1978. Life-forms of phytoplankton as survival alternatives in an unstable environment. *Oceanologica Acta* 1:493–509.

Marinelli, R. L., R. A. Jahnke, D. B. Craven, J. R. Nelson, and J. E. Eckman. 1998. Sediment nutrient dynamics on the south Atlantic bight continental shelf. *Limnology and Oceanography* 43:1305–1320.

Maring, H. B., and R. A. Duce. 1987. The impact of atmospheric aerosols on trace metal chemistry in open ocean surface seawater. *Earth and Planetary Science Letters* 84:381–394.

Marra, J., R. R. Bidigare, and T. D. Dickey. 1990. Nutrients and mixing, chlorophyll and phytoplankton growth. *Deep-Sea Research* 37:127–143.

Martin-Jézéquel, V., M. A. Brzezinski, and M. Hildebrand. 2000. Silicon metabolism in diatoms: Implications for growth. *Journal of Phycology* 36:821–840.

McIntyre, D. C. G., R. J. Geider, and D. C. Miller. 1996. Microphytobenthos: The ecological role of the "secret garden" of unvegetated, shallow-water marine habitats, 1. Distribution, abundance and primary production. *Estuaries* 19:186–201.

McKyes, E., A. Sethi, and R. N. Yong. 1974. Amorphous coatings on particles of sensitive clay soils. *Clays and Clay Minerals* 2:427–433.

McManus, J., W. M. Berelson, D. E. Hammond, T. E. Kilgore, D. J. DeMaster, O. Ragueneau, and R. W. Collier. 1995. Early diagenesis of biogenic silica: Dissolution rates, kinetics, and paleoceanographic implications. *Deep-Sea Research II* 42:871–903.

Meybeck, M. 1982. C, N and P transport by world rivers. *American Journal of Science* 282:401–450.

Michalopoulos, P., and R. C. Aller. 1995. Rapid clay mineral formation in Amazon delta sediments: Reverse weathering and oceanic elemental cycles. *Science* 270:614–617.

Michalopoulos, P., and R. C. Aller. 2004. Early diagenesis of biogenic silica in the Amazon delta: Alteration, authigenic clay formation, and storage. *Geochimica et Cosmochimica Acta* 68(5):1061–1085.

Michalopoulos, P., R. C. Aller, and R. J. Reeder. 2000. Conversion of diatoms to clays during early diagenesis in tropical continental shelf muds. *Geology* 28:1095–1098.

Middelburg, J. J., and K. Soetaert. 2003. The role of sediments in shelf ecosystem dynamics. In *The sea*, edited by A. R. Robinson and K. Brink. Chapter 12, Vol. 13. Cambridge, MA: Harvard University Press.

Migon, C., and V. Sandroni. 1999. Phosphorus in rainwater: Partitioning inputs and impact on the surface coastal ocean. *Limnology and Oceanography* 44(4):1160–1165.

Miralto, A., G. Barone, G. Romano, S. A. Poulet, A. Ianora, G. L. Russo, I. Buttino, G. Mazarella, M. Laabir, M. Cabrini, and M. G. Giacobbe. 1999. The insidious effect of diatoms on copepod reproduction. *Nature* 402:173–176.

Montlucon, D., and S. A. Sanudo-Wilhelmy. 2001. Influence of net groundwater discharge on the chemical composition of a coastal environment: Flanders Bay, Long Island, New York. *Environmental Science and Technology* 35:480–486.

Moriceau, B., M. Garvey, U. Passow, and O. Ragueneau. In press. Reduced biogenic silica dissolution rates in diatom aggregates. *Marine Ecology Progress Series.*

Nedwell, D. B., R. J. Parkes, A. C. Upton, and D. J. Assinder. 1993. Seasonal fluxes across the sediment–water interface, and processes within sediments. *Philosophical Transactions of the Royal Society of London Series A. Physical Sciences and Engineering* 343:519–529.

Nelson, D. M., and Q. Dortch. 1996. Silicic acid depletion and silicon limitation in the plume of the Mississippi River: Evidence from kinetic studies in spring and summer. *Marine Ecology Progress Series* 136:163–178.

Nelson, D. M., and J. J. Goering. 1977a. Near-surface silica dissolution in the upwelling region off northwest Africa. *Deep-Sea Research* 24:65–73.

Nelson, D. M., and J. J. Goering. 1977b. A stable isotope tracer method to measure silicic acid uptake by marine phytoplankton. *Analytical Biochemistry* 78:139–147.

Nelson, D. M., J. J. McCarthy, T. M. Joyce, and H. W. Ducklow. 1989. Enhanced near-surface nutrient availability and new production resulting from the frictional decay of a Gulf Stream warm-core ring. *Deep-Sea Research* 36:705–714.

Nelson, D. M., P. Tréguer, M. A. Brzezinski, A. Leynaert, and B. Quéguiner. 1995. Production and dissolution of biogenic silica in the ocean: Revised global estimates, comparison with regional data and relationship to biogenic sedimentation. *Global Biogeochemical Cycles* 9:359–372.

Newell, C. R., S. E. Shumway, T. L. Cucci, and R. Selvin. 1989. The effects of natural seston particle size and type on feeding rates, feeding selectivity and food resource availability for the mussel *Mytilus edulis* Linnaeus, 1758 at bottom culture sites in Maine. *Journal of Shellfish Research* 8:187–196.

Norkko, A., J. E. Hewitt, S. F. Thrush, and G. A. Funnell. 2001. Benthic–pelagic coupling and suspension-feeding bivalves: Linking site specific sediment flux and biodeposition to benthic community structure. *Limnology and Oceanography* 46(8):2067–2072.

Oberdorfer, J. A., P. J. Hogan, and R. W. Buddemeier. 1990. Atoll island hydrogeology: Flow and fresh water occurrence in a tidally dominated system. *Journal of Hydrology* 120:327–340.

Officer, C. B., and J. H. Ryther. 1980. The possible importance of silicon in marine eutrophication. *Marine Ecology Progress Series* 3:83–91.

Olesen, M., and C. Lundsgaard. 1995. Seasonal sedimentation of autochthonous material from the euphotic zone of coastal ecosystems. *Estuarine, Coastal and Shelf Science* 41:475–490.

Paerl, H. W., and D. R. Whitall. 1999. Anthropogenically derived atmospheric nitrogen deposition, marine eutrophication and harmful algal bloom expansion: Is there a link? *Ambio* 28:307–311.

Pan, Y., D. V. Subba Rao, K. H. Mann, R. G. Brown, and R. Pocklington. 1996a. Effects of silicate limitation on production of domoic acid, a neurotoxin, by the diatom *Pseudo-nitschia multiseries.* I. Batch culture studies. *Marine Ecology Progress Series* 220:225–233.

Pan, Y., D. V. Subba Rao, K. H. Mann, R. G. Brown, and R. Pocklington. 1996b. Effects of silicate limitation on production of domoic acid, a neurotoxin, by the diatom *Pseudo-nitschia multiseries.* II. Continuous culture studies. *Marine Ecology Progress Series* 220:235–243.

Passow, U., R. F. Shipe, A. Murray, D. K. Pak, M. A. Brzezinski, and A. L. Alldredge. 2001. The origin of transparent exopolymer particles (TEP) and their role in the sedimentation of particulate matter. *Continental Shelf Research* 21:327–346.

Patrick, S., and A. J. Holding. 1985. The effect of bacteria on the solubilization of silica in diatom frustules. *Journal of Applied Bacteriology* 59:7–16.

Petersen, C. H., and R. Black. 1987. Resource depletion by active suspension feeders on tidal flat: Influence of local density and tidal elevation. *Limnology and Oceanography* 32:143–166.

Peterson, L. C., G. H. Haug, K. A. Hughen, and U. Röhl. 2000. Rapid changes in the hydrologic cycle of the tropical Atlantic during the last glacial. *Science* 290:1947–1951.

Pondaven, P., C. Fravalo, D. Ruiz-Pino, P. Tréguer, B. Quéguiner, and C. Jeandel. 1998. Modelling the silica pump in the permanently open ocean zone of the Southern Ocean. *Journal of Marine Systems* 17:587–618.

Precht, E., and M. Huettel. 2003. Advective porewater exchange driven by surface gravity waves and its ecological implications. *Limnology and Oceanography* 48(4):1674–1684.

Prins, T. C., and A. C. Smaal. 1994. The role of the blue mussel *Mytilus edulis* in the cycling of nutrients in the Oosterschelde estuary (The Netherlands). Pp. 413–429 in *The Oosterschelde estuary: A case study of a changing ecosystem,* edited by P. H. Nienhuis and A. C. Smaal. Dordrecht, The Netherlands: Kluwer.

Ragueneau, O. 2004. Si-WEBS, a European network for the study of Si fluxes on continental margins. *LOICZ Newsletter* 31:1–4.

Ragueneau, O., L. Chauvaud, and J. Grall. 2004. Benthic–pelagic coupling and eutrophication: The case of the silicate pump. In *Drainage basin nutrient inputs and eutrophication: An integrated approach,* edited by K. Olli and P. Wassmann. Available online at www.ut.ee/~olli/eutr.

Ragueneau, O., L. Chauvaud, A. Leynaert, G. Thouzeau, Y.-M. Paulet, S. Bonnet, A. Lorrain, J. Grall, R. Corvaisier, M. Le Hir, F. Jean, and J. Clavier. 2002a. Direct evidence of a biologically active coastal silicate pump: Ecological implications. *Limnology and Oceanography* 47(6):1849–1854.

Ragueneau, O., L. Chauvaud, B. Moriceau, A. Leynaert, G. Thouzeau, A. Donval, F. Le Loc'h, and F. Jean. 2005a. Biodeposition by an invasive suspension feeder impacts the biogeochemical cycle of Si in a coastal ecosystem (Bay of Brest, France). *Biogeochemistry* doi 10.1007/s10533-004-5677-3.

Ragueneau, O., E. De Blas Varela, P. Tréguer, B. Quéguiner, and Y. Del Amo. 1994. Phytoplankton dynamics in relation to the biogeochemical cycle of silicon in a coastal ecosystem of Western Europe. *Marine Ecology Progress Series* 106:157–172.

Ragueneau, O., M. Gallinari, H. Stahl, A. Tengberg, S. Grandel, R. Lampitt, R. Witbaard,

P. Hall, D. Rickert, and A. Hauvespre. 2001. The benthic silica cycle in the northeast Atlantic: Seasonality, annual mass balance and preservation mechanisms. *Progress in Oceanography* 50:171–200.
Ragueneau, O., C. Lancelot, V. Egorov, J. Vervlimmeren, A. Cociasu, G. Déliat, A. Krastev, N. Daoud, V. Rousseau, V. Popovitchev, N. Brion, L. Popa, and G. Cauwet. 2002b. Biogeochemical transformations of inorganic nutrients in the mixing zone between the Danube River and the northwestern Black Sea. *Estuarine, Coastal and Shelf Science* 54(3):321–336.
Ragueneau, O., B. Quéguiner, and P. Tréguer. 1996. Contrast in biological responses to tidally-induced vertical mixing for two macrotidal ecosystems of Western Europe. *Estuarine, Coastal and Shelf Science* 42:645–665.
Ragueneau, O., N. Savoye, Y. Del Amo, J. Cotten, B. Tardiveau, and A. Leynaert. 2005b. A new method for the measurement of biogenic silica in suspended matter of coastal waters: Using Si:Al ratios to correct for the mineral interference. *Continental Shelf Research* 25:697–710.
Ragueneau, O., and P. Tréguer. 1994. Determination of biogenic silica in coastal waters: Applicability and limits of the alkaline digestion method. *Marine Chemistry* 45:43–51.
Ragueneau, O., P. Tréguer, A. Leynaert, R. F. Anderson, M. A. Brzezinski, D. J. DeMaster, R. C. Dugdale, J. Dymond, G. Fischer, R. François, C. Heinze, E. Maier-Reimer, V. Martin-Jézéquel, D. Nelson, and B. Quéguiner. 2000. A review of the Si cycle in the modern ocean: Recent progress and missing gaps in the application of biogenic opal as a paleoproductivity proxy. *Global and Planetary Change* 26(4):315–366.
Rickert, D. 2000. Dissolution kinetics of biogenic silica in marine environments. *Reports on Polar Research* 351.
Rizzo, W. M. 1990. Nutrient exchanges between the water column and a subtidal benthic microalgal community. *Estuaries* 13(3):219–226.
Robert, R., T. Noel, and R. Galois. 1989. The food value of five unicellular diatoms to the larvae of *Crassostrea gigas* Thunberg. *Special Publication, European Aquaculture Society* 10:215–216.
Rojas de Mendiola, B. 1981. Seasonal phytoplankton distribution along the Peruvian coast. Pp. 330–347 in *Coastal upwelling*, edited by F. A. Richards. Washington, DC: American Geophysical Union.
Romero, O. E., L. Dupont, U. Wyputta, S. Jahns, and G. Wefer. 2003. Temporal variability of fluxes of eolian-transported freshwater diatoms, phytoliths, and pollen grains off Cape Blanc as reflection of land–atmosphere–ocean interactions in northwest Africa. *Journal of Geophysical Research* 108(C5):3153.
Ryther, J. H. 1969. Photosynthesis and fish production in the sea. *Science* 166:72–76.
Sarmiento, J. L., and N. Gruber. 2006. *Ocean biogeochemical dynamics*. Princeton, NJ: Princeton University Press.
Scholln, C. A., F. Gulland, G. J. Doucette, S. Benson, M. Busman, F. P. Chavez, J. Cordaro, R. DeLong, A. DeVogelaere, J. Harvey, M. Haulena, K. Lefebvre, T. Lipscomb, S. Loscutoff, L. J. Lowenstine, R. Marin III, P. E. Miller, W. A. McLellan, P. D. R. Moeller, C. L. Powell, T. Rowles, P. Silvagni, M. Silver, T. Spraker, V. Trainer, and F. M. Van Dolah. 2000. Mortality of sea lions along the central California coast linked to a toxic diatom bloom. *Nature* 403:80–84.
Schultes, S., S. Jansen, and U. Bathmann. In revision. Influence of mesozooplankton grazing on the dissolution rate of Antarctic diatom silica. *Marine Ecology Progress Series.*

Serodio, J., J. M. da Silva, and F. Catarino. 1997. Nondestructive tracing of migratory rhythms of intertidal benthic microalgae using in vivo chlorophyll *a* fluorescence. *Journal of Phycology* 33:545–553.

Shum, K. T., and B. Sundby. 1996. Organic matter processing in continental shelf sediments: The subtidal pump revisited. *Marine Chemistry* 53:81–87.

Shumway, S. E., T. L. Cucci, P. C. Newell, and C. M. Yentsch. 1985. Particle selection, ingestion and absorption in filter-feeding bivalves. *Journal of Experimental Marine Biology and Ecology* 91:77–92.

Sigmon, D. E., and L. B. Cahoon. 1997. Comparative effects of benthic microalgae and phytoplankton on dissolved silica fluxes. *Aquatic Microbial Ecology* 13:275–284.

Slomp, C. P., and P. Van Cappellen. 2004. Groundwater inputs of nutrients to the coastal ocean: Controls and potential impact. *Journal of Hydrology* 295:64–86.

Smayda, T. J. 1990. Novel and nuisance of phytoplankton blooms in the sea: Evidence for a global epimedia. Pp. 29–40 in *Toxic marine phytoplankton*, edited by E. Graneli. New York: Elsevier.

Smetacek, V. S. 1984. The supply of foods to the benthos. Pp. 517–548 in *Flows of energy and materials in marine ecosystems: Theory and practice*, edited by M. J. Fasham. New York: Plenum.

Smetacek, V. S. 1985. Role of sinking in diatom life-history cycles: Ecological, evolutionary and geological significance. *Marine Biology* 84:239–251.

Smetacek, V. 1999. Diatoms and the ocean carbon cycle. *Protist* 150:25–32.

Smith, D. J., and G. J. C. Underwood. 2000. The production of extracellular carbohydrates by estuarine benthic diatoms: The effects of growth phase and light and dark treatment. *Journal of Phycology* 36:321–333.

Soetaert, K., J. Middelburg, P. M. J. Herman, and K. Buis. 2000. On the coupling of benthic and pelagic biogeochemical models. *Earth Science Reviews* 51:173–201.

Srithongouthai, S., Y. Sonoyama, K. Tada, and S. Montani. 2003. The influence of environmental variability on silicate exchange rates between sediment and water in a shallow-water coastal system. *Marine Pollution Bulletin* 47:10–17.

Staats, N., B. De Winter, L. R. Mur, and L. J. Stal. 1999. Isolations and characterization of extrapolysaccharides from the epipelic diatoms *Cylindrotheca closterium* and *Navicula salinarum*. *European Journal of Phycology* 34:161–169.

Steidinger, K. A., and D. G. Baden. 1984. Toxic marine dinoflagellates. Pp. 201–261 in *Dinoflagellates*, edited by D. L. Spector. New York: Academic Press.

Suess, E. 1980. Particulate organic carbon flux in the oceans: Surface productivity and oxygen utilization. *Nature* 288:260–263.

Sundbäck, K. 1991. Influence of sublittoral microphytobenthos on the oxygen and nutrient flux between sediment and water: A laboratory continuous-flow study. *Marine Ecology Progress Series* 74:263–279.

Sundbäck, K., and W. Granéli. 1988. Influence of microphytobenthos on the nutrient flux between sediment and water: A laboratory study. *Marine Ecology Progress Series* 431–432:63–69.

Syrett, P. J. 1981. Nitrogen metabolism of algae. *Canadian Bulletin of Fisheries and Aquatic Sciences* 210:182–210.

Tang, K. W., H. H. Jakobsen, and A. W. Visser. 2001. *Phaeocystis globosa* (Prymnesiophyceae) and the planktonic food web: Feeding, growth and trophic interactions among grazers. *Limnology and Oceanography* 46:1860–1870.

Taniguchi, M., W. C. Burnett, J. E. Cable, and J. V. Turner. 2002. Investigation of submarine groundwater discharge. *Hydrological Processes* 16(11):2115–2129.

Thornton, D. C. O. 2002. Diatom aggregation in the sea: Mechanisms and ecological implications. *European Journal of Phycology* 37:149–161.

Thouzeau, G., L. Chauvaud, J. Grall, and L. Guérin. 2000. Rôle des interactions biotiques sur le devenir du pré-recrutement et la croissance de *Pecten maximus* (L.) en rade de Brest. *Comptes Rendus de l'Académie des Sciences, Paris, Sciences de la Vie/Life Sciences* 323:815–825.

Touratier, F., L. Legendre, and A. Vézina. 1999. Model of copepod growth influenced by the food carbon:nitrogen ratio and concentration, under the hypothesis of strict homeostasis. *Journal of Plankton Research* 21(6):1111–1132.

Tréguer, P., L. Lindner, A. J. Van Bennekom, A. Leynaert, M. Panouse, and G. Jacques. 1991. Production of biogenic silica in the Weddell–Scotia Seas measured with ^{32}Si. *Limnology and Oceanography* 36:1217–1227.

Tréguer, P., D. M. Nelson, A. J. Van Bennekom, D. J. DeMaster, A. Leynaert, and B. Quéguiner. 1995. The silica balance in the world ocean: A reestimate. *Science* 268:375–379.

Van Bennekom, A. J., J. H. F. Jansen, S. J. Van der Gaast, J. M. Van Ieperen, and J. Pieters. 1989. Aluminum-rich opal: An intermediate in the preservation of biogenic silica in the Zaire (Congo) deep-sea fan. *Deep-Sea Research* 36:173–190.

Van Cappellen, P., S. Dixit, and J. van Beusekom. 2002. Biogenic silica dissolution in the oceans: Reconciling experimental and field-based dissolution rates. *Global Biogeochemical Cycles* 16(4):1075.

Wassmann, P. 1991. Dynamics of primary production and sedimentation in shallow fjords and poll of western Norway. *Oceanography and Marine Biology Annual Review* 29:87–154.

Wildish, D., and D. Kristmanson. 1997. *Benthic suspension feeders and flow.* Cambridge: Cambridge University Press.

Wilson, D. L., W. O. Smith Jr., and D. M. Nelson. 1986. Phytoplankton bloom dynamics of the western Ross Sea ice edge. I. Primary productivity and species-specific production. *Deep-Sea Research* 33:1375–1377.

Wollast, R. 2002. Continental margins: Review of geochemical settings. Pp. 15–31 in *Ocean margin systems,* edited by G. Wefer, D. Billett, D. Hebbeln, B. B. Jörgensen, M. Schlüter, and T. J. van Weering. Heidelberg: Springer.

Wong, C. S., and R. J. Matear. 1999. Sporadic silicate limitation of phytoplankton production in the subarctic NE Pacific. *Deep-Sea Research II* 46(11–12):2539–2556.

Wulff, A. S. K., C. Nilsson, L. Carlson, and B. Jonsson. 1997. Effect of sediment load on the microphytobenthic community of a shallow-water sandy sediment. *Estuaries* 203:547–558.

Young, R. W., K. L. Carder, P. R. Betzer, D. K. Costello, R. A. Duce, G. R. DiTullio, N. W. Tindale, E. A. Laws, M. Uematsu, J. T. Merril, and R. A. Feely. 1991. Atmospheric iron inputs and primary productivity: Phytoplankton responses in the North Pacific. *Global Biogeochemical Cycles* 5:119–134.

12

Responses of Coastal Ecosystems to Anthropogenic Perturbations of Silicon Cycling

Olivier Ragueneau, Daniel J. Conley, Aude Leynaert, Sorcha Ni Longphuirt, and Caroline P. Slomp

Changes in phytoplankton dynamics and food webs have been observed in many coastal ecosystems and are related to long-term decreases in riverine Si:N and Si:P nutrient ratios (reviews in Smayda 1990; Conley et al. 1993; Turner et al. 1998). The most obvious explanation for the observed decrease of Si:P and Si:N ratios is eutrophication. Urbanization and agricultural and industrial activities have led to large increases in the delivery of N and P along the land–ocean continuum. On a global basis, the fluxes of these elements to the oceans have increased by a factor of two; it has long been thought that at the same time, DSi fluxes have remained constant because the major source of DSi to rivers is natural silicate rock weathering (Chapter 2, this volume). Numerous reviews have described excessive inputs of N and P compounds to coastal waters (e.g., Nixon et al. 1996; Howarth et al. 1996), their effects on coastal waters (e.g., Cloern 2001), and the subsequent nutrient management strategies needed for sustainable use of coastal zones (Conley 2000). However, Si:N and Si:P ratios also decrease after an anthropogenic reduction of DSi concentrations along the aquatic continuum from land to ocean.

Silica Depletion in Aquatic Systems

Three major mechanisms can contribute to the proposed silica depletion. Two are well known (eutrophication and hydrologic management), and the third one (increasing biodeposition by invasive species) remains to be explored (Ragueneau et al. 2005).

Increased N and P loading—eutrophication—increases aquatic primary production and, in many systems, diatom production. Subsequent increases in the sedimentation rate of diatoms can lead to increasing amounts of diatoms and BSi stored in the sediments, eventually leading to a reduction in DSi supplies to the water column, especially in systems with long residence times (Conley et al. 1993). This nutrient-driven alteration of the biogeochemical Si cycle was first described for the North American Great Lakes (Schelske et al. 1983), but it has also been observed in small lakes (Engstrom et al. 1985; Schelske et al. 1987) and is relevant for the coastal zone, with a number of coastal margin systems showing declines in DSi concentrations that are related to anthropogenically induced eutrophication (review in Conley et al. 1993).

Strong reduction in DSi concentrations in rivers has also been observed after dam construction (the so-called artificial lake effect; Van Bennekom and Salomons 1981), as in the Nile River (Wahbi and Bishara 1979) and the Mississippi River (Turner and Rabalais 1991). Humborg et al. (1997) observed a direct link between dam building, nutrient ratio, and coastal ecosystem structure in the Danube–Black Sea system. In these systems, DSi reductions have been assumed to occur with increased diatom growth after increased residence time of waters in the reservoirs and subsequent retention of Si as diatoms are buried in accumulating sediments behind dams (Figure 12.1). Two other mechanisms to account for DSi reduction with damming are the trapping of phytoliths or other forms of amorphous silica during sedimentation behind dams, removing an important component of particulate Si carried by rivers (Conley 2002), and changes in vegetation associated with hydraulic manipulations that affect the weathering fluxes of DSi, thereby leading to the observed DSi reduction (Humborg et al. 2002). Both mechanisms are discussed in more detail in Chapters 3 and 5.

Another influence on the Si cycle along the land–ocean continuum is increased biodeposition. Indeed, suspension feeders filter enormous amounts of water (Budd et al. 2001); they consume diatoms and produce large quantities of biodeposits enriched in Si relative to C or N because they have no known metabolic need for Si (Ragueneau et al. 2005; see also Tande and Slagstad 1985 or Cowie and Hedges 1996 for an analogy with pelagic grazers). The effects of biodeposition on the Si cycle and ecosystem functioning have been suggested (Chauvaud et al. 2000) and observed at a seasonal scale in terms of phytoplankton dynamics (Ragueneau et al. 2002a; Chapter 11, this volume).

More recently, Ragueneau et al. (2005) explored the consequences of such a mechanism for the Si cycle on longer time scales. Budget calculations demonstrate that in the Bay of Brest, annual Si biodeposition represents nearly 80 percent of DSi river inputs. Some 70 percent of the BSi biodeposited redissolves (Ragueneau et al. 2002a). However, the biodeposition flux is so important and the preservation conditions in *Crepidula fornicata* mats so good (nearly 30 percent of the BSi annually biodeposited eventually gets preserved; Ragueneau et al. 2005) that the annual Si accumulation amounts to about 20 percent of DSi river inputs. Although this retention efficiency is close to the global mean proposed by DeMaster (1981), it is twice as high as the one used at global scale by Tréguer et al. (1995) and probably much higher than it was before *C. fornicata* invaded the ecosystem. In many ecosystems biodeposition increases year after year with the development

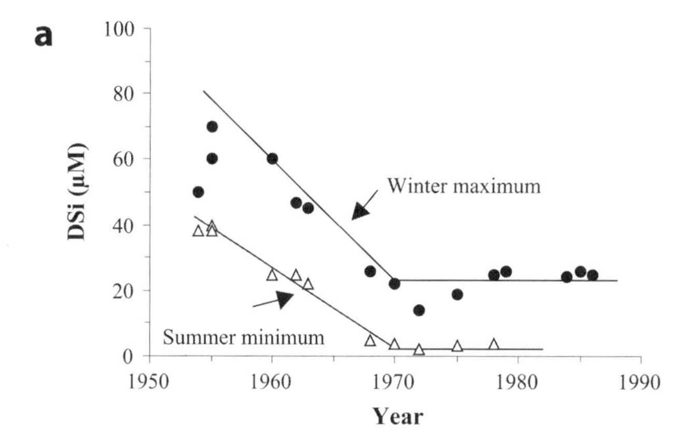

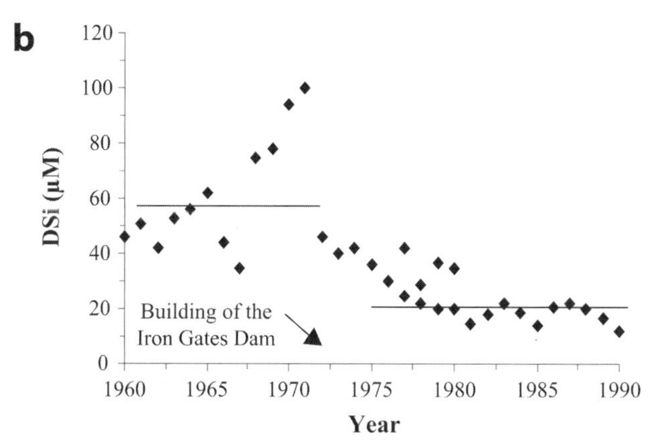

Figure 12.1. Illustration of the DSi depletion hypothesis. *(a)* Decreasing DSi concentrations with eutrophication in Lake Michigan (redrawn from Schelske et al. 1983). *(b)* Declining DSi concentrations at the Danube River mouth after construction of the Iron Gates Dam in the early 1970s (redrawn from Humborg et al. 1997).

of aquaculture and the accidental proliferation of invasive species (Mack et al. 2000; Ruiz et al. 2000). We hypothesize that increasing biodeposition may constitute another route for silica depletion that does not necessarily involve increased diatom production but simply increased BSi preservation. It will be important to test this hypothesis, first in the Bay of Brest ecosystem and then in the many places where biodeposition is increasing. Possible ways of testing this mechanism include looking at changes in BSi accumulation rates in such ecosystems and looking at monitoring data downstream of invasions to search for changes in DSi concentrations.

Anthropogenic DSi Inputs

It was recently recognized that there are significant anthropogenic sources of DSi. Soluble silicates such as sodium silicates (water glass) are some of the largest-volume synthetic chemicals in the world and are used in both industrial and household products, such as laundry detergents (van Dokkum et al. 2004). Recent estimates suggest that the additional load of anthropogenic DSi into rivers from industries and households contributes only 2 percent of the annual DSi in Western Europe (van Dokkum et al. 2004). However, locally, and especially during times of low DSi concentrations or near cities, substantial inputs of DSi may affect local mass balances. Additionally, various forms of soluble silicates are being used as commercial fertilizers, especially on crops such as rice and sugarcane, which have high demands for Si (Datnoff et al. 2001). DSi as a fertilizer is also being used in soilless greenhouse systems. The quantities of the additional sources of DSi are poorly constrained at present.

Ecosystem Response to Decreasing Si:N and Si:P Ratios

Switch from Diatoms to Flagellates: The DSi Paradigm

The growth of diatoms depends on the presence of DSi, whereas the growth of nondiatom phytoplankton does not. Diatoms use DSi, mostly $Si(OH)_4$ (Del Amo and Brzezinski 1999), essentially to build their frustules (Guillard and Kilham 1978). This silica wall may protect them against predators (Hamm et al. 2003) or increase their ability to acquire inorganic carbon (CO_2) from seawater (Milligan and Morel 2002). In the 1970s, many experimental studies were devoted to determining DSi uptake parameters (Officer and Ryther 1980 and references therein), namely the maximum uptake rate (μ_{max}) and the half-saturation constant (K_s). They showed that K_s values ranged between 0.5 and 5.0 μM (i.e., well within the range of DSi concentrations encountered in the field), suggesting that "low silicate concentrations may exert a selective influence on the species composition of phytoplankton populations in the sea" (Paasche 1973). Diatom physiology and DSi kinetic uptake parameters are discussed in more detail by Claquin et al. in Chapter 9.

First Evidence of DSi Limitation

Interestingly, the first evidence of DSi limitation has been reported in lakes. Pearsall (1932) and then Lund (1950) observed a rapid decline of diatoms with DSi concentrations dropping below 8 μM. Kilham (1971) related the seasonal succession of diatoms in eutrophic lakes to changes in ambient DSi concentrations. In the same period, Schelske and Stoermer (1971) related shifts from diatoms to green and blue-green algae in Lake Michigan to DSi depletion, after increased diatom growth and P inputs.

Officer and Ryther (1980) extended this idea to marine waters by reviewing a few case studies in North American estuaries and coastal waters. A sequence from spring diatoms to summer flagellates is a common occurrence in temperate coastal waters (Margalef 1958). Officer and Ryther (1980) related it to DSi availability, which becomes limiting after spring uptake by diatoms and lower subsequent dissolution of diatom Si compared with N and P, which are biologically mediated.

Two Decades of Evidence Sustaining the Paradigm

The Si:N and Si:P ratios are declining at the output of many rivers because of increasing N and P loadings, decreasing DSi concentrations, or both. Year after year, coastal waters move from N to Si limitation (Fransz and Verhagen 1985; Conley and Malone 1992; Ragueneau et al. 1994). Ecosystems move from a natural sequence (from a new production regime during spring, dominated by diatoms, to a regenerated production regime during summer, dominated by flagellates) to a perturbed sequence in which the shift occurs earlier in the season and flagellate blooms become much bigger because they occur under a new production regime, making use of the excess N and P that could not be used by diatoms because of DSi limitation (Billen et al. 1991).

This phenomenon is not confined to a few case studies. Smayda (1990) related the increasing frequency and magnitude of harmful algal blooms in many ecosystems (Baltic Sea, Kattegat, Skagerrak, Dutch Wadden Sea, North Sea, Black Sea) to long-term declines in Si:N and Si:P ratios. Justic et al. (1995) presented two more case studies in the Adriatic Sea and the northern Gulf of Mexico. They noted that if "historically, rivers carried DSi well in excess to DIN and DIP [dissolved inorganic nitrogen and phosphorus, respectively], at present, many world rivers are beginning to experience a stoichiometric nutrient balance or even a DSi deficiency." Perhaps the two best case studies are located on the Louisiana and northwestern Black Sea continental shelves, at the mouths of the Mississippi and Danube rivers, respectively.

Mostly as a result of large increases in N and P fertilizer use, the DSi:DIN ratio has decreased from 3:1 to 1:1 in the twentieth century in Louisiana shelf waters (Turner and Rabalais 1991). Other ecosystem parameters have been affected by these changes in nutrient balance (Figure 12.2): copepod abundance changed from more than 75 percent to less than 30 percent of total mesozooplankton, zooplankton fecal pellets became a minor component of the in situ primary production consumed, and bottom-water oxygen consumption rates became less dependent on fast-sinking (diatom-rich) organic matter packaged mostly as zooplankton fecal pellets (Turner et al. 1998).

This coastal ecosystem can now shift from a food web composed of diatoms and copepods to one with potentially disruptive harmful algal blooms; the shift is controlled by Mississippi River water quality, which is in turn determined by land use practices far inland (Turner et al. 1998).

The second excellent illustration is the northwestern shelf of the Black Sea, which is under the direct influence of the Danube River, which discharges some 70 percent of the total freshwater to the Black Sea (Tolmazin 1985). From the 1960s to the late 1980s, DIN and DIP inputs increased by a factor of five and three, respectively, as a result of

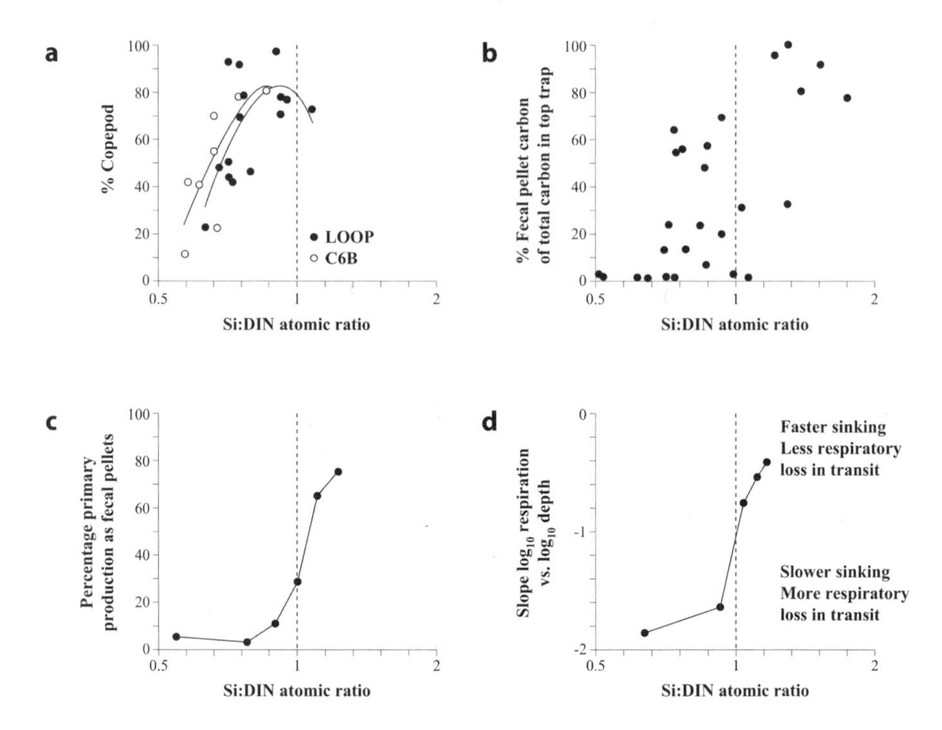

Figure 12.2. Effects of fluctuating DSi:dissolved inorganic nitrogen (DIN) ratios in the Mississippi River on Louisiana shelf plankton food webs (from Turner et al. 1998). *(a)* Effects on copepod abundance, *(b)* effects on the contribution of fecal pellets to the total carbon in a surface sediment trap (5 m), *(c)* effects on the percentage of phytoplankton primary production captured as fecal pellets in a bottom sediment trap (15 m), and *(d)* effects on benthic respiration. LOOP and C6B denote stations from Turner et al. (1998).

human activities in the watersheds of main tributaries and along the shoreline (Cociasu et al. 1996). The consequent changes in DSi:DIN ratios of the Black Sea nutrient load appear to be larger than those caused by eutrophication alone; Humborg et al. (1997) suggested that the damming of the Danube, and in particular the construction of the Iron Gates Dam in the early 1970s, induced a decline in DSi concentrations (Figure 12.1) that could explain in part the observed changes in nutrient balance. These changes are accompanied by a qualitative shift in the oxidation state of nitrogen compounds being delivered to the coastal areas caused by eutrophication in the freshwater reservoirs (Sapozhnikov 1992). Taken together, these changes have stimulated phytoplankton blooms of altered composition (Bodeanu 1992), especially toward nonsiliceous mixotrophic species (Bologa et al. 1995; Bouvier et al. 1998). Cascading effects on pelagic (Bodeanu 1992) and benthic (Gomoiu 1992) food chains have been observed, thought to be at the root of Black Sea ecosystem destabilization (Mee 1992; van Eeckhout and Lancelot 1997).

Mesocosm Evidence Supporting the Paradigm

Mesocosms provide a means to manipulate nutrient concentrations and to study large volumes (several cubic meters) of water containing natural plankton communities. Since the late 1980s and early 1990s, mesocosm experiments have been designed to study the response of phytoplankton populations to varying nutrient concentrations and nutrient balance under different turbulence conditions (Estrada et al. 1988; Svensen et al. 2001), grazing pressure, or both (Alcaraz et al. 1988; Escavarage and Prins 2002). With the increasing evidence linking changes in coastal water phytoplankton composition to changes in river Si:N ratios, mesocosm experiments specifically designed to study the influence of DSi on phytoplankton composition have flourished (Egge and Aksnes 1992; Wassman et al. 1996; Egge and Jacobsen 1997; Svensen et al. 2001; Svensen 2002; Roberts et al. 2003).

However, results from these experiments are somewhat contradictory. But the most cited study sustaining the paradigm we are interested in probably is the one by Egge and Aksnes (1992; Figure 12.3).

Using fourteen enclosures, these authors showed that diatom dominance (more than 80 percent of total cell counts) occurred irrespective of season if DSi concentration exceeded a threshold of approximately 2 μM. They suggested that this could result from a higher maximum growth rate in diatoms at nonlimiting DSi concentrations (see also Martin-Jézéquel et al. 2000). Interestingly, this threshold is close to many K_s values measured in the field (see Chapter 9, this volume). It also corresponds to a threshold above which the BSi:$CaCO_3$ rain ratio increases sharply in the open ocean (Ragueneau et al. 2000). For all these reasons, this value of 2 μM has become the standard value for DSi limitation. But the reality appears to be more complex, as seen from both mesocosm and field studies.

Field and Mesocosm Evidence Against the Paradigm

The paradigm stipulates that DSi availability controls the shift from diatoms to nondiatom species, potentially harmful, with important implications in terms of carbon export toward both deeper waters and higher trophic levels.

The influence of DSi on the export toward higher trophic levels has been discussed indirectly in connection with the importance of diatoms in the food chain (Chapter 11, this volume). In a recent mesocosm study, Roberts et al. (2003) started their experiments with DSi concentrations higher than 2 μM, and initially diatoms dominated, as found by Egge and Aksnes (1992). After Day 8, however, when DSi fell below 2 μM, the relative contribution of diatoms to total autotrophic plankton biomass increased and continued to do so for 5 days. Similarly, in the field (Bay of Brest), Ragueneau et al. (1994) observed several diatom blooms in May and June 1992, while DSi concentrations were

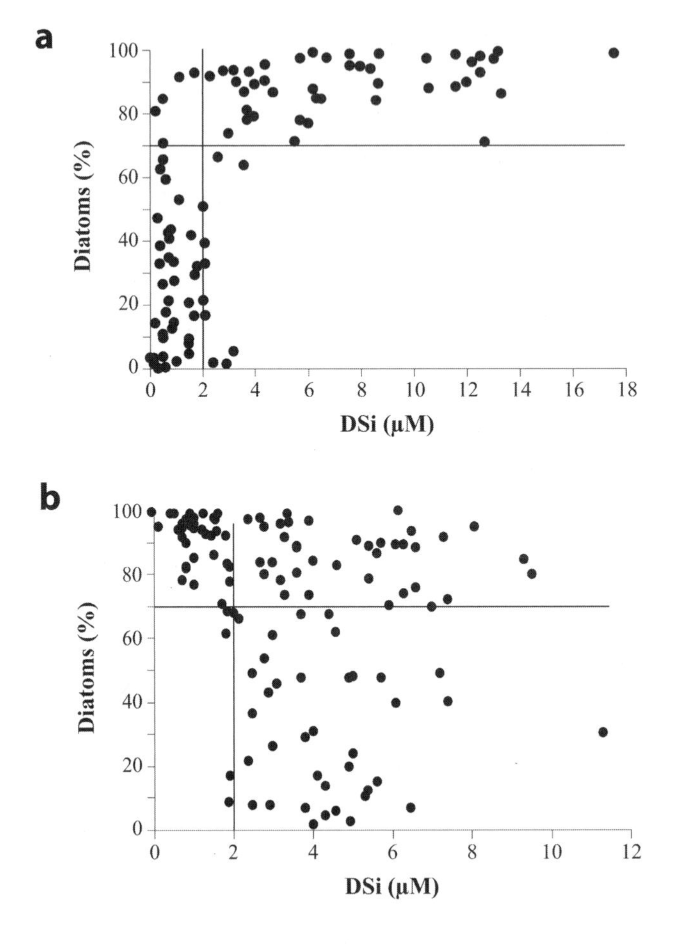

Figure 12.3. Contribution of diatoms to total phytoplankton as a function of DSi concentration. *(a)* Results from mesocosm experiments (Egge and Aksnes 1992). *(b)* Results from the Bay of Brest (France) (Fouillaron et al. submitted).

far below K_s values and even close to zero. Two major reasons can be suggested to account for such observations.

First, the shift from diatoms to nondiatom species can be controlled by silicon regeneration, as suggested by Officer and Ryther (1980). Diatoms may be able to immediately use the DSi that is released from the dissolution of biogenic silica, so that DSi concentrations have no time to build up in the water column. Clearly, as discussed in Chapter 11, BSi production and dissolution fluxes should be measured in addition to DSi concentrations.

Second, the shift in phytoplankton composition may not be immediate because diatoms are known to exhibit some degree of flexibility when they become DSi stressed. Before the shift occurs, Roberts et al. (2003) observed physiological and biochemical changes of the dominant diatom species, such as differences in mean cell volume (MCV), C:MCV ratio, C:N ratio, and glucan concentrations. Laboratory studies have demonstrated that diatoms can sustain high growth rates despite low external DSi concentrations (Olsen and Paasche 1986; Brzezinski et al. 1990). In coastal waters, it has been observed that diatoms adjust to low DSi concentrations by thinning their frustules or decreasing the number and length of spines (Chapter 9, this volume). In the North Sea, Rousseau et al. (2002) observed that the Si:C ratio of the diatoms decreased with declining DSi concentrations during the productive season. This decrease was accompanied by shifts between diatom species. Thus, under increasing DSi stress, if the recycling flux becomes insufficient, diatoms exhibit a gradual response before a complete shift toward nonsiliceous species may occur.

In the summer of 1995, off the Danube River on the northwestern shelf of the Black Sea, the contribution of diatoms to total primary production was negligible, despite the fact that DSi concentrations were greater than 2 μM (Ragueneau et al. 2002b). During that period, N and P limitations were preventing diatoms from blooming. A similar observation has been made by Lancelot et al. (2004) in the North Sea, with DIP again playing a major role in controlling the growth of summer diatoms. In the Bay of Brest and Chesapeake Bay, DSi clearly plays a critical role in spring (Ragueneau et al. 2002a; Conley and Malone 1992). But in summer, DSi builds up again in the water column because it is not being used up by diatoms. In the Bay of Brest in August and September, flagellates dominate while DSi concentrations increase steadily from near zero to more than 4 μM. In this ecosystem, there are many diatoms in spring, when ambient DSi concentrations remain low, and there are no diatoms in late summer, when DSi concentrations build up again in the water column. This is illustrated in Figure 12.3, for direct comparison with the Egge and Aksnes (1992) mesocosm experiments.

Obviously, part of this discrepancy can be resolved by reasoning in terms of fluxes and not concentrations (Fouillaron et al. submitted): DSi concentrations are low in spring because they are being used continuously by diatoms, whose growth rate must be controlled by the efficiency of DSi regeneration, which can be rapid if stimulated by bacterial activity (Bidle and Azam 1999). Diatoms can also acclimate to these low DSi fluxes by modifying their shape or silicification degree or by switching from one diatom species to another. DSi concentrations increase in summer because another factor prevents diatoms from using up the DSi stock. But the discrepancy is only partially relieved: These field observations clearly challenge the concept that diatoms would be better competitors for N and P when DSi concentrations are replete. In the North Sea or off the Danube River, low PO_4 concentrations may explain the observations, because it has been shown that diatoms are poor competitors at low DIP concentrations (Egge

1997), and flagellates can use P from organic sources. But why do diatoms dominate in the Bay of Brest in spring, when both DSi and DIP limit their growth (Del Amo et al. 1997)?

Nutrient Balance and the Development of HABs

In a recent review of harmful algal blooms and eutrophication, Anderson et al. (2002) demonstrated the complexity of the response of coastal waters to nutrient enrichment. DSi is clearly playing a critical role, but it has to be seen as one controlling factor among others. To serve our purposes, a few examples of nutrient effects will be shown here.

Independently of DSi, the balance between N and P also plays an important role. In Tolo Harbor, Hodgkiss and Ho (1997) related the increasing amount of red tides between 1982 and 1989 to declining N:P ratios, from above Redfield to around 10:1. Similar observations were made by Romdhane et al. (1998) in Tunisian lagoons, with blooms of *Gymnodinium aureolumi*, and by Riegman (1995) in Dutch coastal waters, with blooms of *P. pouchetii*. As noted by Anderson et al. (2002), "The nutrient ratio concept has recently been expanded to include the relative abundance of different chemical forms of nutrients, such as organic versus inorganic N and C." The development of HABs has been related to the dissolved organic nitrogen components, particularly urea, as in Chesapeake Bay (Glibert et al. 2001). Several HABs have also been related to elevated ratios of dissolved organic carbon to dissolved organic nitrogen, such as the brown tides in Long Island (Lomas et al. 2001). In fact, it becomes clear that if diatoms are able to exploit nitrate-rich conditions (Goldman 1993; Chapter 11, this volume), microflagellates, including harmful dinoflagellates, are most often associated with low nitrate concentrations, higher ammonium, urea, or dissolved organic nitrogen supply, and consistent physiological preference for reduced N forms (Anderson et al. 2002 and references therein).

Just as some zooplankton species are able to switch from ambush to suspension feeding (Chapter 11, this volume), the ability of many flagellate species, including those forming HABs, to acquire their essential elements from both inorganic and organic (dissolved and particulate) sources provides them with an efficient survival mechanism. Although such mixotrophy has an energetic cost (need for both a photosynthesis apparatus and mechanisms for prey uptake and subsequent digestion; Rothhaupt 1996a, 1996b), it provides them with an enormous advantage when inorganic nutrients limit phototrophic growth (Tittel et al. 2003). With the improvement of methods to measure ingestion and C uptake, the importance of mixotrophy in the development of HABs becomes increasingly recognized (Anderson et al. 2002 and references therein). Note that another important capability of certain HAB species is to acquire some of their nutrients via extracellular oxidation or hydrolysis (Mulholland et al. 1998; Anderson et al. 2002 and references therein).

Conclusion

Among the mechanisms proposed to explain silica depletion, some (e.g., increased diatom production through eutrophication and the artificial lake effect) are well established, whereas others (e.g., trapping of phytoliths behind dams, loss of vegetated soils with damming, and increased BSi retention with increasing biodeposition) clearly deserve more research.

Numerous field and experimental studies (mesocosms) have related the shift from diatoms to nondiatom species, potentially harmful, to declining DSi:DIN and DSi:DIP ratios. But the relationship between DSi availability and diatom importance is not straightforward. A good illustration of this is the response of the Bay of Brest ecosystem to nutrient enrichment and declining DSi:DIN nutrient ratios, without any systematic shift from siliceous to nonsiliceous species. Whether this can be attributed to physical (Le Pape et al. 1996; Le Pape and Menesguen 1997) or biological (Chauvaud et al. 2000; Ragueneau et al. 2002a) factors is still under debate. Specifically related to Si cycling (e.g., in the Bay of Brest ecosystem), the interactions between nutrient enrichment and damming, for example, or between nutrient enrichment and the proliferation of an exotic species (*C. fornicata*) that acts as the biological component of the filter were found to modulate the effect of nutrient enrichment.

Whereas eutrophication is being reduced in Europe and the United States, it is increasing in Southeast Asia with growing population demand. There, major rivers are being dammed at accelerating rates (Milliman 1997), and aquaculture is flourishing to satisfy food demand (Ruiz et al. 2000). Thus, drastic reductions in DSi fluxes to the ocean might occur in this region, which plays a critical role in land–ocean fluxes of dissolved and particulate matter in general (Milliman and Maede 1983) and for DSi in particular (Ittekkot et al. 2000).

It is difficult to determine the extent to which anthropogenic inputs of DSi may compensate for the observed reductions in DSi fluxes. In addition, little is known about changes in the production and weathering of Si with changes in land use (Likens et al. 1970). The long-term effects of disturbance with agriculture and the more recent effects of acid rain on weathering, and thus the production of DSi, also are unknown.

Acknowledgments

This work is part of the Natural and Anthropogenic Modifications of the Si Cycle Along the Land–Ocean Continuum: Worldwide Ecological, Biogeochemical and Socio-economical Consequences (Si-WEBS) Research Training Network, funded by the European Union (HPRN-CT-2002-00218). We are very grateful to V. Ittekkot, C. Humborg, and T. Jennerjahn for inviting us to write this review chapter and organizing the fall 2003 Bremen meeting. Thanks to M. Briand for redrawing the figures.

Literature Cited

Alcaraz, M., E. Saiz, C. Marrasé, and D. Vaqué. 1988. Effects of turbulence on the development of phytoplankton biomass and copepod populations in marine mesocosms. *Marine Ecology Progress Series* 49:117–125.

Anderson, D. M., P. M. Glibert, and J. M. Burkholder. 2002. Harmful algal blooms and eutrophication: Nutrient sources, composition and consequences. *Estuaries* 25(4b):704–726.

Bidle, K. D., and F. Azam. 1999. Accelerated dissolution of diatom silica by marine bacterial assemblages. *Nature* 397:508–512.

Billen, G., C. Lancelot, and M. Meybeck. 1991. N, P, and Si retention along the aquatic continuum from land to ocean. Pp. 20–44 in *Ocean margin processes in global change*, edited by R. F. C. Mantoura, J.-M. Martin, and R. Wollast. Chichester, UK: Wiley.

Bodeanu, N. 1992. Algal blooms and the development of the main phytoplankton species at the Romanian Black Sea littoral in conditions of intensification of the eutrophication process. Pp. 891–906 in *Marine coastal eutrophication: The response of marine transitional systems to human impact—Problems and perspectives for restoration.* Proceedings of the International Conference of Bologna, March 21–24, 1990, edited by R. A. Vollenweider, R. Marchetti, and R. Viviani. Science of the Total Environment, Supplement 1992. Amsterdam: Elsevier.

Bologa, A. S., N. Bodeanu, A. Petran, V. Tiganus, and Y. Zaitsev. 1995. Major modifications of the Black Sea benthic and planktonic biota in the last three decades. *Bulletin de l'Institut Océanographique, Monaco* 15:85–110.

Bouvier, T., S. Becquevort, and C. Lancelot. 1998. Biomass and feeding activity of phagotrophic mixotrophs in the northwestern Black Sea during the summer 1995. *Hydrobiologia* 363:289–301.

Brzezinski, M. A., R. J. Olson, and S. W. Chisholm. 1990. Silicon availability and cell-cycle progression in marine diatoms. *Marine Ecology Progress Series* 67:83–96.

Budd, J. W., T. D. Drummer, T. F. Nalepa, and G. L. Fahrenstiel. 2001. Remote sensing of biotic effects: Zebra mussels (*Dreissena polymorpha*) influence on water clarity in Saginaw Bay, Lake Huron. *Limnology and Oceanography* 46(2):213–223.

Chauvaud, L., F. Jean, O. Ragueneau, and G. Thouzeau. 2000. Long-term variation of the Bay of Brest ecosystem: Pelagic–benthic coupling revisited. *Marine Ecology Progress Series* 200:35–48.

Cloern, J. E. 2001. Our evolving conceptual model of the coastal eutrophication problem. *Marine Ecology Progress Series* 210:223–253.

Cociasu, A., L. Dorogan, C. Humborg, and L. Popa. 1996. Long-term ecological changes in Romanian coastal waters of the Black Sea. *Marine Pollution Bulletin* 32(1):32–38.

Conley, D. J. 2000. Biogeochemical nutrient cycles and nutrient management strategies. *Hydrobiologia* 410:87–96.

Conley, D. J. 2002. Terrestrial ecosystems and the global biogeochemical silica cycle. *Global Biogeochemical Cycles* 16(4):1121.

Conley, D. J., and T. C. Malone. 1992. Annual cycle of dissolved DSi in Chesapeake Bay: Implications for the production and fate of phytoplankton biomass. *Marine Ecology Progress Series* 81:121–128.

Conley, D. J., C. L. Schelske, and E. F. Stoermer. 1993. Modification of the biogeochemical cycle of silica with eutrophication. *Marine Ecology Progress Series* 101:179–192.

Cowie, G. L., and J. I. Hedges. 1996. Digestion and alteration of the biochemical constituents of a diatom (*Thalassiosira weissflogii*) ingested by an herbivorous copepod (*Calanus pacificus*). *Limnology and Oceanography* 41:581–594.

Datnoff, L. E., G. H. Snyder, and G. H. Korndörfer (eds.). 2001. *Silicon in agriculture.* Amsterdam: Elsevier.

Del Amo, Y., and M. A. Brzezinski. 1999. The chemical form of dissolved Si taken up by marine diatoms. *Journal of Phycology* 35:1162–1170.

Del Amo, Y., O. Le Pape, P. Tréguer, B. Quéguiner, A. Menesguen, and A. Aminot. 1997. Impacts of high-nitrate freshwater inputs on macrotidal ecosystems. I. Seasonal evolution of nutrient limitation for the diatom-dominated phytoplankton of the Bay of Brest (France). *Marine Ecology Progress Series* 161:213–224.

DeMaster, D. J. 1981. The supply and accumulation of silica in the marine environment. *Geochimica et Cosmochimica Acta* 45:1715–1732.

Egge, J. K. 1997. Are diatoms poor competitors at low phosphate concentrations? *Journal of Marine Systems* 16:191–198.

Egge, J. K., and D. L. Aksnes. 1992. Silicate as regulating nutrient in phytoplankton competition. *Marine Ecology Progress Series* 83:281–289.

Egge, J. K., and A. Jacobsen. 1997. Influence of silicate on particulate carbon production in phytoplankton. *Marine Ecology Progress Series* 147:219–230.

Engstrom, D. R., E. B. Swain, and J. C. Kingston. 1985. A paleolimnological record of human disturbance from Harvey's Lake, Vermont: Geochemistry, pigments and diatoms. *Freshwater Biology* 15:261–288.

Escavarage, V., and T. C. Prins. 2002. Silicate availability, vertical mixing and grazing control of phytoplankton blooms in mesocosms. *Hydrobiologia* 484:33–48.

Estrada, M., C. Marrasé, and M. Alcaraz. 1988. Phytoplankton response to intermittent stirring and nutrient addition in marine microcosms. *Marine Ecology Progress Series* 48:223–235.

Fouillaron, P., A. Leynaert, P. Huonnic, A. Masson, and S. L'Helguen. Diatom dominance in marine ecosystems: Influence of silicic acid flux rather than concentration. Submitted.

Fransz, H. G., and J. H. G. Verhagen. 1985. Modeling research on the production cycle of phytoplankton in the Southern Bight of the North Sea in relation to riverborne nutrient load. *Netherlands Journal of Sea Research* 19(3–4):241–250.

Glibert, P. M., R. Magnien, M. W. Lomas, J. Alexander, C. Fan, E. Haramoto, M. Trice, and T. M. Kana. 2001. Harmful algal blooms in the Chesapeake and coastal bays of Maryland, USA: Comparison of 1997, 1998 and 1999 events. *Estuaries* 24:875–883.

Goldman, J. C. 1993. Potential role of large oceanic diatoms in new primary production. *Deep-Sea Research I* 40:159–186.

Gomoiu, M. T. 1992. Marine eutrophication syndrome in the north-western part of the Black Sea. Pp. 683–692 in *Marine coastal eutrophication: The response of marine transitional systems to human impact—Problems and perspectives for restoration.* Proceedings of the International Conference of Bologna, March 21–24, 1990, edited by R. A. Vollenweider, R. Marchetti, and R. Viviani. Science of the Total Environment, Supplement 1992. Amsterdam: Elsevier.

Guillard, R. R. L., and P. Kilham. 1978. The ecology of marine planktonic diatoms. Pp. 372–469 in *The biology of diatoms,* edited by D. Werner. Berkeley: University of California Press.

Hamm, C. E., R. Merkel, O. Springer, P. Jurkojc, C. Maier, K. Prechtel, and V. Smetacek. 2003. Architecture and material properties of diatom shells provide effective mechanical protection. *Nature* 421:841–843.

Hodgkiss, I. J., and K. C. Ho. 1997. Are changes in N:P ratios in coastal waters the key to increased red tide blooms? *Hydrobiologia* 852:141–147.

Howarth, R. W., G. Billen, D. Swaney, A. Townsend, N. Jaworski, K. Lajtha, and J. A. Downing et al. 1996. Regional nitrogen budgets and riverine N and P fluxes for the drainages to the North Atlantic Ocean: Natural and human influences. *Biogeochemistry* 35(1):75–139.

Humborg, C., S. Blomqvist, E. Avsan, Y. Bergensund, E. Smedberg, J. Brink, and C.-M. Mörth. 2002. Hydrological alterations with river damming in northern Sweden: Implications for weathering and river biogeochemistry. *Global Biogeochemical Cycles* 16(3):1039.

Humborg, C., V. Ittekkot, A. Cociasu, and B. von Bodungen. 1997. Effect of Danube River dam on Black Sea biogeochemistry and ecosystem structure. *Nature* 386:385–388.

Ittekkot, V., C. Humborg, and P. Schäfer. 2000. Hydrological alterations and marine biogeochemistry: A silicate issue? *BioScience* 50:776–782.

Justic, D., N. Rabalais, R. E. Turner, and Q. Dortch. 1995. Changes in nutrient structure of river-dominated coastal waters: Stoichiometric nutrient balance and its consequences. *Estuarine, Coastal and Shelf Science* 40:339–356.

Kilham, P. 1971. A hypothesis concerning silica and the freshwater planktonic diatoms. *Limnology and Oceanography* 16:10–18.

Lancelot, C., J. Staneva, and N. Gypens. 2004. Modelling the response of coastal ecosystems to nutrient change. *Oceanis* 28(3–4):531–556.

Le Pape, O., Y. Del Amo, A. Menesguen, A. Aminot, B. Quéguiner, and P. Tréguer. 1996. Resistance of a coastal ecosystem to increasing eutrophic conditions: The Bay of Brest (France), a semi-enclosed zone of Western Europe. *Continental Shelf Research* 16:1885–1907.

Le Pape, O., and A. Menesguen. 1997. Hydrodynamic prevention of eutrophication in the Bay of Brest (France): A modeling approach. *Journal of Marine Systems* 12:171–186.

Likens, G. E., F. H. Bormann, N. M. Johnson, D. W. Fisher, and R. S. Pierce. 1970. Effects of forest cutting and herbicide treatment on nutrient budgets in the Hubbard Brook watershed-ecosystem. *Ecological Monographs* 40:23–47.

Lomas, M. W., P. M. Glibert, D. A. Clougherty, D. A. Huber, J. Jones, J. Alexander, and E. Haramoto. 2001. Elevated organic nutrient ratios associated with brown tide blooms of *Aureococcus anophagefferens* (Pelagophyceae). *Journal of Plankton Research* 23:1339–1344.

Lund, J. W. G. 1950. Studies on *Asterionella formosa* Haas, II. Nutrient depletion and the spring maximum. *Journal of Ecology* 38:1–35.

Mack, R., D. Simberloff, V. M. Lonsdale, H. Evans, M. Clout, and F. A. Bazzaz. 2000. Biotic invasion: Causes, epidemiology, global consequences, and control. *Ecological Applications* 10(3):689–710.

Margalef, R. 1958. Temporal succession and spatial heterogeneity in phytoplankton. Pp. 323–349 in *Perspectives in marine biology*, edited by A. A. Buzzati-Traverso. Berkeley: University of California Press.

Martin-Jézéquel, V., M. A. Brzezinski, and M. Hildebrand. 2000. Silicon metabolism in diatoms: Implications for growth. *Journal of Phycology* 36:821–840.

Mee, L. D. 1992. The Black Sea in crisis: A need for concerted international action. *Ambio* 21(4):278–286.
Milligan, A. J., and F. M. M. Morel. 2002. A proton buffering role for silica in diatoms. *Science* 297:1848–1850.
Milliman, J. D. 1997. Blessed dams or damned dams? *Nature* 386:325–326.
Milliman, J. D., and R. H. Maede. 1983. World-wide delivery of river sediments to the oceans. *Journal of Geology* 91:1–21.
Mulholland, M. R., P. M. Glibert, G. M. Berg, L. van Heukelem, S. Pantoja, and C. Lee. 1998. Extracellular amino acid oxidation by micro-plankton: A cross-ecosystem comparison. *Aquatic Microbial Ecology* 15:141–152.
Nixon, S. W., J. W. Ammerman, and S. P. Seitzinger. 1996. The fate of nitrogen and phosphorus at the land–sea margin of the north Atlantic Ocean. *Biogeochemistry* 35:141–180.
Officer, C. B., and J. H. Ryther. 1980. The possible importance of silicon in marine eutrophication. *Marine Ecology Progress Series* 3:83–91.
Olsen, S., and E. Paasche. 1986. Variable kinetics of silicon-limited growth in *Thalassiosira pseudonana* (Bacillariophyceae) in response to changed chemical composition of the growth medium. *Journal of Phycology* 21:183–190.
Paasche, E. 1973. Silicon and the ecology of marine plankton diatoms. II. Silicate uptake kinetics in five diatom species. *Marine Biology* 19:262–269.
Pearsall, W. H. 1932. Phytoplankton of the English lakes, II. The composition of phytoplankton in relation to dissolved substances. *Journal of Ecology* 20:241–262.
Ragueneau, O., L. Chauvaud, A. Leynaert, G. Thouzeau, Y.-M. Paulet, S. Bonnet, A. Lorrain, J. Grall, R. Corvaisier, M. Le Hir, F. Jean, and J. Clavier. 2002a. Direct evidence of a biologically active coastal silicate pump: Ecological implications. *Limnology and Oceanography* 47(6):1849–1854.
Ragueneau, O., L. Chauvaud, B. Moriceau, A. Leynaert, G. Thouzeau, A. Donval, F. Le Loch, and F. Jean. 2005. Biodeposition by an invasive suspension feeder impacts the biogeochemical cycle of Si in a coastal ecosystem (Bay of Brest, France). *Biogeochemistry* 75:19–41.
Ragueneau, O., E. De Blas Varela, P. Tréguer, B. Quéguiner, and Y. Del Amo. 1994. Phytoplankton dynamics in relation to the biogeochemical cycle of silicon in a coastal ecosystem of Western Europe. *Marine Ecology Progress Series* 106:157–172.
Ragueneau, O., C. Lancelot, V. Egorov, J. Vervlimmeren, A. Cociasu, G. Déliat, A. Krastev, N. Daoud, V. Rousseau, V. Popovitchev, N. Brion, L. Popa, and G. Cauwet. 2002b. Biogeochemical transformations of inorganic nutrients in the mixing zone between the Danube River and the northwestern Black Sea. *Estuarine, Coastal and Shelf Science* 54(3):321–336.
Ragueneau, O., P. Tréguer, A. Leynaert, R. F. Anderson, M. A. Brzezinski, D. J. DeMaster, R. C. Dugdale, J. Dymond, G. Fischer, R. François, C. Heinze, E. Maier-Reimer, V. Martin-Jézéquel, D. Nelson, and B. Quéguiner. 2000. A review of the Si cycle in the modern ocean: Recent progress and missing gaps in the application of biogenic opal as a paleoproductivity proxy. *Global and Planetary Change* 26(4):315–366.
Riegman, R. 1995. Nutrient-related selection mechanisms in marine phytoplankton communities and the impact of eutrophication on the planktonic food web. *Water Science and Technology* 32:63–75.
Roberts, E. C., K. Davidson, and L. C. Gilpin. 2003. Response of temperate microplankton communities to N:Si ratio perturbation. *Journal of Plankton Research* 25(12):1–11.

Romdhane, M. S., H. C. Eilertsen, O. Dally Yahia-Kefi, and M. N. Dally Yehia. 1998. Toxic dinoflagellate blooms in Tunisian lagoons: Causes and consequences for aquaculture. Pp. 80–83 in *Harmful algae*, edited by B. Reguera, J. Blanco, M. L. Fernandez, and T. Wyatt. Paris: Xunta de Galicia and IOC of UNESCO.

Rothhaupt, K. O. 1996a. Laboratory experiments with a mixotrophic chrysophyte and obligately phagotrophic and phototrophic competitors. *Ecology* 77:716–724.

Rothhaupt, K. O. 1996b. Utilization of substitutable C- and P-sources by the mixotrophic chrysophyte *Ochromonas sp. Ecology* 77:706–715.

Rousseau, V., A. Leynaert, N. Daoud, and C. Lancelot. 2002. Diatom succession, silicification and silicic acid availability in Belgian coastal waters (southern North Sea). *Marine Ecology Progress Series* 236:61–73.

Ruiz, G. M., P. W. Fofonoff, J. T. Carlton, M. J. Wonham, and A. H. Hines. 2000. Invasion of coastal marine communities in North America: Apparent patterns, processes, and biases. *Annual Review of Ecology and Systematics* 31:481–531.

Sapozhnikov, V. V. 1992. Biohydrochemical causes of the changes of the Black Sea ecosystem and its present condition. *GeoJournal* 2:149–157.

Schelske, C. L., and E. F. Stoermer. 1971. Eutrophication, silica depletion, and predicted changes in algal quality in Lake Michigan. *Science* 173:423–424.

Schelske, C. L., E. F. Stoermer, D. J. Conley, J. A. Robbins, and R. Glover. 1983. Early eutrophication in the lower Great Lakes: New evidence from biogenic silica in sediments. *Science* 222:320–322.

Schelske, C. L., H. Züllig, and L. Boucherle. 1987. Limnological investigation of biogenic silica sedimentation and silica biogeochemistry in Lake St. Moritz and Zürich. *Schweizer Zeitung für Hydrologie* 49:42–50.

Smayda, T. J. 1990. Novel and nuisance of phytoplankton blooms in the sea: Evidence for a global epimedia. Pp. 29–40 in *Toxic marine phytoplankton*, edited by E. Graneli. New York: Elsevier.

Svensen, C. 2002. Eutrophication and vertical flux: A critical evaluation of silicate addition. *Marine Ecology Progress Series* 240:21–26.

Svensen, C., J. K. Egge, and J. E. Stiansen. 2001. Can silicate and turbulence regulate the vertical flux of biogenic matter? A mesocosm study. *Marine Ecology Progress Series* 217:67–80.

Tande, K. S., and D. Slagstad. 1985. Assimilation efficiency in herbivorous aquatic organisms: The potential of the ratio methods using ^{14}C and biogenic silica as markers. *Limnology and Oceanography* 30:1093–1099.

Tittel, J., V. Bissinger, B. Zippel, U. Gaedke, E. Bell, A. Lorke, and N. Kamjunke. 2003. Mixotrophs combine resource use to outcompete specialists: Implications for aquatic food webs. *Proceedings of the National Academy of Sciences of the United States of America* 100(22):12776–12781.

Tolmazin, D. 1985. Changing coastal oceanography of the Black Sea. I: Northwestern shelf. *Progress in Oceanography* 15:217–276.

Tréguer, P., D. M. Nelson, A. J. Van Bennekom, D. J. DeMaster, A. Leynaert, and B. Quéguiner. 1995. The silica balance in the world ocean: A reestimate. *Science* 268: 375–379.

Turner, R. E., N. Qureshi, N. N. Rabalais, Q. Dortch, D. Justic, R. F. Shaw, and J. Cope. 1998. Fluctuating silicate:nitrate ratios and coastal plankton food webs. *Proceedings of the National Academy of Sciences, USA* 95:13048–13051.

Turner, R. E., and N. N. Rabalais. 1991. Changes in Mississippi river water quality this century: Implications for coastal food webs. *BioScience* 41:140–147.

Van Bennekom, A. J., and W. Salomons. 1981. Pathways of nutrients and organic matter from land to ocean through rivers. Pp. 33–51 in *River inputs to ocean systems*, edited by J.-M. Martin, J. D. Burton, and D. Eisma. New York: UNEP, IOC, SCOR, United Nations.

Van Dokkum, H. P., J. H. J. Hulskotte, K. J. M. Kramer, and J. Wilmot. 2004. Emission, fate and effects of soluble silicates (waterglass) in the aquatic environment. *Environmental Science and Technology* 38:515–521.

Van Eeckhout, D., and C. Lancelot. 1997. Modelling the functioning of the north-western Black Sea ecosystem from 1960 to present. Pp. 455–468 in *Sensitivity of North Sea, Baltic Sea and Black Sea to anthropogenic and climatic changes*, edited by E. Ozsoy and A. Mikaelyan. NATO-ASI series. Dordrecht, The Netherlands: Kluwer.

Wahby, S. D., and N. F. Bishara. 1979. The effect of the river Nile on Mediterranean water, before and after the construction of the High Dam at Aswan. Pp. 311–318 in *River inputs to ocean systems*, edited by J.-M. Martin, J. D. Burton, and D. Eisma. Rome: Intergovernmental Oceanographic Commission, Scientific Committee on Oceanographic Research, U.N. Environmental Programme.

Wassmann, P., J. K. Egge, M. Reigstad, and D. L. Aksnes. 1996. Influence of dissolved silicate on vertical flux of particulate biogenic matter. *Marine Pollution Bulletin* 33:1–6.

13

Silicon Isotope–Based Reconstructions of the Silicon Cycle

Christina L. De La Rocha

There is much to be learned about biogeochemical cycles and their relationships with climate through the reconstruction of various aspects of the cycles back through geologic time. This is true not only for the more traditionally studied cycles, such as carbon and nitrogen, but for silicon (Si) as well. Focusing on the Si cycle allows us to study the growth of diatoms in the surface ocean over the last 60–100 million years, for which they have been important. The Si cycle is also of great interest over longer periods of geologic time because dissolved silica (DSi) may be the only major component of seawater whose average concentration in seawater has dropped by two orders of magnitude in the last 600 million years (Siever 1991).

Diatoms control the cycling of Si in the modern ocean and probably have done so since the Eocene (Siever 1991). Diatoms deposit opal (amorphous, hydrated silica) in their cell walls, making them the only major phytoplankton that needs DSi as a major nutrient. Diatoms are also ecologically important in the modern ocean, carrying out the bulk (about 50 percent) of the marine primary production (Nelson et al. 1995). Diatoms also dominate both primary and new production in areas such as the Southern Ocean, where the drawdown of dissolved inorganic carbon is critical to the control of atmospheric concentrations of carbon dioxide (Watson et al. 2000). Diatom production of opal from DSi is the process by which average marine DSi concentrations are as low as 10 μM, with values often approaching zero in surface waters (Tréguer et al. 1995). Before the Mesozoic appearance and subsequent radiation of the diatoms (Tappan and Loeblich 1973), marine DSi concentrations may have been as high as 1,000 μM, based on the types of chert that formed in marine sediments at those times (Siever 1991). It would be very interesting to know when (relative to the continuing evolution of the diatoms), how quickly, and in how many steps this significant drawdown in DSi occurred.

Reconstructing past concentrations of salts in the ocean, or of changes in the balance between fluxes of materials in and out of the ocean, is problematic. One potential means for reconstructing features of the Si cycle is the Si isotopic composition of sedimentary opal ($SiO_2 \cdot nH_2O$). Silicon has three stable isotopes: ^{28}Si at an abundance of 92.23 percent, ^{29}Si at 4.67 percent, and ^{30}Si at 3.10 percent (DeBievre and Taylor 1993). It has long been known that the ratio of these isotopes (e.g., ^{30}Si:^{28}Si) differs slightly between different geologic materials, with opal biominerals showing the greatest range of variation (Tilles 1961a, 1961b; Douthitt 1982; Ding et al. 1996). This suggests that isotopic fractionation occurs during biomineralization. Subsequent studies have worked to quantify the magnitude of Si isotopic fractionation with processes such as opal formation by diatoms, sponges, and land plants (De La Rocha et al. 1997; Ziegler et al. 2000; De La Rocha 2003; Varela et al. 2002) and the extent to which fractionation is influenced by environmental variables such as pH, temperature, and pCO_2 (De La Rocha et al. 1997; Mataliotaki et al. 2003; Milligan et al. 2004). From such studies it does appear that Si isotopes in biogenic opal (biogenic silica [BSi]) record information regarding the state of the silica cycle in the ocean (e.g., De La Rocha et al. 1998; Brzezinski et al. 2003; Wischmeyer et al. 2003) and that this signal may persist in the records for at least tens of millions of years (De La Rocha 2003).

Notation

Variations in the natural abundances of Si isotopes are expressed permil as the isotopic difference between a sample and standard, relative to the standard:

$$\delta^{30}Si = [(R_{sam} - R_{std})/R_{std}] \times 10^3‰, \qquad (1)$$

where R_{sam} and R_{std} are the ^{30}Si:^{28}Si ratios of a sample and standard, respectively. The standard used is NBS-28, a synthetic silica sand whose ^{30}Si:^{28}Si ratio has been characterized as 0.030924/0.922223, or 0.033532 (Coplen et al. 2002).

The degree of isotope fractionation between two materials is expressed in two ways, as α or as ϵ. The fractionation factor, α, is the ratio of the ^{30}Si:^{28}Si ratios (R) of two materials, a and b:

$$\alpha_{a-b} = R_a/R_b. \qquad (2)$$

Generally material a is the phase that is produced from the source reservoir, b. Thus the isotope fractionation between DSi and BSi produced by biomineralization α is calculated as

$$\alpha_{BSi-DSi} = R_{BSi}/R_{DSi}. \qquad (3)$$

When $\alpha = 1$ there is no isotopic fractionation between the two materials. The further away from α one gets in either direction, the greater the degree of isotopic difference between the two materials.

Isotope fractionation may also be expressed in ϵ notation as the difference between the δ values of the two materials, a and b:

$$\epsilon_{a-b} = \delta_a - \delta_b. \qquad (4)$$

In this case the δ value of the source material is subtracted from the δ value of the material produced. For BSi produced from DSi then, the calculation is

$$\epsilon_{BSi-DSi} = \delta^{30}Si_{BSi} - \delta^{30}Si_{DSi}. \quad (5)$$

Fractionation of Silica Isotopes During Opal Biomineralization

The compilation of the $\delta^{30}Si$ of geological materials showed a range of $\delta^{30}Si$ six times greater for BSi than for the igneous silicates that are the ultimate source of Si to the surface of the earth (Douthitt 1982). Subsequent measurements greatly expanding the size of the data set (Ding et al. 1996; De La Rocha et al. 1998, 2000; De La Rocha 2003) confirmed these observations. The average $\delta^{30}Si$ of all of the igneous samples that have been measured is –0.3‰, from a low value of –1.0‰ to a maximum of +0.4‰, a range of 1.4‰ (Figure 13.1).

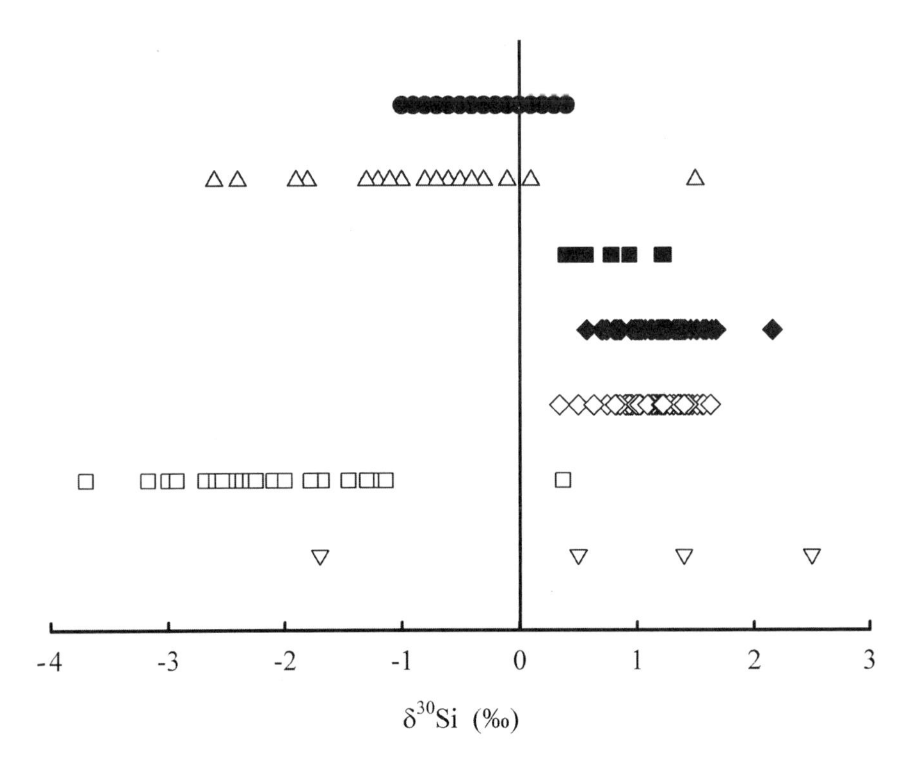

Figure 13.1. $\delta^{30}Si$ of samples of igneous rocks *(filled circles)*, clays *(open triangles)*, river water *(filled squares)*, seawater *(filled diamonds)*, marine diatoms *(open diamonds)*, sponges *(open squares)*, and phytoliths *(open triangles pointing down)*. Data are from Douthitt (1982), Ding et al. (1996), De La Rocha et al. (1998, 2000) and De La Rocha (2003).

By comparison, a handful of sponge samples alone range by 4.1‰, from –3.7‰ to +0.4‰. Diatom and phytolith silica tend to have a more positive $\delta^{30}Si$ signature (Figure 13.1) and bring the upper limit of $\delta^{30}Si$ of BSi measured to date to +2.5‰, for a total range of 6.2‰.

The greater range in $\delta^{30}Si$ of the biogenic samples compared with the igneous samples suggests that Si isotope ratios must be fractionated either during the weathering of Si out of igneous rocks or during opal biomineralization. As a first systematic investigation of this, several species of marine diatom were cultured and the $\delta^{30}Si$ of their opal compared with that of the DSi in their growth media (De La Rocha et al. 1997). Each of the three species yielded an α of 0.9989, corresponding to an ϵ of –1.1, the production of opal that is 1.1‰ more negative than its DSi source. Experiments run at several different temperatures did not yield a systematic variation in α with temperature or growth rate. An ϵ of –1.1‰ for opal biomineralization by diatoms has also been calculated from the slope of the relationship between $\delta^{30}Si$ of DSi and the natural log of the DSi concentration in the euphotic zone in the Monterey Bay upwelling zone during a time of diatom growth (De La Rocha et al. 2000). Recent work suggests that –1.5‰ may be a more accurate value of ϵ (Varela et al. 2002; Mataliotaki et al. 2003; Milligan et al. 2004). This newer estimate is still close enough to the older one to fall within the error bars of the original estimate.

Silicon isotope fractionation during phytolith formation by land plants appears to be on the same order as fractionation by diatoms (Ziegler et al. 2000). A preliminary set of experiments showed an ϵ of –1.5‰ (α = 0.9985) for phytolith production by corn (*Zea mays*) and –0.8‰ (α = 0.9992) for wheat (*Triticum* sp.). However, silicon isotope fractionation by marine sponges is of greater magnitude, with ϵ values ranging from –2.1‰ to –5.0‰ (De La Rocha 2003). The greater magnitude of this isotopic fractionation may be tied to the low affinity the sponges have for DSi compared with organisms such as diatoms.

Rayleigh Distillation of Silica Isotopes

Diatom growth in surface waters of the modern ocean depletes DSi concentrations to low levels. Because diatoms discriminate against the heavier isotopes of Si relative to the lighter ones, they will alter the isotopic composition of the reservoir DSi as they draw it down; the ^{30}Si:^{28}Si ratio (and thus $\delta^{30}Si$ value) of DSi in surface waters should increase as diatoms take it up to form opal (Figure 13.2).

The $\delta^{30}Si$ of the opal produced also increases as the DSi concentrations drop. The opal produced at any given point in time is offset from the $\delta^{30}Si$ of DSi by ϵ (Figure 13.2). However, the material that settles into the sediments probably represents an accumulation of the opal produced over the entire course of the drawdown of DSi.

The $\delta^{30}Si$ of this accumulating opal also increases with the further utilization of DSi (Figure 13.2) but can span only the range equivalent to the value of ϵ (e.g., from the

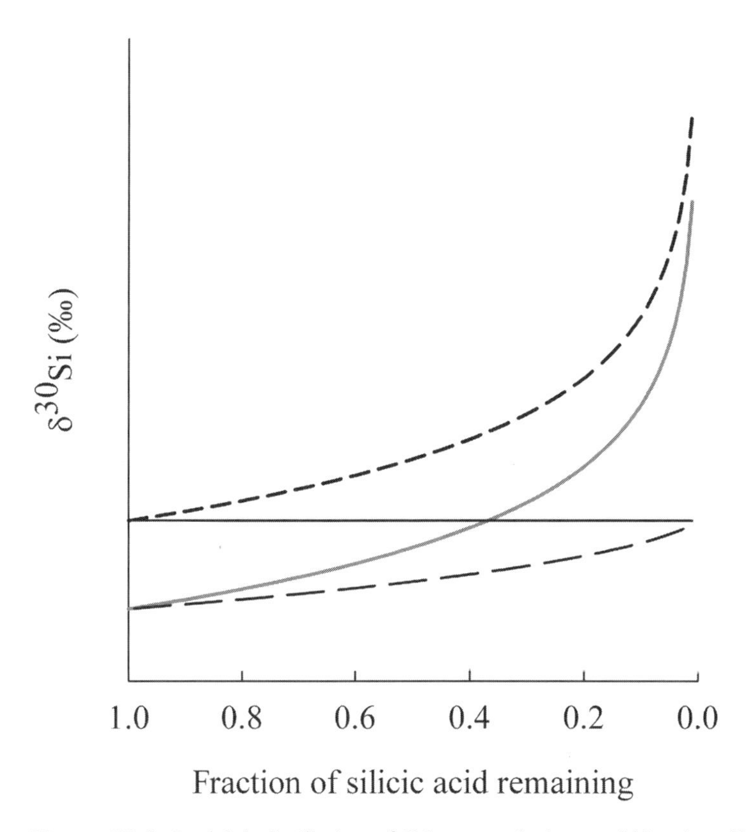

Figure 13.2. Rayleigh distillation of Si isotopes during opal biomineralization from a DSi reservoir of finite size. As the DSi concentration drops because of the conversion of DSi to biogenic opal, the δ^{30}Si of DSi *(bold dashed line)* increases. Paralleling this increase (but offset by ϵ) is the increase in the δ^{30}Si produced at a given moment in time during the drawdown of DSi *(solid gray line)*. The δ^{30}Si of all the opal produced from DSi *(dashed line)*, representing, for example, opal accumulating on the sea floor in the ocean, also increases as DSi concentrations drop, but only up to the initial δ^{30}Si value of DSi.

initial DSi δ^{30}Si plus ϵ to the initial DSi δ^{30}Si). In the case of marine diatoms, then, the expected range of δ^{30}Si in sedimentary opal at a given site is only about 1‰ or 1.5‰, depending on the exact value of the fractionation factor. Over this 1–1.5‰ possible range of variation the δ^{30}Si of sedimentary opal carries information concerning the degree of utilization of DSi by diatoms over a certain period of time, such as a bloom or a growing season. In general, the more positive the δ^{30}Si of marine diatoms is, the greater the extent to which they used up the DSi available to them.

Silica Isotopes in the Glacial Southern Ocean

The initial reconstructions of δ^{30}Si of sedimentary opal have focused on sites in the Southern Ocean (De La Rocha et al. 1998; Brzezinski et al. 2003), south of the present-day Antarctic Polar Front. Reconstruction of primary production and nutrient utilization in this region are particularly critical for understanding the ocean's role in the approximately 90 ppm drop in atmospheric concentrations of CO_2 between interglacial and glacial times (Knox and McElroy 1984; Sarmiento and Toggweiler 1984; Siegenthaler and Wenk 1984). Additionally, in this region diatoms are important primary producers, accounting, for example, for more than 90 percent of the productivity during ice edge blooms (Nelson and Smith 1986).

Four Southern Ocean sediment cores have yielded the same general patterns of δ^{30}Si over the last interglacial–glacial cycle (De La Rocha et al. 1998; Brzezinski et al. 2003): E50-11 and RC11-94 in the Indian sector of the Southern Ocean and RC13-269 and RC13-259 in the Atlantic sector of the Southern Ocean. As illustrated for E50-11 in Figure 13.3, in diatom opal from these cores the most positive values of δ^{30}Si occur near the top of the core, in the early Holocene or present interglacial.

The most negative values of δ^{30}Si occur at the last glacial maximum, the maximal extent of the continental and polar ice sheets, about 20,000 years ago. Intermediate values of δ^{30}Si occur in the lower but still glacial levels of the core. This pattern is repeated back at least through the beginning of the penultimate glacial cycle (Brzezinski et al. 2003).

The patterns of δ^{30}Si from Southern Ocean sediments have been interpreted to indicate a drop in DSi uptake during glacial times, south of the present-day Antarctic Polar Front. This interpretation is supported by the observation of lower opal accumulation rates in this area at this time relative to the interglacials (François et al. 1997; Frank et al. 2000). The export of organic carbon to the sediments at this time is also thought to be lower (François et al. 1997; Frank et al. 2000). Taken together, most of the proxy evidence for levels of export production suggests that during glacial times that DSi utilization was low, primary production was also at a minimum.

The one proxy whose results conflict with those of δ^{30}Si in the Southern Ocean is the nitrogen isotopes (δ^{15}N) measured on organic matter encapsulated in the silica of diatom frustules. The δ^{15}N of this diatom frustule organic matter implies that when levels of DSi utilization, export production, and opal accumulation are low, the utilization of nitrate is high. Increased stratification of the water column during glacial times could account for the decoupling between the extent of nitrate utilization and primary production (François et al. 1997; Sigman and Boyle 2000) but not for the decoupling between the two nutrient proxies.

It is difficult to reconcile how the utilization of one major nutrient might increase while the other decreases, and it is tempting to consider that one of the proxies may not be faithfully recording its nutrient cycle. But the antiphasing of the diatom records of δ^{15}N and δ^{30}Si could be tied (Brzezinski et al. 2003) to the amelioration of the Fe-limited conditions of the interglacials caused by increased input of Patagonian dust dur-

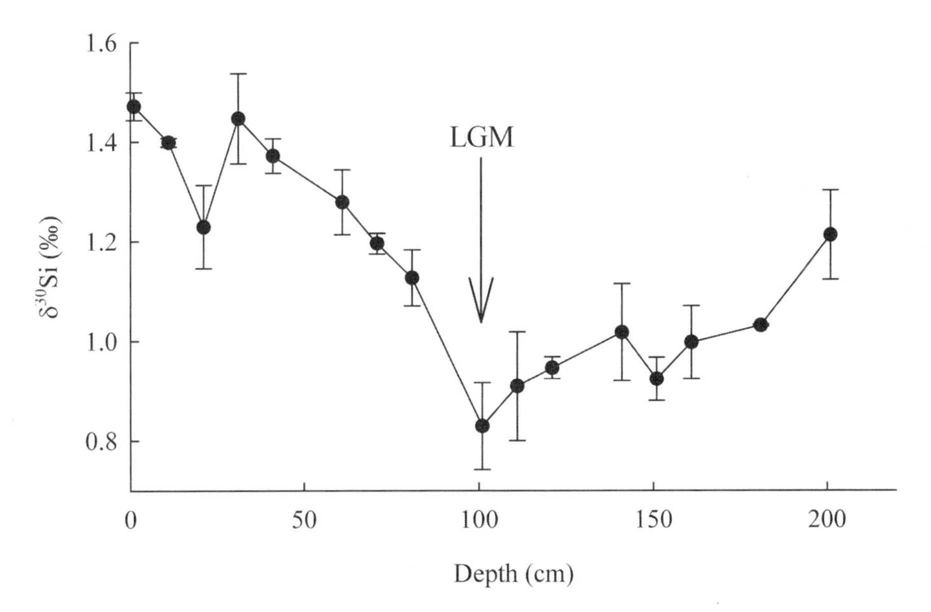

Figure 13.3. Typical pattern of δ^{30}Si of diatom opal over the last glacial cycle in the Southern Ocean, south of the present-day polar front, illustrated in core E50-11 (55°56′S, 104°57′E). The arrow indicates the location of the last glacial maximum in this core, as identified from the abundance of *Cycladophora davisiana.* Data are from De La Rocha et al. 1998.

ing the glacials (Petit et al. 1990). Because the N:Si uptake ratio of Fe-limited diatoms is 1:2 or 1:3 (Takeda 1998; Hutchins and Bruland 1998), compared with values of 1:1 for diatoms growing under nutrient-replete conditions (Brzezinski 1985), a shift toward Fe limitation would cause a drop in nitrate utilization and an increase in DSi utilization for a given level of carbon fixation.

Taking all of the proxy evidence of opal accumulation, export production, nitrate utilization, and DSi utilization together yields a picture of the glacial Southern Ocean as having equal or lower productivity than that of the modern day, diatoms less silicified than those currently growing there, and nitrate levels completely drawn down by a combination of Fe fertilization and a decrease in the upwelling flux of nutrients to the euphotic zone.

This interpretation is based on the results of only a handful of sediment cores and warrants further validation. The approximately 15,000-year residence time of Si in the ocean means that an imbalance between inputs and outputs to the ocean could result in shifts in average marine δ^{30}Si over the 100,000 years of a glacial–interglacial cycle. However, it is unlikely that changes in the silica budget of the ocean over glacial–interglacial cycles contribute more than a minor amount of the observed signal of δ^{30}Si

(De La Rocha et al. 1998; De La Rocha and Bickle 2005). Modeling exercises further suggest that regional variations in the δ^{30}Si of DSi upwelled to the euphotic zone are unlikely to contribute to the variations observed in the Southern Ocean sediments (Wischmeyer et al. 2003).

Potential for Reconstruction of Long-Term Changes in the Silica Cycle

The Si isotopic signal recorded in silica may be robust enough to allow δ^{30}Si to be reconstructed much further back in time than the few hundred thousand years that have been looked at by De La Rocha et al. (1998) and Brzezinski et al. (2003). Although work directly assessing the effects of diagenesis of opal-*a* (amorphous hydrated silica, as is produced during biomineralization) to opal-*ct* (a mixture of cristobalite and tridymite, metastable crystalline phases of silica) and quartz on δ^{30}Si remains to be carried out, measurements suggest that Si isotope signatures may last for tens to hundreds of millions of years. For example, eighteen samples of radiolarian chert from the Holocene back through the Cambrian show variations in δ^{30}Si of 1.4‰, averaging +0.1 ± 0.5‰ (Ding et al. 1996). Although there is no way of knowing the δ^{30}Si signature of the radiolarians before their diagenesis to chert, the range of values measured on the chert is similar to the values of +0.2 to +0.3‰ measured on modern radiolarians collected from the Pacific (Ding et al. 1996). Sponge spicules from the Eocene–Oligocene boundary also show δ^{30}Si signatures very similar to those of sponges growing in modern seawater (De La Rocha 2003). Even more convincingly, Precambrian silicic rocks, including banded iron formations, show δ^{30}Si signatures greater than 2‰ and 3‰ (Ding et al. 1996; P. Knauth and C. L. De La Rocha, unpublished data), far above the average igneous rock value of –0.3‰.

If δ^{30}Si signatures are preserved in siliceous materials over at least the last 100 million years, they may prove useful in dating the drawdown of the DSi concentrations of ocean waters associated with the rise of diatoms to dominance over the silica cycle in the ocean (Maliva et al. 1989; Siever 1991; Racki and Cordey 2000). Before about 60 million years ago, the distribution and texture of chert produced in the geologic record suggest that marine DSi concentrations were on the order of 1,000 µM (Siever 1991). After the radiation of the diatoms and their rise to dominance over the marine silica cycle, DSi concentrations declined, resulting in the thinning of radiolarian tests (Racki and Cordey 2000) and sponge spicules (Maldonado et al. 1999). As the DSi concentration of the ocean dropped over millennia through removal by diatoms, the δ^{30}Si of the deep ocean should have also declined by 1–2‰ (De La Rocha and Bickle 2005). The exact isotopic shift and its duration should reflect the duration over which there was net removal of Si from the ocean and the amount by which export fluxes of Si exceeded the input fluxes. The presence of singular or multiple isotopic excursions might also indicate whether the drop in DSi concentration to modern levels occurred in one or many steps between the Cretaceous and Eocene. Such information might be gleaned from the δ^{30}Si of deep-sea sponges, which should be recording the average δ^{30}Si of seawater DSi.

Control of the Silicon Isotopic Composition of Silicon Inputs to the Ocean

The δ^{30}Si of DSi in the deep ocean could vary over geologic history for other reasons than the excessive removal of Si by the sedimentation of BSi. The δ^{30}Si of DSi entering the ocean might vary significantly over time. Many processes could change this value. The most important candidate is the removal of DSi in estuaries. A large portion of the DSi in river waters is stripped out by BSi production in estuaries (Conley and Malone 1992; Tréguer et al. 1995), resulting in an increase, due to Rayleigh distillation, in the δ^{30}Si of the DSi input to the ocean (De La Rocha, unpublished data). The global ratio of estuarine area relative to river influx to the ocean is not fixed in time and may influence the proportion of riverine DSi that never reaches the sea. As a result, climatic features such as sea level transgression and regression and rainfall patterns may have a profound impact on the δ^{30}Si of the major Si input to ocean waters.

The δ^{30}Si of DSi in rivers is also quite variable, on the order of 2–3‰ (De La Rocha et al. 2000; Ding et al. 2004). This may be due to the uptake of Si into phytoliths or biogenic silica produced by land plants, or it may be linked to processes of weathering and clay formation (De La Rocha et al. 2000; Ziegler et al. 2005a, 2005b) or the formation of silica cements (Basile-Doelsch et al. 2005). Both phytolith production and clay formation appear to fractionate Si isotopes (Ziegler et al. 2000, 2005a, 2005b). Phytoliths and clays form large repositories for Si weathered out of rocks (Conley 2002), potentially serving as reservoirs of Si isotopically fractionated such that riverine Si has δ^{30}Si values higher than those of igneous rocks (De La Rocha et al. 2000; Ding et al. 2004).

The data available for the δ^{30}Si of clays (Douthitt 1982; Ding et al. 1996) and phytoliths (Douthitt 1982), which are shown in Figure 13.1, suggest that clays, not phytoliths, control the isotopic composition of river water. Clays (open triangles in Figure 13.1) have a lower average δ^{30}Si (–0.7‰) than both igneous rocks (–0.3‰) and river water (+0.8‰) and thus could serve as the accumulation of low δ^{30}Si silicon needed in the terrestrial environment to shift up the δ^{30}Si between igneous and riverine Si. Recent work on the isotopic composition of basaltic soils of various ages and within a granitic saprolite demonstrates that a combination of primary mineral dissolution, secondary mineral formation, and biogenic cycling of Si controls the δ^{30}Si of local stream waters (Ziegler et al. 2005a, 2005b).

The δ^{30}Si of rivers should vary with the intensity of weathering, which in turn is a function of temperature, terrain, and rainfall, factors that vary both regionally and over time and are linked with climate. Because Si isotopes appear to be fractionated during the formation of clays, the δ^{30}Si of the DSi in river water and groundwater is a function of both the fractionation factor and the proportion of weathered Si that is locked up in clays (De La Rocha et al. 2000). When a smaller proportion of the Si released from rocks is sequestered in clays, the δ^{30}Si of river waters should be more negative (assuming an ϵ of about −1‰; Ziegler et al. 2000). Conversely, when a greater proportion of the weathered Si becomes incorporated into clay minerals, the river water δ^{30}Si should be more positive. If the proportion of weathered Si that is sequestered into clays varies over time, as with changing climate, this could result in a shift in the δ^{30}Si of deep ocean seawater.

Unless the weathering-related shifts in δ^{30}Si are large, they will be difficult to distinguish from δ^{30}Si variations caused by nutrient utilization, shifts in the average value for Si isotope fractionation during biomineralization, or imbalances between inputs and outputs of Si from the ocean. For those of us who work in the ocean, weathering-related variations in δ^{30}Si of river waters will be a nagging possibility of a few tenths of permil of any variations observed over a few thousand years of record. But in the terrestrial system Si isotopes may be useful for looking at local changes in weathering. Analysis of δ^{30}Si of diatoms from lake sediments, especially when coupled with analyses of Ge:Si ratios (Filippelli et al. 2000), may prove a useful indicator of weathering intensity in specific catchment areas over climate cycles. They might also provide information about the introduction of crops such as wheat and rice.

Conclusion

Silicon isotopes have only started to show their utility in the reconstruction of past Si cycling. As more measurements are made over the next few years, the picture of Si cycling in the Southern Ocean over the Quaternary should fill in, and the first isotope-based records of Si cycling at lower latitudes will be produced. Investigation of the distribution and behavior of Si isotopes in small catchments, soils, and rivers will help to pin down the controls on the δ^{30}Si of Si inputs to the ocean and their potential range of variation over time. This in turn will promote the understanding of longer-term variations in δ^{30}Si in the sedimentary record.

Literature Cited

Basile-Doelsch, I., J. D. Meunier, and C. Parron. 2005. Another continental pool in the terrestrial silicon cycle. *Nature* 433:399–402.

Brzezinski, M. A. 1985. The Si:C:N ratio of marine diatoms: Interspecific variability and the effect of some environmental variables. *Journal of Phycology* 21:347–357.

Brzezinski, M. A., C. J. Pride, V. M. Franck, D. M. Sigman, J. L. Sarmiento, K. Matsumoto, N. Gruber, G. H. Rau, and K. H. Coale. 2003. A switch from $Si(OH)_4$ to NO_3^- depletion in the glacial Southern Ocean. *Geophysical Research Letters* 29:10.1029/2001GL014349.

Conley, D. J. 2002. Terrestrial ecosystems and the global biogeochemical silica cycle. *Global Biogeochemical Cycles* 16:10.1029/2002GB001894.

Conley, D. J., and T. C. Malone. 1992. Annual cycle of dissolved silicate in Chesapeake Bay: Implications for the production and fate of phytoplankton biomass. *Marine Ecology Progress Series* 81:121–128.

Coplen, T. B., J. A. Hopple, J. K. Böhlke, H. S. Peiser, S. E. Rieder, H. R. Krouse, K. J. R. Rosman, T. Ding, R. D. Vocke Jr., K. M. Révész, A. Lamberty, P. Taylor, and P. De Bievre. 2002. *Compilation of minimum and maximum isotope ratios of selected elements in naturally occurring terrestrial materials and reagents.* USGS Water-Resources Investigations Report 01-4222. Reston, VA: U.S. Department of the Interior and U.S. Geological Survey.

DeBievre, P., and P. D. P. Taylor. 1993. Table of the isotopic composition of the elements. *International Journal of Mass Spectrometry and Ion Physics* 123:149.

De La Rocha, C. L. 2003. Silicon isotope fractionation by marine sponges and the reconstruction of the silicon isotope composition of ancient deep water. *Geology* 31:423–426.

De La Rocha, C. L., and M. J. Bickle. 2005. Sensitivity of silicon isotopes to whole-ocean changes in the silica cycle. *Marine Geology* 217:267–282.

De La Rocha, C. L., M. A. Brzezinski, and M. J. DeNiro. 1997. Fractionation of silicon isotopes by marine diatoms during biogenic silica formation. *Geochimica et Cosmochimica Acta* 61:5051–5056.

De La Rocha, C. L., M. A. Brzezinski, and M. J. DeNiro. 2000. A first look at the distribution of the stable isotopes of silicon in natural waters. *Geochimica et Cosmochimica Acta* 64.2467–2477.

De La Rocha, C. L., M. A. Brzezinski, M. J. DeNiro, and A. Shemesh. 1998. Silicon-isotope composition of diatoms as an indicator of past oceanic change. *Nature* 395:680–683.

Ding, T., S. Jiang, D. Wan, Y. Li, J. Li, H. Song, Z. Liu, and X. Yao. 1996. *Silicon isotope geochemistry.* Beijing: Geological Publishing House.

Ding, T., D. Wan, C. Wang, and F. Zhang. 2004. Silicon isotopic compositions of dissolved silicon and suspended matter in the Yangtze River, China. *Geochimica et Cosmochimica Acta* 68:205–216.

Douthitt, C. B. 1982. The geochemistry of the stable isotopes of silicon. *Geochimica et Cosmochimica Acta* 46:1449–1458.

Filippelli, G. M., J. W. Carnahan, L. A. Derry, and A. Kurtz. 2000. Terrestrial paleorecords of Ge/Si cycling derived from lake diatoms. *Chemical Geology* 168:9–26.

François, R., M. A. Altabet, E.-F. Yu, D. M. Sigman, M. P. Bacon, M. Frank, G. Bohrman, G. Bareille, and L. D. Labeyrie. 1997. Contribution of Southern Ocean surface–water stratification to low atmospheric CO_2 concentrations during the last glacial period. *Nature* 389:929–935.

Frank, M., R. Gersonde, M. Rutgers van der Loeff, G. Bohrmann, C. C. Nürnberg, P. W. Kubik, M. Suter, and A. Mangini. 2000. Similar glacial and interglacial export bioproductivity in the Atlantic sector of the Southern Ocean: Multiproxy evidence and implications for glacial atmospheric CO_2. *Paleoceanography* 15:642–658.

Hutchins, D., and K. W. Bruland. 1998. Iron-limited growth and Si:N uptake ratios in a coastal upwelling regime. *Nature* 393:561–564.

Knox, F., and M. B. McElroy. 1984. Changes in atmospheric CO_2: Influence of the marine biota at high latitude. *Journal of Geophysical Research* 89:4629–4637.

Maldonado, M., M. C. Carmona, M. J. Uriz, and A. Cruzado. 1999. Decline in Mesozoic reef-building sponges explained by silicon limitation. *Nature* 401:785–788.

Maliva, R. G., A. H. Knoll, and R. Siever. 1989. Secular change in chert distribution: A reflection of evolving biological participation in the silica cycle. *Palaios* 4:519–532.

Mataliotaki, I., C. L. De La Rocha, U. Passow, and D. A. Wolf-Gladrow. 2003. Impact of the pH-dependent speciation of silicic acid on the silicon isotope composition of diatoms. *Geophysical Research Abstracts* 5:09375.

Milligan, A. J., D. E. Varela, M. A. Brzezinski, and F. M. M. Morel. 2004. Dynamics of silicon metabolism and silicon isotopic discrimination in a marine diatom as a function of pCO_2. *Limnology and Oceanography* 49:322–329.

Nelson, D. M., and W. O. Smith Jr. 1986. Phytoplankton bloom dynamics of the western Ross Sea ice edge. II. Mesoscale cycling of nitrogen and silicon. *Deep-Sea Research* 33:1389–1412.

Nelson, D. M., P. Tréguer, M. A. Brzezinski, A. Leynaert, and B. Quéguiner. 1995. Production and dissolution of biogenic silica in the ocean: Revised global estimates, comparison with regional data and relationship to biogenic sedimentation. *Global Biogeochemical Cycles* 9:359–372.

Petit, J. R., L. Mounier, J. Jouzel, Y. S. Korotkevich, V. I. Kotlyakov, and C. Lorius. 1990. Paleoclimatological and chronological implications of the Vostok core dust record. *Nature* 343:56–58.

Racki, G., and F. Cordey. 2000. Radiolarian palaeoecology and radiolarites: Is the present the key to the past? *Earth Science Reviews* 52:83–120.

Sarmiento, J. L., and J. R. Toggweiler. 1984. A new model for the role of the oceans in determining atmospheric pCO_2. *Nature* 308:621–624.

Siegenthaler, U., and T. H. Wenk. 1984. Rapid atmospheric CO_2 variations and ocean circulation. *Nature* 308:624–626.

Siever, R. 1991. Silica in the oceans: Biological–geochemical interplay. Pp. 287–295 in *Scientists on Gaia,* edited by S. H. Schneider and P. J. Boston. Cambridge, MA: MIT Press.

Sigman, D. M., and E. A. Boyle. 2000. Glacial/interglacial variations in atmospheric carbon dioxide. *Nature* 407:859–869.

Takeda, S. 1998. Influence of iron availability on nutrient consumption ratio of diatoms in oceanic waters. *Nature* 393:774–777.

Tappan, H., and A. P. Loeblich Jr. 1973. Evolution of the oceanic plankton. *Earth Science Reviews* 9:207–240.

Tilles, D. 1961a. Natural variations in isotopic abundances of silicon. *Journal of Geophysical Research* 66:3003–3013.

Tilles, D. 1961b. Variations of silicon isotope ratios in a zone pegmatite. *Journal of Geophysical Research* 66:3015–3020.

Tréguer, P., D. M. Nelson, A. J. Van Bennekom, D. J. DeMaster, A. Leynaert, and B. Quéguiner. 1995. The silica balance in the world ocean: A reestimate. *Science* 268:375–379.

Varela, D. E., C. J. Pride, M. A. Brzezinski, and M. J. DeNiro. 2002. Natural variations in silicon isotope abundances as indicators of silica production in the Southern Ocean. *2002 Ocean Sciences Meeting Abstracts,* OS21l-08.

Watson, A. J., D. C. E. Bakker, A. Ridgwell, P. W. Boyd, and C. S. Law. 2000. Effect of iron supply on Southern Ocean CO_2 uptake and implications for glacial atmospheric CO_2. *Nature* 407:730–733.

Wischmeyer, A. G., C. L. De La Rocha, E. Maier-Reimer, and D. A. Wolf-Gladrow. 2003.

Control mechanisms for the oceanic distribution of silicon isotopes. *Global Biogeochemical Cycles* 17:10. 1029/2002GB002022.
Ziegler, K., O. A. Chadwick, M. A. Brzezinski, and E. F. Kelly. 2005a. Natural variations of δ^{30}Si ratios during progressive basalt weathering, Hawaiian Islands. *Geochimica et Cosmochimica Acta* 69:4597–4610.
Ziegler, K., O. A. Chadwick, E. F. Kelly, and M. A. Brzezinski. 2000. Silicon isotope fractionation during weathering and soil formation: Preliminary experimental results. *Journal of Conference Abstracts* 5:1135.
Ziegler, K., O. A. Chadwick, A. F. White, and M. A. Brzezinski. 2005b. δ^{30}Si systematics on a granitic saprolite, Puerto Rico. *Geology* 33:817–820.

14

Long-Term Oceanic Silicon Cycle and the Role of Opal Sediment

Christoph Heinze

Biogeochemical forcing of the long-term marine Si cycle ultimately starts on land through the erosion of siliceous particles from dry areas and the subsequent input of silicic acid to the ocean through the dissolution of particles deposited on the surface and through river runoff, mainly in dissolved form as silicic acid. The input of silicic acid through river loads is estimated to be larger than the eolian input by a factor of about 10 (Tréguer et al. 1995). However, the amount of Si, which does not reach the open ocean but rather is sedimented in estuaries and shelf seas, has not yet been precisely quantified. The ocean acts as an interface between this Si input from external sources and output through sediment accumulation. The marine "biological particle pump" together with early diagenetic processes perform the conversion from Si input to Si output. This pump is overridden by a series of complex biogeochemical processes. Si as a nutrient is essential for the formation of siliceous shell material of diatoms (phytoplankton) and radiolarians (zooplankton). Si is not a biolimiting nutrient like N and P. Life in the ocean would also be present without large Si supplies. However, lack of Si can slow down biological production in ecological systems that otherwise favor the existence of diatoms. Wherever $Si(OH)_4$ in significant amounts is pumped into the euphotic zone, siliceous primary producers thrive if enough light is available and no ice cover reduces plant growth.

In the paleocean, the strength of the biological pump and hence the amount of biogenic silica (opal) (BSi) in sediments is determined by the Si supplies from surface layers and laterally through Si input from land and through vertical internal supply by ocean currents.

The concentration of BSi in marine sediments has potential as a paleoclimate tracer (Ragueneau et al. 2000). However, the space–time pattern of BSi in the sediment results from a complex interplay of physical, chemical, and biological factors. This makes the use of opal as a quantitative paleoclimate tracer difficult.

What reflects the opal sediment primarily; that is, for what is it a useful tracer in a qualitative sense? In today's ocean and in the past, high concentrations of BSi are found in the sediment in oceanic upwelling regions (Leinen et al. 1986) and other regions with vertical water movements. Pronounced opal sediment weight percentages prevail in the eastern and central equatorial Pacific, the Southern Ocean, the Namibian upwelling regime, and to a lesser extent in the equatorial Atlantic and along the northwest sides of the North Atlantic and North Pacific oceans. In these areas, nutrients—and also silicic acid—are brought to the sea surface from deeper layers and induce significant biological productivity in the euphotic zone and a corresponding export flux of particles to below the euphotic zone. Accordingly, the remnants of diatoms and radiolarians sink to the ocean floor at these sites. The nature of the vertical movements and of the biological production mode can be different. Higher upward current velocities and related biological export production occur in wind-induced upwelling areas but also at deep water production sites, areas of large seasonal mixed layer depth variations, and areas of intense eddy mixing activity. No simple correlation can be established between annually averaged biological export production and concentration of BSi in the sediment. For example, in the Southern Ocean, large amounts of opal dominate the top sediment, although the annually averaged biological BSi production in this high nutrient, low chlorophyll region probably is not at its potential maximum. On the other hand, deposition of the opaline material on the ocean floor is strongly seasonally pulsed around Antarctica (Fischer et al. 1988; Beucher et al. 2004). This seasonality effect therefore might contribute more to strong opal sedimentation than to a high average export production rate.

Uncertainties in the Use of BSi as a Paleoclimate Tracer

These physical and biological influences combine with a series of chemical factors, which determine the rates of dissolution or preservation of opal in the water column and the sediment (Van Cappellen et al. 2002). Altogether at least nine factors influence the dissolution kinetics—or, conversely, the degree of opal preservation—in the water column and the top sediment layer: organic coatings on the particles, degree of undersaturation in surrounding seawater, temperature, presence of dissolved aluminum (or further contaminants), specific surface area of particulate material, settling velocity, the ratio of detrital material to opal, bioturbation, and aging. The top sediment layer is defined here as the bioturbated zone or sediment mixed layer, that is, the zone where porewater tracers deviate significantly from the asymptotic concentration achieved in deeper sediment layers, where diffusive property transport is only negligibly small. All these factors have a qualitatively similar effect on biogenic opal: They either increase or decrease rates of dissolution. As long as no clear quantitative laws for the effects of the different processes on opal dissolution rates can be made, it is very difficult to estimate the contributions of these different processes to opal degradation from a fit of models to observations. Respective systems for inverse computations of the governing parameters are not well conditioned, and only a subset of parameters can be uniquely determined (Heinze et al. 2003).

Also confusing the issue are the many different ways in which the opal content of sediments are reported in the literature. Usually the weight percentage of opal is given either relative to total dry weight of sediment or relative to dry weight on a calcite-free basis, that is, excluding any $CaCO_3$ contributions to the total sediment weight. Furthermore, sometimes opal accumulation rates are reported, that is, flux rates of solid opal out of the sediment mixed layer down into the lithosphere. From sediment core analysis, these flux rates are more difficult to determine than the weight percentage data because quantifications of the sedimentation rate or deposition rate onto the ocean bottom have to be made. These can be made on the basis of ^{230}Th measurements, under the assumption that ^{230}Th, which is produced locally in a water column above a unit area of sediment, is locally completely washed out by particles through scavenging. This assumption includes potential errors of up to 100 percent of the sedimentation rate, as shown by Henderson et al. (1999).

A further uncertainty complicates the interpretation of opal data: sediment focusing. Locally sediment can be transported slowly or quickly into topographic lows and be eroded from topographic highs. This can lead to large errors in the opal accumulation rate estimates.

However, the uncertainties associated with the general degradation of biogenic opal in the oceans are not large enough to exclude opal as a useful paleoclimate tracer. It is certain that opaline sediments clearly mark oceanic upwelling systems, with the exception of very sluggish basinwide upwelling such as in the North Pacific Ocean at the end of the ocean conveyor belt circulation.

Observations on BSi Sediment and Sediment Porewater

Our knowledge on the behavior of BSi as an oceanic tracer is based largely on observations from the modern ocean. Measurements of porewater silicic acid are not as numerous as core top solid sediment opal weight percentages. However, the available $Si(OH)_4$ porewater data are numerous enough to show the basic distributions and their variability. The apparent solubility values of opal in sediment can be deduced from the asymptotic $Si(OH)_4$ porewater concentrations to which the actual silicic acid depth profiles in the sediment approach. These k_D values differ from site to site and reveal potential inhibiting factors on opal solubility such as admixture of aluminum from continental sources (e.g., Kongo river plume; Van Bennekom et al. 1989) and also as yet unresolved factors (e.g., Southern Ocean; Rabouille et al. 1997). So far the wide variation in the solubility of opal, as documented by the different asymptotic concentrations of silicic acid in porewater at different sites, poses more questions than answers on the processes influencing the solubility and dissolution kinetics of BSi in sediment. These questions were discussed in Archer et al. (1993).

The core top sediment distribution is summarized at present in three different data sets, which build on each other. The first comprehensive surface sediment (*surface* here means the "surface" of the sediment column, which can lie several kilometers deep on the ocean floor) data set for opal on a calcite-free basis was published by Leinen et al. (1986). This data set, which was transferred into digitized form by David Archer, added an opal core top data set relative to total sediment weight. The data set by Archer was then extended by Dittert et al. (2002). It now includes previously unpublished data and provides not only a general core top data subset but also data with more detailed sampling depth levels down in the sediment cores (Figures 14.1 and 14.2).

In general, all surface sediment samples, especially from regions with low sedimentation rates, have to be considered carefully with respect to the age of the sample. The most recent part of the sediment core can get lost during sampling or be removed by erosion by bottom currents, and the top sediment layer does not always represent the real modern ocean but rather can reflect the sedimentary situation several thousand years ago. The dating of opal sediment samples is a major problem not only for core top samples but downcore as well. Age models are missing in most cases where downcore data sets are reported in publicly accessible databases. In sediments where opal dominates and $CaCO_3$ material is essentially lacking, dating of the opal material itself is more difficult because there is a shortage of material on which ^{14}C or ^{18}O measurements can be carried out straightforwardly. In the future, standardized publicly accessible databases for paleoceanographic data are necessary for merging data sets and model results. Still, we are a long way from databases in which it is possible to directly synchronize downcore opal data with those of other sedimentary and porewater components. The determination of age models for single cores can be supported by simulations of synthetic sediment cores, where the model delivers a consistent age model relative to the downcore sediment distributions (Heinze 2001). However, relying on the correctness of sediment deposition and accumulation rates alone will not be sufficient, and the resulting age models will have to be checked at least by a number of reliable direct age determinations of the downcore sediment distribution.

Because of the age determination and synchronization difficulties, data sets for past time slices such as the last glacial maximum (21 kyr BP) or the Miocene ocean with open Panama Isthmus are scarce and still associated with uncertainties (e.g., Keller and Barron 1983; Heinze and Dittert 2005). The full potential of opal as a paleoclimate tracer might be used once all available data are made accessible to all scientists together with the best possible information on sample ages. A worldwide data policy and agreement between scientists following rules of good scientific practice will allow a more straightforward use of potentially accessible BSi data sets with benefits for all (Dittert et al. 2001).

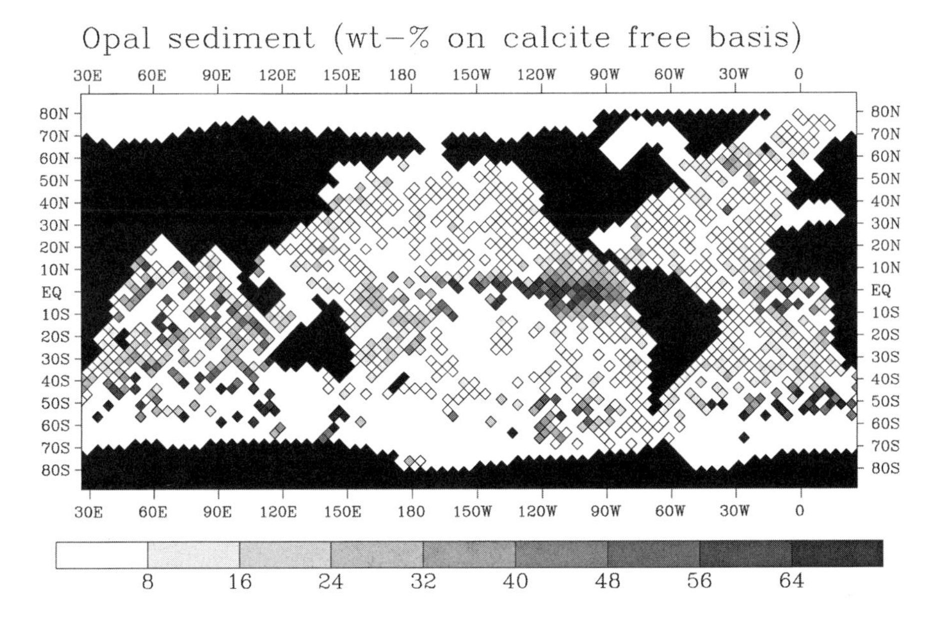

Figure 14.1. Core top BSi concentrations on a calcite-free basis as given in the data set from Dittert et al. (2002). This data set also includes the opal data sets as compiled by Leinen et al. (1986).

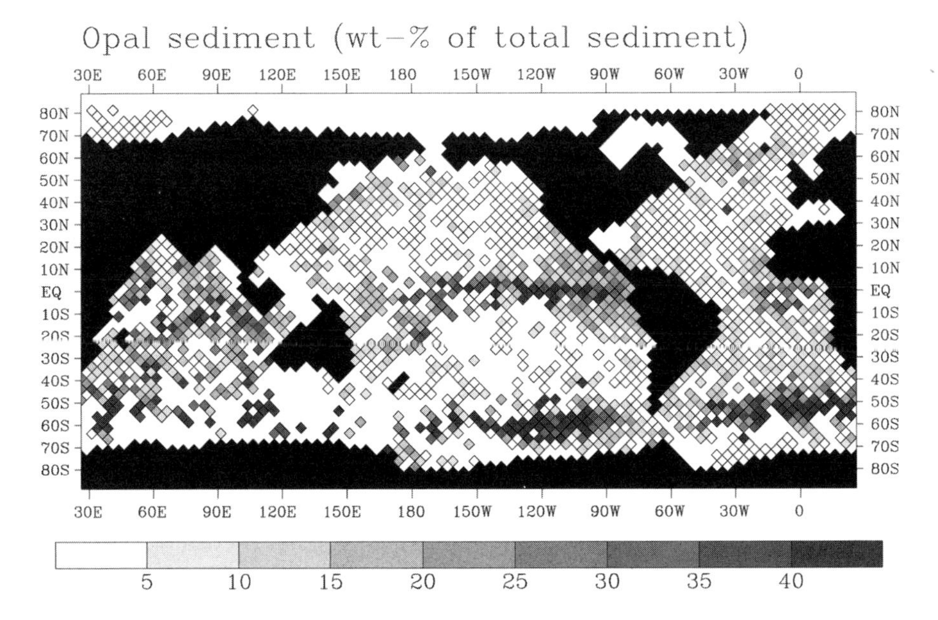

Figure 14.2. Core top BSi concentrations relative to total sediment as given in the data set from Dittert et al. (2002). This data set also includes the opal data sets as compiled by Leinen et al. (1986).

Modeling the Silica Sediment

Modeling the sediment always must have a clear scientific reason and should not be done for its own sake, as pointed out by Boudreau (1997:4): "You must have a well-defined problem/question, and one for which modeling is useful, if not necessary, in providing an answer." That means that the goal of the modeling exercise must be clear before the correct model can be formulated and applied.

First of all, why are models necessary for quantifications of the marine opal sediment as a function of space and time? In general, five reasons can be named for the necessity of modeling next to observational or experimental work:

- Some sedimentary processes are so slow that they cannot be directly observed easily, and it would take more than a lifetime to obtain usable measurements.
- Measurement systems perturb the systems (e.g., sampling of sedimentation with bottom sediment traps or direct measurements of porewater silicic acid concentrations with electrodes).
- Methods for analysis of sediment core samples and corresponding dating and for porewater measurements are laborious and expensive, so a close sampling network in space and time probably will never be feasible.
- As compared with the atmosphere, oceanic processes work on small spatial scales and long time scales. Therefore, synoptically reliable data sets are difficult to achieve.
- Predictions cannot be measured per se.

A classic sediment model taking into account at least two different reactive sediment species was developed by Archer (1991) and further developed by Archer et al. (1993) to include the silicon cycle. The fundamental differential equations in this model for the solid material components and the porewater concentrations are

$$\frac{dS}{dt} = D_B \frac{\partial^2 S}{\partial z^2} - \frac{\partial}{\partial z}(w \cdot S) - \frac{R \cdot M}{\rho \cdot (1 - \Phi)}, \quad (1)$$

and

$$\frac{dP}{dt} = \frac{\partial}{\partial z}(D_W \frac{\partial P}{\partial z}) + \frac{R}{\Phi}, \quad (2)$$

where

t = time
z = depth
S = solid component (in weight fraction of total sediment)
P = dissolved porewater component
D_B = diffusion coefficient for bioturbation
W = vertical advection velocity of solid compounds
R = reaction rate

M = molecular weight of solid component
Φ = density of bulk sediment
Φ = porosity
D_W = diffusion coefficient for porewater diffusion.

In the work of Archer et al. (1993) the system of differential equations is solved for the steady state using a relaxation method. Conceptually, the diagenetic model of Archer et al. (1993) was introduced into a three-dimensional biogeochemical ocean general circulation model (BOGCM). At first $CaCO_3$ and organic material (Archer and Maier-Reimer 1994) were included and later $CaCO_3$, organic carbon, BSi, and "clay" (an inert solid fraction) were included by Heinze et al. (1999). The BOGCM is the Hamburg ocean carbon cycle circulation (HAMOCC) model, based on the original model version by Maier-Reimer (1993). The HAMOCC model can be configured for different purposes. For long-term integrations, an annually averaged model is available. It is computationally very efficient and can be integrated over hundreds of thousands of years (Heinze et al. 1999, 2003). The model usually is initialized from constant water column tracer concentrations and clay-only sediment (clay being treated as an inert solid fraction that serves primarily as a diluter of the biogenic sediment). When we integrate the model forward from this initial condition by applying temporally constant sources of $Si(OH)_4$ (and also TCO_2, total alkalinity, PO_4^{3-}, and O_2), the model gradually builds up its own dedicated biogenic sediment cover consisting of opal, $CaCO_3$, and organic carbon compounds, which all are diluted by the clay fraction. After several thousand years, the model approaches equilibrium conditions (steady state). Then global input as prescribed and global output through sediment accumulation balance each other. The global sediment accumulation rate in this case is not a prognostic model variable but a prescribed value through specification of the external source strengths, which have been kept constant in time. However, the spatial sediment distribution is a prognostic model variable because it depends on the flow field and the biogeochemical process parameterizations applied in the model world.

The BSi export production as simulated by the model closely follows the vertical velocity distribution (Figures 14.3 and 14.4).

The structure of the BSi sediment mirrors the export production closely regardless of the basis (calcite-free or total sediment) on which the opal sediment fraction is reported (Figures 14.5 and 14.6).

Case Study: Variable Delivery of Silica to the World Ocean

Now we will address the question of how the external inputs of Si to the sea surface and outputs through sediment accumulation and hence loss to the lithosphere balance each other. In complete equilibrium, inputs and outputs would balance each other. Although this ideal state may never completely be fulfilled in nature, it is easy to realize such a situation in an ocean model.

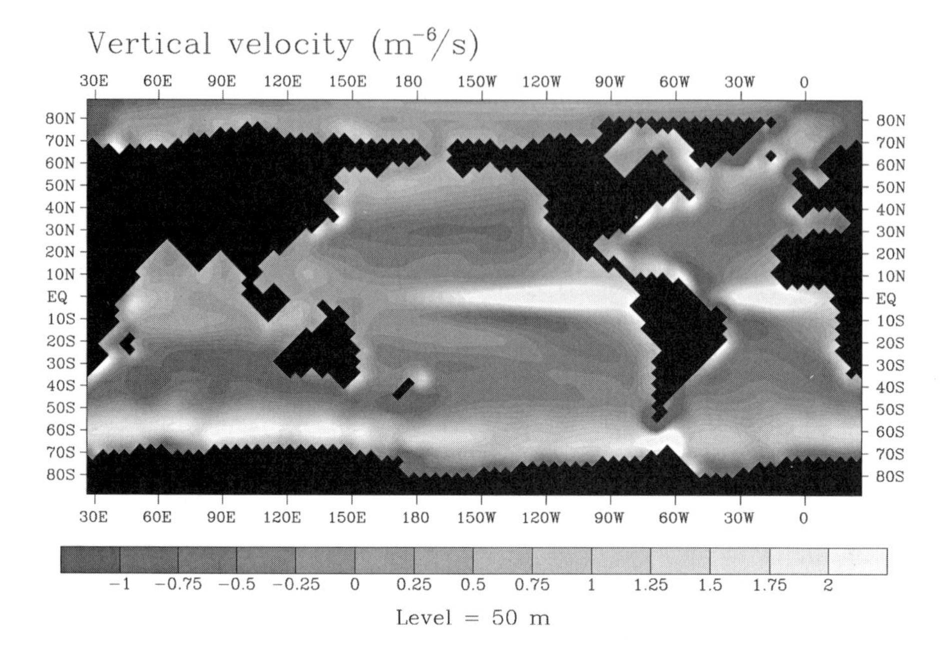

Figure 14.3. Vertical velocity of model circulation at 50 m according to the preindustrial velocity field as computed by Winguth et al. (1999) and averaged over one annual cycle.

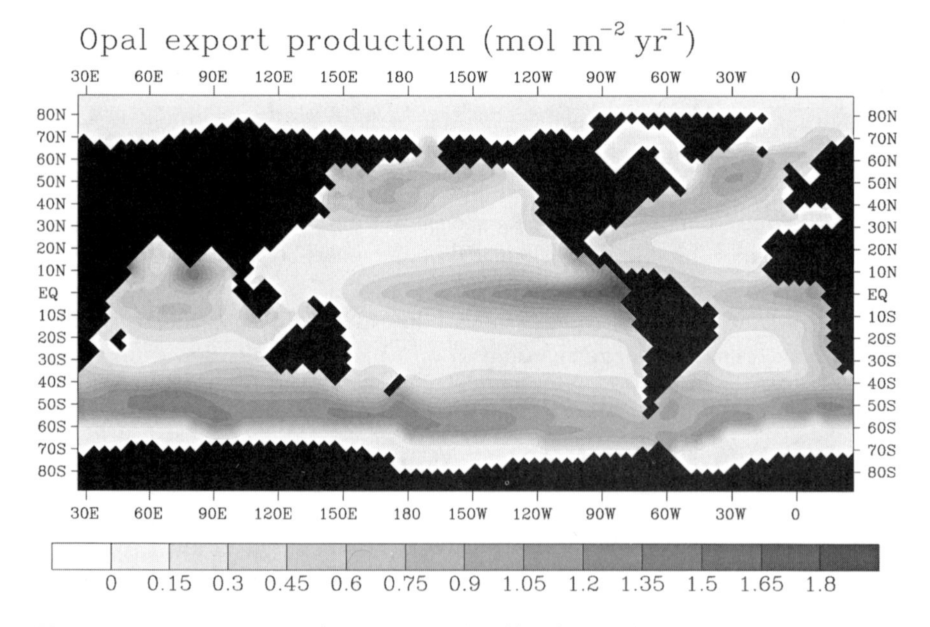

Figure 14.4. BSi export production as simulated by the Hamburg ocean carbon cycle circulation model (HAMOCC model) (after Heinze et al. 2003).

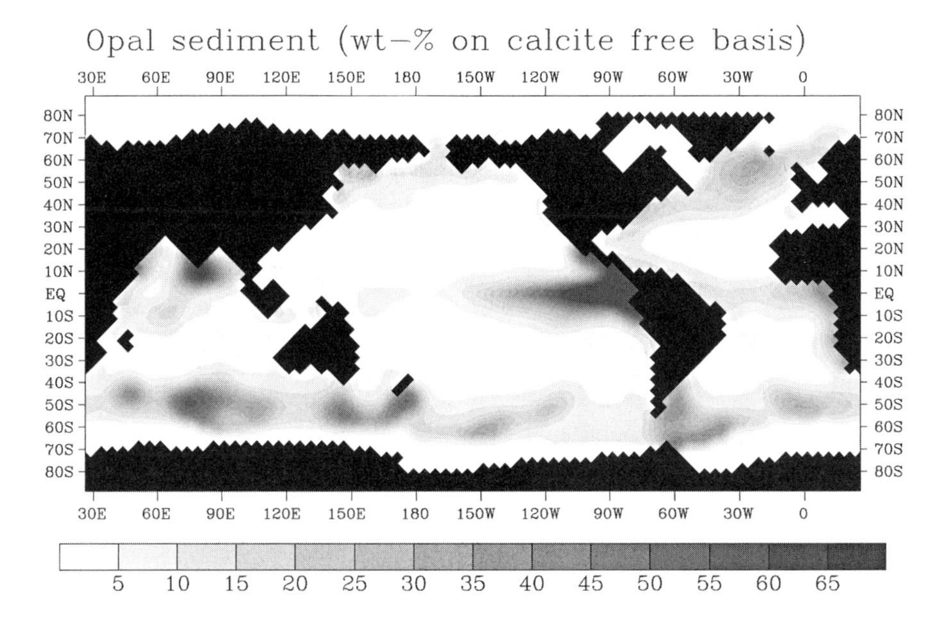

Figure 14.5. Modeled BSi sediment reported on a calcite-free basis (HAMOCC model, after Heinze et al. 2003).

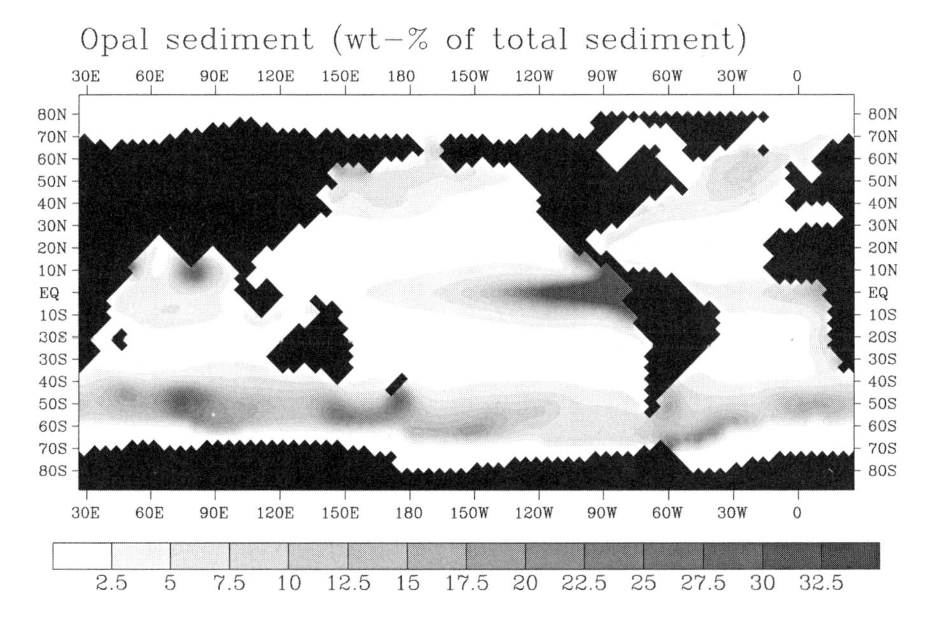

Figure 14.6. Modeled BSi sediment reported relative to total sediment (HAMOCC model, after Heinze et al. 2003).

What happens if this ideal equilibrium between input and output of Si is perturbed? Candidates for perturbing factors are in principle all parameters that govern the marine Si cycle. However, several factors may be particularly efficient in changing the marine Si cycle and hence marine biogeochemistry in general, such as climate-induced weathering changes, mountain uplift and stronger erosion, closure and opening of ocean gateways, wind stress strength and curl changes, ecological changes, changes in ocean stratification, changes in seasonality, and admixture of Al.

Because it is ultimately the biogeochemical flux from land that fuels the oceanic silicon cycle, we now test how and on which time scale the ocean biogeochemical system reacts to variations in Si delivery. We will carry out four very simple sensitivity experiments relative to an established control run. This control run is the optimized Si cycle simulation from Heinze et al. (2003) with the circulation field for the interglacial ocean as provided by Winguth et al. (1999). The corresponding Si budget for the global ocean is shown in Figure 14.7 together with the respective numbers as given by Tréguer et al. (1995).

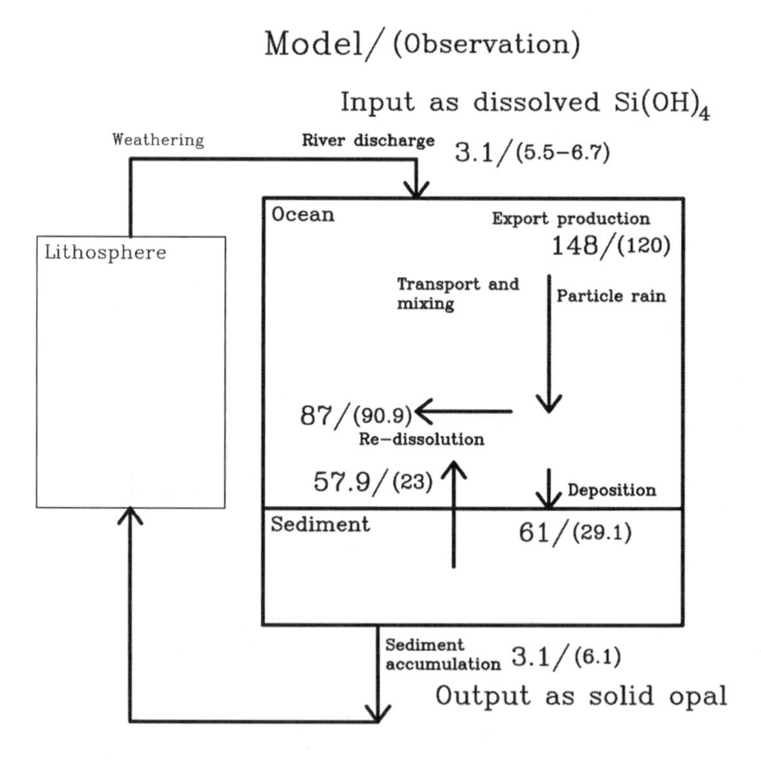

Figure 14.7. Si budget of the global ocean following Heinze et al. (2003) *(large numbers).* For comparison, the fluxes according to Tréguer et al. (1995) are shown in parentheses *(small numbers).* All numbers are fluxes of 10^{12} mol Si per year.

The following sensitivity experiments with respect to the control run where carried out:

Run 1: Si input reduced to 25 percent.
Run 2: Si input increased by 400 percent.
Run 3: Si input set to zero.
Run 4: Step function. Si input increased by a factor of ten over 20 kyr, then reduced to zero over 20 kyr and finally set back to control run value.

All runs were restarted from control run conditions; that is, they were not spun up separately but rather used the control run fields as initial conditions.

The first two runs mirror each other and address strong decreases and increases in external Si delivery to the world ocean. These changes theoretically can result from large-scale retention of Si in freshwater reservoirs or from increased weathering caused by mountain uplift. Run 3 is an extreme scenario for an instantaneous stop of Si input from external sources. Finally, Run 4 shows a strong step function with a strong peak in Si delivery, a complete stop of Si delivery after 20,000 years, and a return to normal conditions after another 20,000 years. Experiments 1–3 show that after a sudden change in external Si input, which is then kept at a new constant level, it takes about 50,000 years until the sediment accumulation and the external Si input again balance each other globally (Figures 14.8–14.10). The corresponding e-folding time is consistent with the estimated average residence time of 20,000 years for Si in the present ocean (Broecker and Peng 1982).

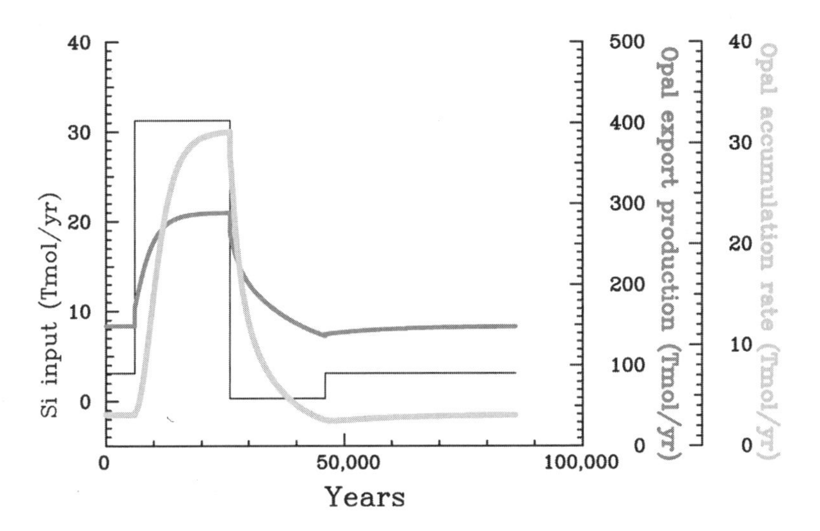

Figure 14.8. Model experiment Run 1: After about 50,000 years, the sediment accumulation has achieved the new equilibrium value, which equals the decreased external Si input rate.

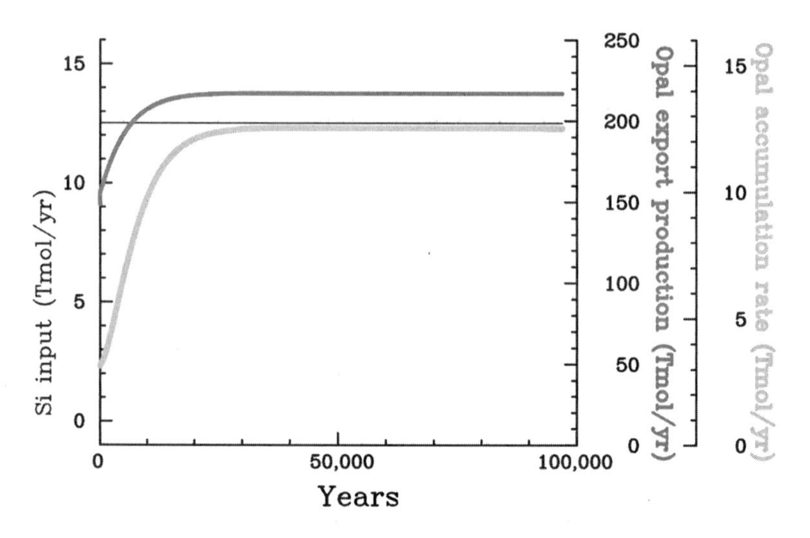

Figure 14.9. Model experiment Run 2: Mirror image experiment to Run 1. After about 50,000 years, the sediment accumulation has achieved the new equilibrium value, which equals the highly increased external Si input rate.

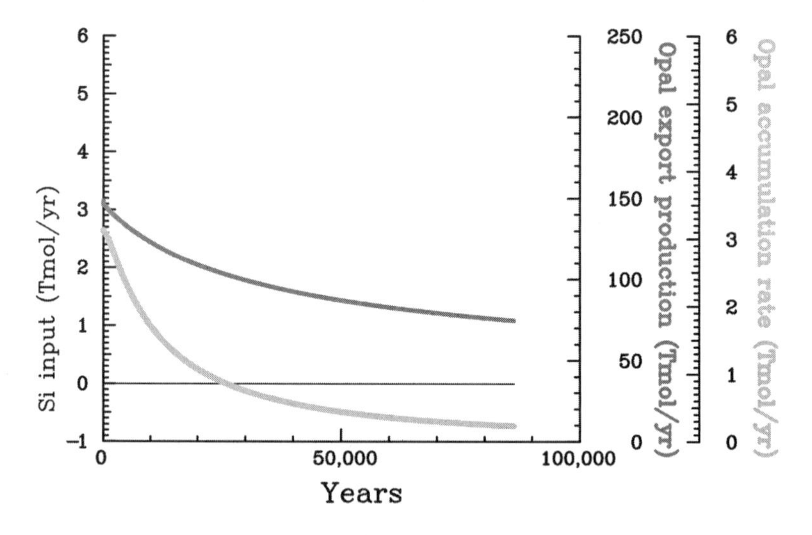

Figure 14.10. Model experiment Run 3: The external Si input is stopped completely. After 80,000 years, the sediment accumulation is almost zero, but the opal export production is still operating at about 50% of the control run value.

In all cases, it is interesting to see that the reaction of the opal export production is much smaller than the reaction of the sediment accumulation rates. Even after 80,000 years and a complete stop in Si input to the pelagic ocean, the global biogenic silica export production is still operating at about 50 percent of its present value (which corresponds to the control run export production rate). Also, the transient experiment (Run 4, Figure 14.11) illustrates the smaller relative change of the opal export production relative to the sediment accumulation.

Because the total ocean throughput of Si is only a small fraction of the total ocean Si inventory, this small change in opal export production is no surprise.

The strong Si input, as prescribed in Run 2, may be unrealistic in view of co-limitation by P and N, which have not been increased relative to the Si delivery in that experiment.

After a perturbation of the Si input rates, the model always tries to adjust so that input and output rates balance each other. Such an equilibration is reassuring because one would also expect such an equilibration to take place in the real world. On the other hand, errors in the formulation of the porewater chemistry or export production can lead to a wrong inventory of Si in the water column. An attempt to obtain reliable paleoceanographic reconstructions of the silicic acid concentration in the water column and a corresponding model optimization based on these data could efficiently reduce the uncertainties.

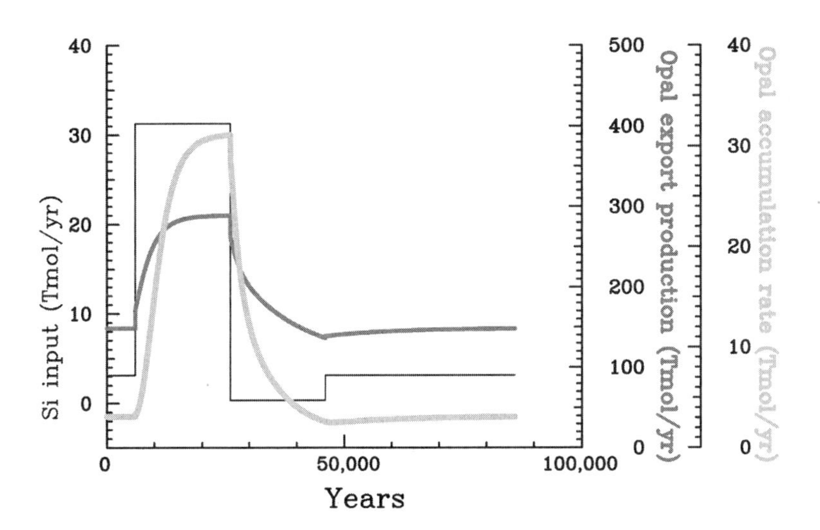

Figure 14.11. Model experiment Run 4: The external Si input is prescribed as a sudden peak, a sudden drop, and a return to the control run value. The relative changes in opal export production are smaller than those for sediment accumulation.

Conclusion

In case of a change in external Si delivery to the ocean, the large opal redissolution rate from the sediment works as a negative feedback. The percentage of the opal particles, which enter the bioturbated sediment zone from above, increases in case of a decrease in Si delivery and vice versa. The Si system of the open ocean adjusts slowly to changes in the external Si supply because of the large Si inventory in the open ocean water column. Therefore, a drastic change in Si delivery caused by water reservoir and dam building or by a change in Si supply at the last deglaciation will have no significant impact on the open ocean Si system. Of course, this may in no way hold for shallow coastal oceans, where changes in the continental Si delivery may have severe consequences for marine ecology. Because of the stability of the oceanic Si system, variability in the opal sediment on shorter time scales than the reaction time scale to external input changes may reliably reflect internal changes in productivity and circulation of the open ocean.

Acknowledgments

Support for this work from European Union grants HPRN-CT-2002-00218 Si-WEBS and EVK2-CT-2001-00100 ORFOIS is gratefully acknowledged. Thanks to Christina De La Rocha for very useful comments. This is publication No. A 107 of the Bjerknes Centre for Climate Research.

Literature Cited

Archer, D. 1991. Modeling the calcite lysocline. *Journal of Geophysical Research* 96(C9):17037–17050.

Archer, D., M. Lyle, K. Rodgers, and P. Froelich. 1993. What controls opal preservation in tropical deep-sea sediments? *Paleoceanography* 8:7–21.

Archer, D., and E. Maier-Reimer. 1994. Effect of deep-sea sedimentary calcite preservation on atmospheric CO_2 concentration. *Nature* 367:260–263.

Beucher, C., P. Tréguer, A. M. Hapette, R. Corvaisier, N. Metzl, and J. J. Pichon. 2004. Intense summer Si-recycling in the surface Southern Ocean. *Geophysical Research Letters* 31(9):L09305.

Boudreau, B. P. 1997. *Diagenetic models and their implementation.* Berlin: Springer Verlag.

Broecker, W. S., and T.-H. Peng. 1982. *Tracer in the sea.* Palisades, NY: Eldigio Press.

Dittert, N., M. Diepenbroek, and H. Grobe. 2001. Scientific data must be available to all. *Nature* 414:393.

Dittert, N., M. Diepenbroek, C. Heinze, and O. Ragueneau. 2002. Managing (pale-)oceanographic data sets using the PANGAEA information system: The SINOPS example. *Computers & Geosciences* 28:789–798.

Fischer, G., D. Fütterer, R. Gersonde, S. Honjo, D. R. Ostermann, and G. Wefer. 1988. Seasonal variability of particle flux in the Weddell Sea and its relation to ice cover. *Nature* 335:426–428.

Heinze, C. 2001. Towards the time dependent modelling of sediment core data on a global basis. *Geophysical Research Letters* 28(22):4211–4214.
Heinze, C., and N. Dittert. 2005. Impact of paleocirculations on the silicon redistribution in the world ocean. *Marine Geology* 214(1–3):201–213.
Heinze, C., A. Hupe, E. Maier-Reimer, N. Dittert, and O. Ragueneau. 2003. Sensitivity of the marine biospheric Si cycle for biogeochemical parameter variations. *Global Biogeochemical Cycles* 17(3):1086.
Heinze, C., E. Maier-Reimer, A. M. E. Winguth, and D. Archer. 1999. A global oceanic sediment model for longterm climate studies. *Global Biogeochemical Cycles* 13:221–250.
Henderson, G. M., C. Heinze, R. F. Anderson, and A. M. E. Winguth. 1999. Global distribution of the ^{230}Th flux to ocean sediments constrained by GCM modelling. *Deep-Sea Research I* 46:1861–1893.
Keller, G., and J. A. Barron. 1983. Paleoceanographic implications of Miocene deep-sea hiatuses. *Geological Society of America Bulletin* 94:590–613.
Leinen, M., D. Cwienk, G. R. Heath, P. E. Biscaye, V. Kolla, J. Thiede, and J. P. Dauphin. 1986. Distribution of biogenic silica and quartz in recent deep-sea sediments. *Geology* 14:199–203.
Maier-Reimer, E. 1993. Geochemical cycles in an ocean general circulation model: Preindustrial tracer distributions. *Global Biogeochemical Cycles* 7:645–677.
Rabouille, C., J.-F. Gaillard, P. Tréguer, and M.-A. Vincendeau. 1997. Biogenic silica recycling in surficial sediments across the Polar Front Zone of the Southern Ocean (Indian sector). *Deep Sea Research II* 44:1151–1176.
Ragueneau, O., P. Tréguer, A. Leynaert, R. F. Anderson, M. A. Brzezinski, D. J. DeMaster, R. C. Dugdale, J. Dymond, G. Fischer, R. François, C. Heinze, E. Maier-Reimer, V. Martin-Jézéquel, D. M. Nelson, and B. Quéguiner. 2000. A review of the Si cycle in the modern ocean: Recent progress and missing gaps in the application of biogenic opal as a paleoproductivity tracer. *Global and Planetary Change* 26:317–365.
Tréguer, P., D. M. Nelson, A. J. Van Bennekom, D. J. DeMaster, A. Leynaert, and B. Quéguiner. 1995. The balance of silica in the world ocean: A re-estimate. *Science* 268:375–379.
Van Bennekom, A. J., J. H. F. Jansen, S. J. van der Gaast, J. M. van Iperen, and J. Pieters. 1989. Aluminium-rich opal: An intermediate in the preservation of biogenic silica in the Zaïre (Congo) deep-sea fan. *Deep-Sea Research* 36:173–190.
Van Cappellen, P., S. Dixit, and J. van Beusekom. 2002. Biogenic silica dissolution in the oceans: Reconciling experimental and field-based dissolution rates. *Global Biogeochemical Cycles* 16(4):1075.
Winguth, A. M. E., D. Archer, J.-C. Duplessy, E. Maier-Reimer, and U. Mikolajewicz. 1999. Sensitivity of paleonutrient tracer distributions and deep sea circulation to glacial boundary conditions. *Paleoceanography* 14:304–323.

15

The Perturbed Silicon Cycle

Venugopalan Ittekkot, Daniela Unger, Christoph Humborg, and Nguyen Tac An

Silicon probably is the only element that so clearly shows the link between rock and life. Previous chapters describe this link by tracing its movement from rocks and minerals to life and then back to minerals and rock. On its way it exerts control over the distribution and fate of carbon and other nutrient elements, thereby affecting the four planetary spheres. Here we summarize the information from previous chapters and describe the major findings and recommendations from the project based on deliberations at the regional workshops (Ittekkot 2000; Ittekkot et al. 2000, 2003; Tac An et al. 2002).

Chemical weathering of silicate rocks and minerals is the natural source of dissolved silica (DSi) in rivers (White and Brantley 1995). The weathering process uses up carbon dioxide and water and depends on temperature and precipitation. Because of their geological and climatic settings, tropical river basins such as those of Asia appear to play a major role in chemical weathering and the transfer of DSi and alkalinity to rivers and oceans (Chapters 2 and 4, this volume).

The acidification of rain and continental surface waters is one of the anthropogenic perturbations that affect the chemical weathering of silicates and the corresponding DSi inputs to water bodies. The increase of CO_2 in the atmosphere also has an effect, but this effect is believed to occur in the long term. However, recent studies suggest that the increase in alkalinity observed from long-term monitoring records (about 50 years) in some of the major world rivers such as the Mississippi could have resulted from an acceleration of chemical weathering, which in turn is linked to land use changes (Raymond and Cole 2003), urbanization (Chapter 4, this volume), and increased atmospheric CO_2 (Ittekkot 2003).

Except perhaps for processes such as weathering and the related transfer of DSi to rivers, information on the terrestrial cycling of Si has been scarce. This is especially true for Si in the terrestrial biosphere (Chapter 3, this volume). For example, amorphous silica in phytoliths and soils is an important Si reservoir and could have an impact on the global Si cycle by affecting the transfer of Si from terrestrial to aquatic systems. The processes by which this transfer occurs are not understood as well as the effects anthropogenic perturbation might have on terrestrial biogeochemical cycling of Si.

The riverine input of DSi to the ocean shows strong geographic variability. Tropical rivers have higher concentrations of DSi than nontropical rivers (Chapter 4, this volume). They contribute about 70 percent of the total annual DSi inputs by rivers into the ocean, primarily through climatic, geological, and geomorphological factors rather than anthropogenic factors such as river regulation, waste disposal, and land use change. On a global scale, the rivers of Asia and Oceania are the most efficient sources of DSi because of optimum weathering and transport conditions such as young geology and tectonic activity as well as high precipitation and runoff. Si storage in sediments and phytoliths in the floodplains leads to lower DSi concentrations and yields in large Asian rivers than in small Asian rivers. Conversely, large African rivers reveal high DSi contents relative to small rivers because of their high vegetation and the soils in their drainage basins.

Boreal and arctic rivers, on the other hand, have DSi concentrations that are 30–50 percent lower than the global average of 150 μM (Chapter 5, this volume). Physical denudation, temperature, and changing terrestrial vegetation cover in high-latitude watersheds appear to control riverine DSi input to the northern seas. Because boreal and arctic systems are weathering limited, they are especially affected by hydrological alterations not only in the context of retention but also in changing the weathering regimes of entire watersheds. This is of particular interest in possible future changes in Si fluxes to the oceans because global warming is believed to be especially pronounced at high latitudes of the Northern Hemisphere (IPCC 2001). A climate-induced change in vegetation could alter quite rapidly the biogeochemistry of river catchments and, as a consequence, land–ocean fluxes of Si along the coasts of the Arctic Ocean.

Rivers exhibit not only regional but also strong temporal variability in the concentration and fluxes of DSi, as Zhang et al. (Chapter 6) report for the lower reaches of the Yangtze River. The variation is related to extensive soil erosion and mobilization of Si in the 1960s and to the possible influence of changing hydrological regimes in the drainage basins caused by climate change. This is seen from the observed correlation, though weak, between DSi fluxes and the Southern Oscillation Index.

The fate of fluvially delivered DSi in estuaries and coastal zones is determined by physical and hydrodynamic properties of the respective estuarine system (Chapter 8, this volume). In large stratified estuaries, DSi is almost quantitatively transferred to the coastal zone because high turbidity and low concentrations of N and P nutrients inhibit biological uptake. DSi may be consumed by diatoms in the outer estuary and the adjacent coastal zone when turbidity is reduced. Si:N and Si:P ratios are much larger in pristine, undisturbed rivers than the ratio needed for the development of diatoms so that complete consumption of DSi requires additional inputs of N and P nutrients from upwelling or vertical mixing of nutrient-rich oceanic water. In smaller estuaries that are tidally well mixed and that are often strongly disturbed by anthropogenic activities, the concentrations of dissolved N and P species are very high and do not limit DSi uptake.

Diatoms are the major phytoplankton group to need silicic acid as a major nutrient. Marine diatoms in particular are often limited by Si, whereas river diatoms suffer from Si limitation only occasionally under high anthropogenic inputs of N and P (Chapter 9, this volume). Although data on riverine diatom assemblages are scarce, the physiological differences between marine and freshwater diatoms appear generally to reflect the abundance of DSi in ambience. In rivers, discharge also plays an important role in bloom development. Details of the relationship between Si content, Si metabolism, and environmental factors such as salinity, pH, and osmotic pressure are scarce and warrant further research.

Ragueneau et al. (Chapters 11 and 12) show that Si for diatoms in coastal waters is delivered from both internal and external sources. Recycling within the water column, at the sediment–water interface, and in sediments constitutes the internal source in coastal waters. In certain areas coastal upwelling is an additional source. Major external sources are the rivers that supply Si in the dissolved and particulate form. Groundwater inputs appear to be of minor importance, especially at the global scale. A similar conclusion can be drawn for atmospheric inputs, with most of the atmospheric inputs fertilizing open ocean areas (Chapter 7, this volume). Estimates show that the highest amounts of dust Si are deposited in the tropical North Atlantic and the western North Pacific. The eolian Si input to the ocean is equivalent to about 3–9 Tmol Si per year. Assuming a solubility of particulate Si between 1 percent and 10 percent, 0.03–1 Tmol of dissolved Si is added by deposited soil dust to the oceans each year.

Although human activities have increased N and P fluxes in many rivers, the flow of DSi from the land to sea is increasingly reduced by Si retention in reservoirs and lakes with increased diatom productivity and by possible loss of vegetation and soils (Chapter 5, this volume). Anthropogenic inputs of DSi (e.g., in the form of sodium silicate or Si fertilizers; Chapters 11 and 12, this volume) appear to contribute little to overall river DSi fluxes (Chapter 10, this volume) but might be important locally. At present, it is difficult to determine the extent to which such anthropogenic inputs may compensate for the observed reductions in DSi fluxes.

The changes in DSi fluxes in rivers have an impact by shifting the ratios of available nutrients for plankton growth in water bodies. Under nutrient-replete conditions, diatoms, for example, have N:Si ratios of about 1:1 (Brzezinski 1985), and in pristine lakes and rivers there is an excess of Si over N and P relative to the need of diatoms. In many systems, however, human perturbation has resulted in a decline in the ratio of Si:N:P to 1:1 or less, with severe impacts on the quality and structure of aquatic ecosystems (Chapters 5, 11, and 12, this volume).

Numerous field and experimental studies (mesocosms) have linked the shift from diatoms to nondiatom species, potentially harmful, to declining ratios of DSi to dissolved inorganic nitrogen and dissolved inorganic phosphorus (Chapters 11 and 12, this volume). But the relationship between DSi availability and diatom importance is not as straightforward as previously thought, and it emerges that there is a need to incorporate organic nutrients and heterotrophic potential in both monitoring and modeling of harmful algal bloom population dynamics.

Modeling is an important tool for investigating past and future changes of the Si cycle. An example of the modeling approach, the RIVERSTRAHLER model, is given by Garnier et al. (Chapter 10, this volume). The model offers a general framework for the study of the biogeochemical functioning of river systems by considering input from the watershed and the biological uptake and transformation processes within them. In three case studies (the Seine and Danube rivers in Europe, and the Red [Hong] River in Vietnam) it is used to explore the effect of hydrological regime, climate, and degree of human activities on nutrient fluxes in the respective river networks. Anthropogenic impact leads to Si limitation relative to P or to N in the Seine and Danube. With concentration of DSi about twice those of the other rivers, DSi is generally not limiting in the Red River.

In another modeling approach, Heinze (Chapter 14, this volume) shows that the Si system of the open ocean adjusts slowly to changes in the external Si supply because of the large Si inventory in the open ocean water column. Therefore, past and present changes in riverine DSi fluxes will have no significant impact on the open ocean Si system, whereas the same changes may have severe consequences for the ecology in coastal marine regions. Because of the stability of the oceanic Si system, variability in opal sedimentation on shorter time scales rather than the time scale of the reaction to external input changes may reliably reflect the internal changes in productivity and circulation of the open ocean.

A promising approach to studies on Si cycling is the use of stable Si isotopes, which have just begun to show their utility in such studies (Chapter 13, this volume). Isotopic fractionation of Si occurs during chemical weathering, precipitation, and biomineralization. Recently it has been shown that the continental precipitation of siliceous cements, which are strongly depleted in ^{30}Si, and the associated fractionation might represent a major process accounting for the enrichment of dissolved river and sea ^{30}Si relative to igneous rocks (Basile-Doelsch et al. 2005). Investigation of the distribution and behavior of Si isotopes in small catchments, soils, and rivers will help to pin down the controls on the δ^{30}Si of Si inputs to the ocean and their potential range of variation over time. This in turn will promote the understanding of longer-term variations in δ^{30}Si in the sedimentary record.

Cycles of Silicon and Carbon

Interactions between the cycles of Si and other nutrient elements (i.e., N and P) were described mainly by focusing on relative changes in their availability and associated consequences for phytoplankton speciation in estuaries and coastal seas. However, there is also a fundamental coupling between the cycling of Si and carbon (see also Ittekkot et al. 2003). Both elements have been closely linked through geological times, as in silica (clay minerals)–organic complexes implicated during the early development of life (Cairns-Smith 1985; Hanczyk et al. 2003) or during biomineralization, during which organic molecules mediate the formation of Si-containing frustules of diatoms and radiolarians (Hecky et al. 1973; Degens 1976).

Today, silicon–carbon interactions are of special interest against the background of rising atmospheric CO_2 as both chemical weathering of silicates (Chapter 2, this volume) and the sequestration of atmospheric CO_2 through the biological carbon pump affect atmospheric CO_2 content. Perturbation of the Si cycle affects both these processes, with major adverse consequences in the short and long term.

DSi inputs to the sea stimulate the production of diatoms, which play a crucial role in the biological uptake of CO_2 in the ocean by the so-called biological pump (Dugdale et al. 1995; Smetacek 1998), a process by which CO_2 is incorporated into organic matter (the organic carbon pump) and calcium carbonate (the carbonate pump). The formation of organic matter during photosynthesis decreases partial pressure of CO_2 in the surface layers, leading to a net CO_2-uptake, whereas carbonate production increases the partial pressure of CO_2 at the ocean surface. Thus, for each mole of carbonate formed, a mole of CO_2 is released. Key players in the functioning of the biological carbon pump are silica-secreting diatoms and carbonaceous coccolithophorids. The efficiency of the biological pump in the short term is determined by the relative abundance of the two species: Diatoms are more efficient at carbon sequestration than coccolithophorids because the latter secrete carbonaceous shells. Any alterations in the supply of DSi to the productive surface layers of the oceans either by changing river inputs or upwelling intensities and the accompanying potential shifts in species composition could have a significant impact on carbon cycling, especially on the ocean's capacity to store CO_2.

Outlook and Recommendations

There is no doubt that anthropogenic perturbation of the silicon cycle could unsettle a host of regulatory and socioeconomic functions of the world's water bodies. The ability of these water bodies to sustain economically important fisheries will be reduced; severe perturbations can be expected in the biogeochemical cycling of elements, with consequences for the ability of coastal seas to act as sinks for anthropogenic gases such as CO_2. Because Si is also implicated in the long-term sequestration of organic C into marine sediments, the largest long-term sink on Earth, human alterations of Si delivery to the oceans could have profound effects on carbon burial and ultimately atmospheric oxygen.

Changing Si:N ratios may lead to drastic changes in the planktonic community, causing severe perturbations of food webs including the development of harmful algal blooms affecting fisheries. The size of phytoplankton and the grazing regime in turn determine carbon fluxes to the deep sea. A shift from diatoms to other species would enhance recycling of organic matter in the upper water column because diatoms are very effective in carbon sequestration (Dugdale et al. 1995; Ittekkot et al. 2000). Elevated recycling rates could bring about anoxic conditions (Justić et al. 2002), although this depends also on the fluxes of carbon and the ventilation of the water column. Modified emissions of trace gases such as dimethylsulfide, organohalogens, CO_2, CO, and N_2O could also arise from changing plankton composition, with consequences for climate and atmospheric acidity, raising the possibility of feedback on weathering rates. These impacts, though reported from only a few areas, justify serious concern over the continued human perturbations of the Si cycle and call for future research efforts to better understand it and to mitigate the impact of future interventions.

From the few published long-term data sets on riverine nutrient fluxes it emerges that severe changes have resulted from anthropogenic action in rivers and adjacent seas (e.g., Humborg et al. 1997; Justić et al. 1995a, 1995b; Raymond and Cole 2003). The need to identify these changes of Si and associated nutrient fluxes and the underlying causes makes it necessary to explore long-term data that appear to exist in many countries and that to date have not been analyzed adequately. Available historical data sets from direct measurements should be supplemented by information retrieved from sites at reservoirs, rivers, and lakes and from marine sediments and corals. For this purpose both DSi measurements and analysis of planktonic remains could be of use.

It is also essential that future river monitoring programs include DSi measurements. Monitoring procedures should follow common standards to ensure high-quality data enabling direct comparison of regional data sets. Monitoring activities must be extended to fluxes at reservoir inlets and outlets to refine our knowledge of their impact on riverine DSi fluxes so that the pattern of change arising from existing perturbations such as large-scale hydrological alterations could be described fully. Furthermore, Si budgets in freshwater lakes and reservoirs, spanning a range of water residence times, should be compiled from existing data sets on a global scale, together with representative determinations of Si fluxes in the coastal zone. Especially in coastal seas affected by large inputs of suspended matter (e.g., in Southeast Asia), sampling strategies in estuaries could be refined when remote sensing methods were used to identify sampling sites. A few temperate and tropical rivers, both small and large, affected by variable anthropogenic perturbations should be selected for comprehensive case studies. These should be accomplished by the investigation and monitoring of land use changes that might distinctly alter the transfer of Si from the terrestrial to the aquatic system. In combination, these efforts would help to identify and bridge gaps in data sets, permitting us to track changes over time.

Factors controlling uptake, storage, and recycling of Si in coastal systems must be known in detail in order to assess the reaction of different coastal ecosystems to changes in silicate inputs and to develop Si budgets for systems with different physical settings, such as large and small estuary regions and high- and low-energy coastal zones, ranging from polluted to pristine systems.

However, budget calculations will remain unsatisfactory until the remarkably large deficiency of analytical tools and quantitative knowledge of several key biogeochemical processes, such as Si uptake and dissolution in the water column and in sediments, has been compensated. Against this background it is particularly important to develop and apply new techniques in Si biogeochemistry, including the improvement of biogenic silica measurements in sediments or in suspended material, which are currently based mainly on controlled dissolution experiments. The selectivity of this method is questionable and the sensitivity poor, especially in riverine and estuarine environments where a huge amount of lithogenic silica is present. Significant progress in Si uptake and dissolution kinetics and in identification of the sources and sinks of Si in the environment could be achieved by the application of suitable radioisotope tracers. Furthermore, the combined application of Si and H nuclear magnetic resonance will allow the study of the interactions between DSi and organic molecules such as humic and fulvic acids and proteins.

Apart from improved techniques and data assimilation, the application of mathematical models offers another tool to elucidate Si cycling. Models that include Si-related processes or describe the Si cycle as a whole are rare, be they empirical or theoretical ones, process-oriented or system descriptive approaches. Therefore, Si-related variables should be included in existing hydrodynamic and biogeochemical models. Hydrobiogeochemical models should be developed for aquatic continua around the world to test scenarios of human impacts in watersheds and the consequences for coastal zones. The impact of changes in the hydrological regime, such as those induced by climate changes, should be explored in collaboration with the community of scientists developing general circulation models.

Finally, more international cooperation and exchanges of information are needed to resolve the transnational issues emerging, for example, from the damming of large rivers or from climatically induced changes in the hydrological regimes and their consequences for the receiving rivers and coastal waters. Adequate dissemination of the obtained knowledge is also necessary to ensure its incorporation into river and coastal zone management strategies so that they can respond effectively to the potential adverse consequences of an anthropogenically perturbed global Si cycle.

Literature Cited

Basile-Doelsch, I., J. D. Meunier, and C. Parron. 2005. Another continental pool in the terrestrial silicon cycle. *Nature* 433:399–402.

Brzezinski, M. A. 1985. The Si:C:N ratio of marine diatoms: interspecific variability and effects of some environmental variables. *Journal of Phycology* 21:347–357.

Cairns-Smith, A. G. 1985. The first organisms. *Scientific American* 252:90–100.

Degens, E. T. 1976. Molecular mechanisms on carbonate, phosphate and silica deposition in the living cell. *Topics in Current Chemistry* 64:1–112.

Dugdale, R. C., F. P. Wilkerson, and H. J. Minas. 1995. The role of a silicate pump in driving new production. *Deep-Sea Research* 42:697–719.

Hanczyk, M. M., S. M. Fujikawa, and J. W. Szostak. 2003. Experimental models of primitive cellular compartments: Encapsulation, growth, and division. *Science* 302:618–622.

Hecky, R. E., K. Mopper, P. Kilham, and E. T. Degens. 1973. The amino acids and sugar composition of diatom cell walls. *Marine Biology* 19:323–331.

Humborg, C., V. Ittekkot, A. Cociasu, and B. von Bodungen. 1997. Effect of Danube River dam on Black Sea biogeochemistry and ecosystem structure. *Nature* 386:385–388.

IPCC. 2001. *Climate change 2001: The scientific basis.* Cambridge: Cambridge University Press.

Ittekkot, V. 2000. Perturbed silicon cycle discussed. *EOS, Transactions, American Geophysical Union* 81:198, 200.

Ittekkot, V. 2003. A new story from the Ol' Man River. *Science* 301:56–57.

Ittekkot, V., C. Humborg, L. Rahm, and N. Tac An. 2003. Carbon–silicon interactions. Pp. 311–322 in *Interactions of the major biogeochemical cycles: Global change and human impact,* edited by J. M. Mellilo, C. B. Field, and B. Moldan. SCOPE 61. Washington, DC: Island Press.

Ittekkot, V., C. Humborg, and P. Schäfer. 2000. Large-scale hydrological alterations and marine biogeochemistry: A silicate issue. *BioScience* 50:776–782.

Justić, D., N. N. Rabalais, and R. E. Turner. 1995a. Stoichiometric nutrient balance and origin of coastal eutrophication. *Marine Pollution Bulletin* 30:41–46.

Justić, D., N. N. Rabalais, and R. E. Turner. 2002. Modelling impacts of decadal changes in riverine nutrient fluxes on coastal eutrophication near the Mississippi River delta. *Ecological Modelling* 152:33–46.

Justić, D., N. N. Rabalais, R. E. Turner, and Q. Dortch. 1995b. Changes in nutrient structure of river-dominated coastal waters: Stoichiometric nutrient balance and its consequences. *Estuarine, Coastal and Shelf Science* 40:339–356.

Raymond, P., and J. Cole. 2003. Increase in the export of alkalinity from North America's largest river. *Science* 301:88–91.

Smetacek, V. 1998. Diatoms and the silicate factor. *Nature* 391:224–225.

Tac An, N., B. von Bodungen, V. Ittekkot, R. E. Turner, and workshop participants. 2002. Summary and conclusions. In *SCOPE Workshop on Land–Ocean Nutrient Fluxes: The Silica Cycle,* Nha Thrang, Vietnam, September 25–27, 2000. *Collection of Marine Research Works* 12 Supplementary Issue, Proceedings, 5–10.

White, A. F., and S. L. Brantley (eds.). 1995. Chemical weathering rates of silicate minerals. *Reviews in Mineralogy* 31:583. Washington, DC: Mineralogical Society of America.

Contributors

Seno Adi
Directorate of Technology for Land Resources and Region Management, Agency for the Assessment and Application of Technology
Jl. M.H. Thamrin 8
10340 Jakarta
Indonesia

Gilles Billen
UMR Sisyphe—CNRS
Université Pierre et Marie Curie
4, Place Jussieu
F–75252 Paris
France

Gregg J. Brunskill
Australian Institute of Marine Science
PMB 3, Townsville MC
4810 Queensland
Australia

Lei Chou
Laboratoire d'Océanographie Chimique et Géochimie des Eaux
Université Libre de Bruxelles
Campus Plaine—CP 208
B-1050 Brussels
Belgium

Pascal Claquin
Laboratoire de Biologie et de Biotechnologies Marines
Université de Caen Basse-Normandie
Esplanade de la Paix
F–14032 Caen Cedex
France

Daniel J. Conley
Department of Marine Ecology
National Environmental Research Institute
DK-4000 Roskilde
Denmark
and
Department of Marine Ecology
University of Aarhus
DK-8000 Aarhus
Denmark

Åsa Danielsson
Department of Water and Environmental Studies
Linköping University
SE-581 83 Linköping
Sweden

Christina L. De La Rocha
Alfred-Wegener-Institut für Polar- und Meeresforschung
Columbusstrasse
D–27568 Bremerhaven
Germany

Hans Dürr
UMR Sisyphe—CNRS
Université Pierre et Marie Curie
4, Place Jussieu
F–75252 Paris
France

Josette Garnier
UMR Sisyphe—CNRS
Université Pierre et Marie Curie
4, Place Jussieu
F–75252 Paris
France

Christoph Heinze
Geophysical Institute & Bjerknes Centre for Climate Research
University of Bergen
Allegaten 70
N-5007 Bergen
Norway

Christoph Humborg
Institute of Applied Environmental Research
Stockholm University
SE-106 91 Stockholm
Sweden

Venugopalan Ittekkot
Center for Tropical Marine Ecology
Fahrenheitstrasse 6
D-28359 Bremen
Germany

Tim C. Jennerjahn
Center for Tropical Marine Ecology
Fahrenheitstrasse 6
D-28359 Bremen
Germany

Danuta Kaczorek
Department of Soil Environment Sciences
Agricultural University of Warsaw (SGGW)
Nowoursynowska 159
02-776 Warsaw
Poland

Bastiaan A. Knoppers
Department of Geochemistry
Federal University of Fluminense
Morro do Valonguinho s/n
24020-007 Niterói, RJ
Brazil

Karen E. Kohfeld
School of Earth and Environmental Sciences
Queens College, City University of New York
New York
USA

Aude Leynaert
Laboratoire des Sciences de l'Environnement Marin, IUEM
Université de Bretagne Occidentale
F–29280 Plouzané
France

Rui Xiang Li
First Institute of Oceanography
State Oceanic Administration
6 Xianxialing Road
266061 Qingdao
China

Su Mei Liu
College of Chemistry and Chemical Engineering
Ocean University of China
5 Yushan Road
266003 Qingdao
China

Natarajan Madhavan
School of Environmental Sciences
Jawaharlal Nehru University
New Delhi
India

Jean Dominique Meunier
CEREGE, Europôle Méditerranéen de l'Arbois
BP 80
F-13545 Aix-en-Provence Cedex 4
France

Michel Meybeck
UMR Sisyphe—CNRS
Université Pierre et Marie Curie
4, Place Jussieu
F–75252 Paris
France

Carl-Magnus Mörth
Department of Geology and Geochemistry
Stockholm University
SE-106 91 Stockholm
Sweden

Sorcha Ni Longphuirt
Laboratoire des Sciences de l'Environnement Marin, IUEM
Université de Bretagne Occidental
F-29280 Plouzané
France

Xiao Hong Qi
College of Chemistry and Chemical Engineering
Ocean University of China
5 Yushan Road
266003 Qingdao
China

Olivier Ragueneau
Laboratoire des Sciences de l'Environnement Marin, IUEM
Université de Bretagne Occidentale
F-29280 Plouzané
France

Lars Rahm
Department of Water and Environmental Studies
Linköping University
SE-581 83 Linköping
Sweden

Loredana Saccone
Department of Marine Ecology
National Environmental Research Institute

P.O. Box 358
DK-4000 Roskilde
Denmark

Agata Sferratore
UMR Sisyphe—CNRS
Université Pierre et Marie Curie
4, Place Jussieu
F–75252 Paris
France

E. Ivan L. Silva
Institute of Fundamental Studies
Hantana Rd
Kandy
Sri Lanka

Caroline P. Slomp
Faculty of Geosciences
Utrecht University
3508 TA Utrecht
The Netherlands

Erik Smedberg
Department of Systems Ecology
Stockholm University
SE-106 91 Stockholm
Sweden

Michael Sommer
Leibniz-Centre for Agricultural Landscape and Land Use Research (ZALF), Institute of Soil Landscape Research
Eberswalder Str. 84
D-15374 Müncheberg
Germany

Weber F. Landim de Souza
Laboratório de Análises Inorgânicas
Instituto Nacional de Tecnologia
A.V. Venezuela, 82 sl 210
20081-312 Rio de Janeiro
Brazil

Vaidyanatha Subramanian
School of Environmental Sciences
Jawaharlal Nehru University
New Delhi
India

Nguyen Tac An
Institute of Oceanography
01 Cau Da, Nha Trang
Vietnam

Ina Tegen
Leibniz-Institute for Tropospheric Research
Permoserstrasse 15
D-04318 Leipzig
Germany

Daniela Unger
Center for Tropical Marine Ecology
Fahrenheitstrasse 6
D-28359 Bremen
Germany

Roland Wollast
Laboratoire d'Océanographie Chimique et Géochimie des Eaux
Université Libre de Bruxelles
Campus Plaine—CP 208
B-1050 Brussels
Belgium

Ying Wu
State Key Laboratory of Estuarine and Coastal Research
East China Normal University
3663 Zhongshan Road North
200062 Shanghai
China

Guo Sen Zhang
College of Chemistry and Chemical Engineering
Ocean University of China
5 Yushan Road
266003 Qingdao
China

Jing Zhang
State Key Laboratory of Estuarine and Coastal Research
East China Normal University
3663 Zhongshan Road North
200062 Shanghai
China

SCOPE Series List

SCOPE 1–59 are now out of print. Selected titles from this series can be downloaded free of charge from the SCOPE Web site (http://www.icsu-scope.org).
SCOPE 1: *Global Environment Monitoring,* 1971, 68 pp
SCOPE 2: *Man-made Lakes as Modified Ecosystems,* 1972, 76 pp
SCOPE 3: *Global Environmental Monitoring Systems (GEMS): Action Plan for Phase I,* 1973, 132 pp
SCOPE 4: *Environmental Sciences in Developing Countries,* 1974, 72 pp
SCOPE 5: *Environmental Impact Assessment: Principles and Procedures,* Second Edition, 1979, 208 pp
SCOPE 6: *Environmental Pollutants: Selected Analytical Methods,* 1975, 277 pp
SCOPE 7: *Nitrogen, Phosphorus and Sulphur: Global Cycles,* 1975, 129 pp
SCOPE 8: *Risk Assessment of Environmental Hazard,* 1978, 132 pp
SCOPE 9: *Simulation Modelling of Environmental Problems,* 1978, 128 pp
SCOPE 10: *Environmental Issues,* 1977, 242 pp
SCOPE 11: *Shelter Provision in Developing Countries,* 1978, 112 pp
SCOPE 12: *Principles of Ecotoxicology,* 1978, 372 pp
SCOPE 13: *The Global Carbon Cycle,* 1979, 491 pp
SCOPE 14: *Saharan Dust: Mobilization, Transport, Deposition,* 1979, 320 pp
SCOPE 15: *Environmental Risk Assessment,* 1980, 176 pp
SCOPE 16: *Carbon Cycle Modelling,* 1981, 404 pp
SCOPE 17: *Some Perspectives of the Major Biogeochemical Cycles,* 1981, 175 pp
SCOPE 18: *The Role of Fire in Northern Circumpolar Ecosystems,* 1983, 344 pp
SCOPE 19: *The Global Biogeochemical Sulphur Cycle,* 1983, 495 pp
SCOPE 20: *Methods for Assessing the Effects of Chemicals on Reproductive Functions, SGOMSEC 1,* 1983, 568 pp
SCOPE 21: *The Major Biogeochemical Cycles and their Interactions,* 1983, 554 pp
SCOPE 22: *Effects of Pollutants at the Ecosystem Level,* 1984, 460 pp
SCOPE 23: *The Role of Terrestrial Vegetation in the Global Carbon Cycle: Measurement by Remote Sensing,* 1984, 272 pp

SCOPE 24: *Noise Pollution,* 1986, 466 pp
SCOPE 25: *Appraisal of Tests to Predict the Environmental Behaviour of Chemicals,* 1985, 400 pp
SCOPE 26: *Methods for Estimating Risks of Chemical Injury: Human and Non-Human Biota and Ecosystems, SGOMSEC 2,* 1985, 712 pp
SCOPE 27: *Climate Impact Assessment: Studies of the Interaction of Climate and Society,* 1985, 650 pp
SCOPE 28: *Environmental Consequences of Nuclear War*
Volume I: Physical and Atmospheric Effects, 1986, 400 pp
Volume II: Ecological and Agricultural Effects, 1985, 563 pp
SCOPE 29: *The Greenhouse Effect, Climatic Change and Ecosystems,* 1986, 574 pp
SCOPE 30: *Methods for Assessing the Effects of Mixtures of Chemicals, SGOM-SEC 3,* 1987, 928 pp
SCOPE 31: *Lead, Mercury, Cadmium and Arsenic in the Environment,* 1987, 384 pp
SCOPE 32: *Land Transformation in Agriculture,* 1987, 552 pp
SCOPE 33: *Nitrogen Cycling in Coastal Marine Environments,* 1988, 478 pp
SCOPE 34: *Practitioner's Handbook on the Modelling of Dynamic Change in Ecosystems,* 1988, 196 pp
SCOPE 35: *Scales and Global Change: Spatial and Temporal Variability in Biospheric and Geospheric Processes,* 1988, 376 pp
SCOPE 36: *Acidification in Tropical Countries,* 1988, 424 pp
SCOPE 37: *Biological Invasions: a Global Perspective,* 1989, 528 pp
SCOPE 38: *Ecotoxicology and Climate with Special References to Hot and Cold Climates,* 1989, 432 pp
SCOPE 39: *Evolution of the Global Biogeochemical Sulphur Cycle,* 1989, 224 pp
SCOPE 40: *Methods for Assessing and Reducing Injury from Chemical Accidents, SGOMSEC 6,* 1989, 320 pp
SCOPE 41: *Short-Term Toxicity Tests for Non-genotoxic Effects, SGOMSEC 4,* 1990, 353 pp
SCOPE 42: *Biogeochemistry of Major World Rivers,* 1991, 356 pp
SCOPE 43: *Stable Isotopes: Natural and Anthropogenic Sulphur in the Environment,* 1991, 472 pp
SCOPE 44: *Introduction of Genetically Modified Organisms into the Environment,* 1990, 224 pp
SCOPE 45: *Ecosystem Experiments,* 1991, 296 pp
SCOPE 46: *Methods for Assessing Exposure of Human and Non-human Biota SGOMSEC 5,* 1991, 448 pp
SCOPE 47: *Long-Term Ecological Research. An International Perspective,* 1991, 312 pp

SCOPE 48: *Sulphur Cycling on the Continents: Wetlands, Terrestrial Ecosystems and Associated Water Bodies,* 1992, 345 pp
SCOPE 49: *Methods to Assess Adverse Effects of Pesticides on Non-target Organisms, SGOMSEC 7,* 1992, 264 pp
SCOPE 50: *Radioecology after Chernobyl,* 1993, 367 pp
SCOPE 51: *Biogeochemistry of Small Catchments: a Tool for Environmental Research,* 1993, 432 pp
SCOPE 52: *Methods to Assess DNA Damage and Repair: Interspecies Comparisons, SGOMSEC 8,* 1994, 257 pp
SCOPE 53: *Methods to Assess the Effects of Chemicals on Ecosystems, SGOMSEC 10,* 1995, 440 pp
SCOPE 54: *Phosphorus in the Global Environment: Transfers, Cycles and Management,* 1995, 480 pp
SCOPE 55: *Functional Roles of Biodiversity: a Global Perspective,* 1996, 496 pp
SCOPE 56: *Global Change, Effects on Coniferous Forests and Grasslands,* 1996, 480 pp
SCOPE 57: *Particle Flux in the Ocean,* 1996, 396 pp
SCOPE 58: *Sustainability Indicators: a Report on the Project on Indicators of Sustainable Development,* 1997, 440 pp
SCOPE 59: *Nuclear Test Explosions: Environmental and Human Impacts,* 1999, 304 pp
SCOPE 60: *Resilience and the Behavior of Large-Scale Systems,* 2002, 287 pp
SCOPE 61: *Interactions of the Major Biogeochemical Cycles: Global Change and Human Impacts,* 2003, 384 pp
SCOPE 62: *The Global Carbon Cycle: Integrating Humans, Climate, and the Natural World,* 2004, 526 pp
SCOPE 63: *Alien Invasive Species: A New Synthesis,* 2004, 352 pp.
SCOPE 64: *Sustaining Biodiversity and Ecosystem Services in Soils and Sediments,* 2003, 308 pp
SCOPE 65: *Agriculture and the Nitrogen Cycle,* 2004, 320 pp

SCOPE Executive Committee 2005–2008

Index

Island Press Board of Directors